Join us on the web at
agriculture.delmar.com

COMPANION ANIMALS
in Society

COMPANION ANIMALS
in Society

Stephen Zawistowski, Ph.D.

Australia • Brazil • Canada • Mexico • Singapore • Spain • United Kingdom • United States

Companion Animals in Society
Stephen Zawistowski, Ph.D.

Vice President, Career Education Strategic Business Unit:
Dawn Gerrain

Director of Learning Solutions:
John Fedor

Acquisitions Editor:
David Rosebaum

Managing Editor:
Robert Serenka

Product Manager:
Christina Gifford

Editorial Assistant:
Scott Royael

Director of Production:
Wendy Troeger

Production Manager:
Mark Bernard

Senior Content Project Manager:
Kathryn B. Kucharek

Technology Project Manager:
Sandy Charette

Director of Marketing:
Wendy Mapstone

Channel Manager:
Gerard McAvey

Marketing Coordinator:
Jonathan Sheehan

Art Director:
David Arsenault

Cover and Front Matter Images:
Goldfish and Hamster—Getty Images, Inc.; Cat and Dog—Copyright © Image Source Limited

Cover Design:
Judi Orozco

© 2008 Thomson Delmar Learning, a part of the Thomson Corporation. Thomson, the Star logo, and Delmar Learning are trademarks used herein under license.

Printed in Canada
1 2 3 4 5 XXX 11 10 09 08 07

For more information contact Delmar Learning,
5 Maxwell Drive, PO Box 8007, Clifton Park, NY 12065-2919.

Or you can visit our Internet site at
http://www.delmarlearning.com.

ALL RIGHTS RESERVED. No part of this work covered by the copyright hereon may be reproduced or used in any form or by any means—graphic, electronic, or mechanical, including photocopying, recording, taping, Web distribution or information storage and retrieval systems—without written permission of the publisher.

For permission to use material from this text or product, submit a request online at http://www.thomsonrights.com
Any additional questions about permissions can be submitted by email to thomsonrights@thomson.com

Library of Congress Cataloging-in-Publication Data

Zawistowski, Stephen.
    Companion animals in society/Stephen Zawistowski. — 1st ed.
      p. cm.
    Includes index.
    ISBN-13: 978-1-4180-1370-7 (alk. paper)
    1. Pets. 2. Pets—Social aspects. I. Title.
    SF411.5.Z39 2007
    636.088'7—dc22
                                  2007043747

NOTICE TO THE READER

Publisher does not warrant or guarantee any of the products described herein or perform any independent analysis in connection with any of the product information contained herein. Publisher does not assume, and expressly disclaims, any obligation to obtain and include information other than that provided to it by the manufacturer.

The reader is expressly warned to consider and adopt all safety precautions that might be indicated by the activities herein and to avoid all potential hazards. By following the instructions contained herein, the reader willingly assumes all risks in connection with such instructions.

The publisher makes no representation or warranties of any kind, including but not limited to, the warranties of fitness for particular purpose or merchantability, nor are any such representations implied with respect to the material set forth herein, and the publisher takes no responsibility with respect to such material. The publisher shall not be liable for any special, consequential, or exemplary damages resulting, in whole or in part, from the readers' use of, or reliance upon, this material.

CONTENTS

PREFACE ix
ABOUT THE AUTHOR xiii
ACKNOWLEDGMENTS xv

CHAPTER 1
INTRODUCTION TO COMPANION ANIMALS 1
What Is a Companion Animal? / 1
Introduction to Pet Keeping / 3
Companion Animal Demographics / 8

CHAPTER 2
DEVELOPMENT OF COMPANION ANIMALS 14
Domestication/Development of Companion Animals / 17

CHAPTER 3
DEVELOPMENT OF ANIMAL PROTECTION 53
Early Examples / 53
Other Opinions / 63
Modern Animal Rights / 65

CHAPTER 4
ANIMAL SHELTERS AND RESCUE 71
Animal Shelter Organizations / 75
Animal Shelter Developments / 78
Animal Shelter Programs / 79
Support and Professional Organizations / 84
No-Kill Movement / 88
Shelter Design / 90

Shelter Medicine / 92
Behavior Programs / 93
Technology / 99
Asilomar Accords / 100
Annual Live Release Rate Formulas / 105

CHAPTER 5
COMPANION ANIMALS, LAW, AND ANIMAL CRUELTY 112
Federal Laws / 114
State and Local Laws / 119
The Cycle of Violence / 130

CHAPTER 6
PET CARE INDUSTRY 134
Pet Products / 135
Pet Industry Organization / 137
Pet Foods / 138
Kitty Litter / 142
Veterinary Medicine / 143
Animal Behavior Services / 149
Dog Grooming / 153
Boarding, Day Care, and Pet Sitting / 154
Media / 156
Charities and Nonprofit Organizations / 158

CHAPTER 7
COMPETITIONS 164
Shows (Conformation) / 166
Agility / 170
Earthdog Events / 172
Flyball / 173
Lure Coursing / 174
Herding / 175
Obedience Trials / 176
Iditarod: *The Last Great Race on Earth* / 176
Weight Pulling / 178
Canine Freestyle / 179
Cat Shows / 180
Rats and Mice / 181

CHAPTER 8
ASSISTANCE DOGS 183
  Guide Dogs / 189
  Service Dogs / 197
  Bomb-Sniffing Dogs / 199
  Schutzhund / 200
  Military / 201
  Conservation Dogs / 204

CHAPTER 9
COMPANION ANIMALS IN THE HOME 206
  How Pets Are Chosen / 208
  Children and Animals / 210
  Children and Pets / 214
  Animals and Human Health / 216
  Companion Animal Death and Grief / 217
  Conclusions / 220

CHAPTER 10
HOT BUTTON ISSUES 224
  Owner or Guardian? / 224
  Exotic or Wild Animals as Pets? / 225
  Breed-Specific Legislation (BSL) / 226
  Declawing Cats / 227
  Docking and Cropping of Dogs / 228
  Outdoor Cats / 228
  Horse Slaughter / 229
  Pet Store Sales of Dogs / 230

CHAPTER 11
COMPANION ANIMAL CARE GUIDES 232
  Dogs / 232
  Cats / 236
  Ferrets / 239
  Hamsters / 240
  Gerbil / 243
  Guinea Pig / 245
  Mice / 247

Rats / 249
Rabbits / 251
Chinchillas / 253
Birds / 255
Fish / 257

APPENDIX AMERICAN DOG AGILITY ORGANIZATIONS 265
GLOSSARY 267
INDEX 273

PREFACE

The *why* for this book might easily be explained through a simple thought experiment, or *gedanken*. Close your eyes, and starting with getting out of bed this morning, trace your day until this point. Did you wake up with a pet sleeping on your bed? Do you kiss them good-morning? Did your morning include serving breakfast to the pets in your home? Were they waiting for you when you came home? Chances are that you answered *yes* to at least one of these questions. It is estimated that 63% of American households, or just about 69 million households, have a pet (American Pet Products Manufacturers Association, 2005–2006). As you had your morning cup of coffee, did you watch TV or listen to the radio and hear a commercial for a pet product or service? If you are a baseball fan, did you check the box scores, and did your team play in a ball park named for a major pet retail chain? Did you make a note to pick up pet food, treats, or toys that day, contributing to the nearly $36 billion that will be spent on pets this year (Guthrie, 2005)? On your way to school or work did you see people walking their dogs or dumping the cat litter? Did you see stray dogs or cats on the way? Were they bold or furtive in their movements seeking food and shelter? If you stopped in someone's office, did they have a picture of their family on the wall or desk, complete with family pet? What about your wallet or purse, does it hold a picture of your pet?

By now it should be clear that companion animals are an intimate part of our culture and environment. Whether you are part of the majority that currently shares your home with a pet, or part of the minority that does not, companion animals affect your life. You may consider your

Animals are sensitive to the passage of time and subtle environmental cues. As a result they will often be waiting for their human companions at the end of the day. *(Courtesy Darlene Larson)*

The pet industry is growing in revenue and visibility. The San Diego Padres play in PETCO Park. *(Courtesy San Diego Padres)*

pets part of your family and close companions. Or you may think that other people's pets are a waste of time and a nuisance (the New York City 311 non-emergency complaint line lists barking dogs as one of the top nuisance calls). Stray dogs and cats may anger you by damaging your lawn or garden, or bring you to tears with their plight and stimulate you to contribute to one of the many humane groups that care for homeless pets.

This text will be a broad introduction to the role and impact of *Companion Animals in Society.* Each chapter could easily be expanded into a book on that topic, and in many cases that has already been done. I have attempted to distill an enormous amount of information into a manageable package that will provide a gateway into the complex and expanding field of companion animal studies. Twenty-five years ago this book would have been unlikely. Academic interest in companion animals was generally limited to the nation's veterinary colleges, where the primary focus was on their health and medicine. It is true that there was the unusual professor here and there who might study dogs, cats, or other pets, or the relationships between humans and animals. Their work stands out and is important because it was rare, and helped to establish a base from which subsequent scholars have built a substantial body of new research. If nothing else, I expect that readers will be surprised by the wide range of topics covered, once again, reflecting the extent to which companion animals have been integrated into our society. Biology, psychology, sociology, history, anthropology, law, economics, medicine, and other disciplines have contributions to make in this endeavor.

Weaving these various themes into a coherent narrative is a challenge, but necessary.

If nothing else, we need these different perspectives to help us recognize, if not always understand, the depth of the connection between companion animals and people, and the striking inconsistencies in their treatment. How do we explain the long running broadcast popularity of the Westminster Kennel Club Dog Show when juxtaposed to the abandonment of millions of dogs each year at animal shelters? Cats have endured a roller-coaster ride of human attention, worshipped in ancient Egypt, vilified and killed as evil in medieval Europe, and now the most common and popular companion animal in America. Over 70 million cats are kept as pets, but millions more live as unsocialized ferals on the fringes of our communities. In 2005, the first cloned cat came with a $50,000 price tag (Eisenberg, 2005), while at the same time, the state of Wisconsin considered making it legal to hunt unowned stray cats (<http://www.animallaw.info/statues/stuswi2005question62.htm>).

It is unlikely that the subsequent topics in this book will resolve these contradictions, but they should help to chart some of the history, biology, and other elements that trace the presence and nature of these issues. The topics that an author decides to include or not include in a book are always a combination of many different things. These decisions are often easier in a well-established field where previous works have helped to delimit the boundaries of the field, and the range of expected knowledge has been established. While pet keeping is an ancient human practice, companion animal studies is still a developing discipline. It is a special challenge since it synthesizes the information from a variety of other academic disciplines. History, sociology, law, genetics, animal behavior, veterinary medicine, philosophy, and economics all have something to say about companion animals in society. My choices for this book have been based on my own history and academic background, several decades working with companion animals and with a wide range of organizations and institutions that affect companion animals and the people associated with them. The introduction provided by this text is broad, but it may not be complete. That remains to be seen. However, it should provide a foundation to participate in the dialogue about *Companion Animals in Society*.

ABOUT THE AUTHOR

Stephen L. Zawistowski, Ph.D., joined the ASPCA in 1988 as Vice President of Education, and is currently Executive Vice President for National Programs and Science Advisor. A well-known speaker on education, animal behavior, and animal welfare issues, Dr. Zawistowski is frequently quoted in the media, and in early 2000 he appeared as the host of ASPCA Pet Check segments on PBS.

Dr. Zawistowski is a certified applied animal behaviorist and chairs the Animal Behavior Society's Board of Professional Certification from 1998–2007. Published in a number of scientific journals, he is the founding co-editor of the *Journal of Applied Animal Welfare Science* (*JAAWS*). Since joining the ASPCA, he has authored and consulted on more than 20 books on animals and pet care, including *Animal Shelter Medicine for Veterinarians and Staff,* co-edited with Dr. Lila Miller, which was published by Blackwell Publishing in spring 2004.

Dr. Zawistowski's deep concern for animal welfare grew out of his work as a caretaker for laboratory animals while earning his B.A. in biology from Canisius College in 1977. He was then awarded a graduate fellowship at the University of Illinois to study behavior/genetics in the Department of Psychology, where he completed his master's degree in 1979, and Ph.D. in 1983. Later that year, while teaching biology at Indiana University as a visiting assistant professor, the National Science Foundation awarded Dr. Zawistowski a post-doctoral fellowship in environmental biology. He left Indiana University in 1985 to become assistant professor of psychology at St. John's University, where he taught until joining the ASPCA.

Dr. Zawistowski was a founding board member and past president of the National Council on Pet Population Study and Policy. He serves on numerous boards and committees, including the Harmony Institute Community Advisory Board, the Scientific Advisory Committee of Humane Farm Animal Care, and the Scientific Advisory Panel of the World Society for the Protection of Animals. He is chairman of the board of the Alliance for the Contraception of Cats and Dogs. In 2003–04 Dr. Zawistowski served on the National Academy of Sciences–National Research Council Committee on a Review of the Smithsonian Institution's National Zoological Park, and in 2006 on the Institute for Laboratory Animal Resources committee to update the report on the Recognition and Alleviation of Pain and Distress in Laboratory Animals.

Dr. Zawistowski is also an adjunct professor at the University of Illinois College of Veterinary Medicine.

In 1989 Psychologists for the Ethical Treatment of Animals (PSYeta) named Dr. Zawistowski "Psychologist of the Year" for outstanding contributions to animal welfare science. St. John's University presented him with the Patrick Daly Memorial Award for a career in education marked by compassion and commitment in spring 2002. He is also listed in *Who's Who in America* and *Who's Who in Science and Engineering*.

Dr. Zawistowski lives on Staten Island, N.Y., with his wife Jane, a beagle, two tabby cats, and some fish.

ACKNOWLEDGMENTS

An endeavor of this sort requires a cadre of prompters, cajolers, helpers, and partners. I'd like to start by thanking the folks at Delmar Learning. Gerald O'Malley "sold" me on the idea of doing the book. Christina Gifford has been calm and patient throughout as my editor, as one crisis after another seemed to erupt in my "day job" at the ASPCA. This was especially true when I left a message in the midst of Katrina rescue efforts (which also coincided with the time frame for the first samples to be done), that I was going to be out of contact for the next couple of months . . . "Bye."

I'm not sure if I would have contemplated writing the book if I had not been at the ASPCA. For what has now been almost two decades, the ASPCA has been my home. Much of this book is based on my experiences working here. The organization and my colleagues here have been wonderfully supportive and open in sharing their experience and ideas. It would be hard to list everyone who helped or contributed, I know that I would miss many. However, there are some folks who deserve special mention because we have worked together for so long, and have spent so many hours talking about animal stuff. Julie Morris, who taught me about shelters; Lila Miller, who was my professor of shelter medicine; Steve Musso, who has been confidant and friend; Jacque Schultz, who knows bunches of stuff and will talk to you about it; Lisa Weisberg, who can really help you to understand the laws; Mark MacDonald, who will show you how to enforce those laws; and Stephanie LaFarge, who will always make you blink twice and think about something three times.

This book would not have been possible without the help of Emily Manos. She found the pictures, kept things in order, and kept people off my back when needed. A special thanks to my boss, Ed Sayres. He talks the talk and walks the walk when it comes to using knowledge to help animals. He granted me the freedom to work on the book and access to the ASPCA resources for research.

Linda Koebner stepped in when I needed some help, researching and drafting Chapters 7 and 8. Lynda Pope provided research help and comments on early drafts, and Andrea Pace helped with the legalese of Chapter 5. Friends and colleagues have spent many years helping me to learn about animals, people, and how they go together. Thank you to Randy Lockwood (whom I met so long ago we don't remember when or where, but he's now at the ASPCA), Carter Luke, Tom Dent, all of the folks from the National Council on Pet Population and Study (especially Mo Salman, John New, and Jan Scarlett), friends from the Interdisciplinary Forum in Applied Animal Behavior (Pam Reid, whom I met there and is now at the ASPCA), Gary Patronek, Ken Shapiro, Amy Marder, Sam Ross, Mike Kaufman, and it certainly could go on for many pages. If you were on one of those many pages not written, I'm sorry, but I am grateful nonetheless.

I'd also like to thank the folks at Green Chimneys, Harmony, Purina, Iams, Morris Animal Foundation, the Cat Fanciers' Association, Women's Humane Society, Massachusetts SPCA, SPCA Serving Erie County, McKissick Museum at the University of South Carolina, Sergeant's Pet Care Products, and so many others for access to their archives and materials.

It can be trite to thank your parents, but they were fundamental to my interest and approach to animals. They were tolerant but tough as I collected a small menagerie of pets. They gave me the two gifts most needed to survive and succeed in this field, passion and compassion.

Finally, thank you to Jane, my friend and partner and wife. We may reclaim the back bedroom (lucky the kid got married and went off to grad school, but thanks, Matt, for taking care of all the pets while you were growing up), and three corners of the living room where stacks of books, reprints, and files have accumulated. I'm not sure we'll ever reclaim the middle of the bed from the dog, or the corners from the cats. I know that we'll never reclaim our hearts from the many animals that we have welcomed into our home.

Delmar Learning and the author would also like to thank the following individuals for taking the time to review the manuscript:

Linda Case
University of Illinois

Colleen Brady
Purdue University

Kimberly Ange-van Heugten
North Carolina State University

Sarah Hurley
Parkland College

Stuart Porter
Blue Ridge Community College

Janice Siegford
Michigan State University

Introduction to Companion Animals

CHAPTER 1

Children and dogs are as necessary to the welfare of the country as Wall Street and railroads.

–Harry S. Truman

KEY TERMS

species	companion animal	households (HHs)
pet	natural history	birth rates
survey	hobbyists	death rates

WHAT IS A COMPANION ANIMAL?

Diamond (1999) and Caras (1996) have both considered the significant role that the domestication of animals has played in the development of human civilizations. Diamond points out the important advantages that human populations enjoyed if they lived in areas where the biological resources included animals and plants that lent themselves to domestication. Domestication provided these groups with readily available sources of food, fiber, and power. Freed from the daily struggle of finding sustenance as hunter-gatherers, they were able to invest time and effort into the development of crafts, science, and social structures that are the basis of civilizations. Diamond's thesis is that the development of dominant civilizations did not arise from some intrinsic superiority of a particular group of people, but rather was the result of the surrounding resources. He points out that many simple cultures have an exquisite knowledge of the animals and plants that surround them. He also points out, however, if these animals and plants were not amenable to domestication, the people were never able to divorce themselves from the uncertainties of the hunter-gatherer lifestyle

(Diamond, 1999). It is remarkable that given the extraordinary diversity of **species** on the planet, domestication of animals has seldom occurred and only in a limited number of places. Caras (1996) argues that the domestication of animals was critical to the development of human culture as it now exists. "We arrived at the human state incomplete, willing but unable to take giant steps. At each turn in our evolution as culture makers, animals enabled us to move on to the next level" (p. 19). Diamond's premise is that these events are the stepping-stones on a path that led some civilizations to dominance and left others to either wither or be vanquished. Within this context of civilizations rising and falling, how is it that of the several dozen species that have been domesticated, only a few have moved beyond their role as hunter, herder, mouser, meal, or some other role and taken up the mantle of companion? Perhaps it is as simple as Darwin's finches, adapting and radiating to fill the available ecological niches on the Galapagos Islands. Humans are social and desire companionship, and a small cadre of animals, already domesticated for other purposes, was ready to fill the role of companions. In retrospect, companionship is the primary reason why these animals continue to share our homes. Mugford (1980) indicates that pet owners most frequently cited affiliation and self-esteem as reasons for having a **pet**. In a **survey**, *Psychology Today* (Horn & Meer, 1984) may have found an even more fundamental reason for the success of companion animals: pet owners are more likely to feel satisfied with their lives than non-pet owners. The people responding to the survey apparently understood and experienced a phenomenon that research scientists are still attempting to evaluate.

Defining Companion Animals

What is a **companion animal**? How is it different from livestock or a wild animal? Is it the same as a pet? Let us begin by answering the last question. Companion animals and pets essentially are the same. K. C. Grier (2006) notes that *pet* is likely derived from the French word *petit*, meaning small. It was first applied to indulged or spoiled children, but by the mid-sixteenth century included animals as well. These animals were often runts of family livestock, commonly lambs, and were raised by hand with special attention. The animals provided companionship and playmates for children on remote farms who were otherwise isolated from neighboring children. These pets were frequently given names and generally avoided slaughter, further distinguishing them from other animals the family might own.

Companion animal has become a more common term in recent years. Often it is used to avoid the possible pejorative sense of the word *pet*. It also reflects accurately the sense of companionship that people believe they share with animals that live with them. Obviously, different animals fulfill the role of companion in various ways. A tank of tropical fish may not seem to be much in the way of companions. However, their presence can have a profound effect on people. They have a more significant impact on someone than inanimate objects in the same room. Simply watching the fish can help lower blood pressure. On the other hand, a person who is blind and relies on a guide dog has a more intimate relationship with their companion. In this case the dog is more than a companion; the dog is an essential partner in accomplishing daily tasks.

For most of us, however, our pets are companions in very simple yet important ways. They greet us when we come home from school or work, they keep us company when we read or watch television, they give us a reason to get out of the house and go for a walk. In the end, we know they are companions because when they die, we miss them, we may grieve, and we notice that they are gone.

INTRODUCTION TO PET KEEPING

Grier's *History of Pets in America* is an important contribution to understanding the role that pets have played in American society (2006). While Native Americans had developed complex relationships with their dogs, it was European colonists who brought their pet keeping habits to America and established the traditions that we continue to follow in many ways. Dogs and cats came with the Europeans. Spanish conquistadors brought war dogs to attack and intimidate the indigenous peoples they found and conquered. Colonists brought dogs to hunt, herd, and guard. Cats were the constant companions for sailors, protecting ships' stores from the inevitable on-board rats. Cats came ashore as well and took up their traditional role protecting food and supplies from rodents. Europeans had a long history of keeping small birds in cages. In the days before recorded music, birds' songs and calls provided spontaneous gaiety to otherwise silent rooms, while their bright colors and movement enlivened the space. These were often wild birds captured in nearby fields or woods or caught elsewhere and imported. North American birds provided a new spectrum of color and sound for people to keep as pets. They also found a wide variety of small animals. Chipmunks, squirrels, and flying squirrels would all prove popular.

By the second half of the nineteenth century, many of the basic pet keeping habits of America were established, including a wide range of still commonly kept animals. In 1893 Marshall Saunders authored *Beautiful Joe*, an important contribution to humane education literature. In 1908, she wrote *My Pets, Real Happenings in My Aviary*. In the dedication of *My Pets*, she indicates to the boys and girls who are her friends that, while other stories she has written are "partly true," this is the story of her aviary and the pets that live in it (Saunders, 1908). The story was taken from her diaries; the birds and other animals described in it were real and girls and boys were welcome to visit them. Her pets included dogs, cats, guinea pigs, rabbits, rats, pigeons, and canaries. There were also owls, cardinals, sparrows, and grosbeaks. In addition to anticipating the breadth of species kept as pets, as the author of a classic book that sensitized children to the treatment of animals, Saunders anticipated some of the concerns about the welfare of companion animals that occupy us a century later. She also described the enormous trade in canaries from Germany to the United States and how the birds were shipped in tiny wicker cages. The voyage is difficult and "There is often great suffering among them" (p. 183). She continues, "I believe more caged winged creatures die from monotony than anything else..." (p. 187).

Wild animals as pets are also a theme in the 1939 book, *Animals as Friends, and How to Keep Them* by Margaret Shaw and James Fisher. Originally written for a British audience, it reflects what was a long-standing connection between

natural history study, collecting, and pet keeping. Nature hobbies were common at this time. Collecting rocks and fossils as well as amateur geology practice was appreciated by both the young and the old. Having a foreword written by the eminent biologist Julian Huxley reflects the high esteem that pet keeping held as a hobby. His comments include, "The keeping of pets is a very desirable habit...it tends to make people more humane, to make them interested in animals.... On the other hand, it is not always a simple art, and many pets undoubtedly suffer from misplaced kindness and mere ignorance" (Shaw & Fisher, p. xi). It would not be surprising to see similar sentiments expressed in the forward of a current pet care book.

The range of animals considered includes hedgehogs, bats, otters, squirrels, deer, a variety of birds, fish (including salt water fish), amphibians, reptiles, dogs, and cats. In a number of cases information is provided on how to catch young wild animals. Each entry includes some *don'ts*. The dog section concludes with, "*Don't* keep a dog unless you are prepared to make him one of the family" (Shaw & Fisher, p. 189). The snake section concludes with some remarkable information. Expecting that some people "are set on keeping poisonous snakes," the authors provide information on how to treat a viper bite, including the address for the acquisition of snake antivenin! (pp. 131–32).

Pet keeping was in many ways a do-it-yourself hobby from the mid-nineteenth to mid-twentieth centuries. Pet care books most often provided diagrams to build cages, recipes to prepare foods, and a variety of home remedies to treat sick pets. The rudiments of the pet supply industry were developing during this period, but did not blossom until the years following World War II. The availability of plastics and other new fabrication materials, combined with more sophisticated electronics to provide light, heat, and aeration helped to fuel an expansion of products that made pet keeping easier. When these new products met the economic expansion in the post-war years, they rapidly became fixtures in the new tract homes that drew families out of the city and into suburbs.

A Pet of Your Own by Georg Zappler and Paul Villiard (1981) was promoted as being five complete books in one. It was in fact a compilation of several books published from 1969 to 1974 and written about different types of pets ranging from reptiles and amphibians to birds and wild mammals. In some ways *A Pet of Your Own* bridges the transition period from the earlier *catch what you can, and build your own cages for keeping it* approach, to more modern concepts in caring for pets. Readers are encouraged to scout the local woods and ponds for reptiles and amphibians. In an echo of *Animals as Friends, and How to Keep Them*, the authors of *A Pet of Your Own* also caution the reader in keeping poisonous snakes, but accept the fact that some readers will try (Zappler & Villiard, 1981). While information is provided on how to catch and care for wild mammals such as squirrels, raccoons, and groundhogs, they do not advise catching wild birds as pets, sticking with canaries, parakeets, finches, and other birds that can be purchased. On the other hand, they do suggest trapping sparrows and other small birds to feed to larger lizards and snakes.

The fish-keeping sections are particularly instructive of the advances in the field from the pre–World War II books, and how much was to come in this area.

FIGURE 1-1

After dogs and cats, freshwater fish are the next most common pets in homes. Cats in particular seem to enjoy watching the fish swim about as much as people do. *(Courtesy Dr. Stephen Zawistowski)*

They dismiss the concept of the "balanced aquarium" where plants provide fish with oxygen and the fish provide the plants with needed carbon dioxide as an "old" idea (Zappler & Villiard, 1981). Aerators were now widely available, along with a wider range of filters, including the introduction of under-gravel filters. Trying to balance the need to place a fish tank in a location to get enough sunlight to promote plant growth, but not too much to cause temperature fluctuations is no longer necessary (Figure 1-1). GRO-LUX® fluorescent lights are now available and they not only promote plant growth, but also enhance viewing the colors of the fish in the tank (1981).

Another interesting theme that comes up in early pet keeping books is breeding pets. Sophisticated **hobbyists** and fanciers prided themselves on the ability to breed various species in captivity. Dogs and cats had been kept and bred for centuries, attempting to develop new breeds or types, or improve upon the available breeds. In other cases, however, where the pets were recently collected from the wild, extensive efforts would be made to determine the conditions required to breed them successfully. This was especially true in the developing tropical fish hobby in the 1950s and 60s. New fish were being discovered regularly and introduced to the aquarium trade. Hobbyists were able to play an active part in helping to develop and understand the biology of these new specimens. Breeding a new species could bring both recognition and potential economic return if the young could be raised and then sold. The important role that amateurs could play in ichthyology is shown in the introductory chapters to the *Handbook of Tropical Aquarium Fishes* by Axelrod and

Schultz (1990). First published in 1955, it includes sections on how scientific names are assigned, details on the collection of fishes from the wild, and how to send an unidentified fish to an ichthyologist for identification (1990). Over time, a sophisticated system of aquaculture has developed to supply the commercial pet market with a wide range of domestically bred specimens. Aquarium dealers, hobbyists, and fish importers may still be an important resource for biologists studying the taxonomy and developmental biology of fishes (Pennisi, 2006).

The Pet Keepers

History would certainly seem to confirm that pet keeping has been a universal human habit. Grier (2006) describes the broad range of individuals who have left documentation of their pet keeping habits. While finances and social status will influence the types of pets someone may keep, and the care provided, pet keeping is more about interest than income. Nonetheless, society is strongly influenced by the habits of the famous, and may emulate their behaviors. Coren (2005) tells the story that Ramses the Great had a hound named Kami who slept with the Pharaoh and is mentioned in the carvings on Ramses' tomb. Alexander the Great went into battle with his beloved greyhound, Peritas. Peritas died in battle saving Alexander's life. In some cases the link between a personage and their pets is so significant, it has been memorialized in the name of the breed. King Charles II was so taken with a type of small spaniel that he conferred a hereditary title upon the dogs and all the descendents of that breed (Figure 1-2). Hence, Cavalier King Charles spaniels are entitled access to the House of Lords.

America's rulers have been no less in love with their pets, especially dogs. George Washington was well known for the quality of his foxhounds (Grier, 2006). Calvin Coolidge had a number of dogs as did Teddy Roosevelt and other presidents (Kirkwood, 2005). Franklin D. Roosevelt's dog, Fala, sparked one of the most colorful pet events in presidential history. During the 1944 presidential campaign, a rumor surfaced that during a trip, Fala was left by mistake in the Aleutian Islands and FDR had sent a naval ship back to retrieve the dog at significant taxpayer expense. FDR turned the tables on his critics, however, by denying the event, and making a joke out of it. In a speech to the Teamsters he indicated that not only had he not sent a ship back for Fala, but since Fala was a Scottish terrier, even thinking of such an expense would be an

FIGURE 1-2

The Cavalier King Charles spaniel is a small dog with growing popularity. *(Courtesy Kelly Cunningham)*

offense to his Scotch soul. Franklin D. Roosevelt and Fala would not be the last president and pet to be embroiled in a controversy. Richard M. Nixon used the family cocker spaniel, Checkers, as part of his response to critics while running for vice president. He had been questioned about inappropriate gifts that he may have received. In a classic television moment, he admitted that Checkers had been a gift, and because he was a beloved family member, would not be leaving the Nixon home. Lyndon B. Johnson was excoriated for being photographed lifting one of his beagles into the air by his ears. George H. W. Bush and William J. Clinton, though from different political persuasions and parties, both dealt with criticism from the animal rights and animal welfare groups. Millie was Bush's English springer spaniel. Millie was already famous as an "author," after "dictating" a book to Barbara Bush, when the first family decided to breed her in 1989. Animal shelter groups' reactions around the country ranged from disappointment to outrage. At a time when millions of homeless animals were being euthanized, they felt that this was a poor example to set for the nation (personal recollection). One of the puppies, Spot, went to the president's son, George W. Bush, and eventually returned to live in the White House. Bill Clinton and his family arrived at the White House without a dog. Their cat, Socks, became quite the personality. Being the first *first cat* in some time, Socks received substantial fan mail and attention. When the Clintons decided to acquire a dog, they also attracted the displeasure of animal shelter groups. Despite a concerted public campaign urging them to adopt a homeless dog from an animal shelter, the first family settled on a chocolate Labrador retriever named Buddy after a favorite uncle (personal recollection). Buddy and Socks would "co-author" their own book, *Dear Socks, Dear Buddy*, a collection of letters from school children.

Movie stars, entertainers, and athletes have all been connected to pets as a way to enhance their own images, and they have influenced the image of pets as well. Publications that focus on the pet market will frequently feature articles on celebrities and their pets. In other cases mainstream magazines such as *People* or *Sports Illustrated* will include photos of people with their pets. Many times these celebrities may lend their names to one or another pet cause, appealing to the public to be responsible pet owners.

No less appealing than the famous, and perhaps more so, are the stories of the simple families that love and care for their pets. Grier (2006) provides numerous examples of both the well-off with their pets and those of modest means. In many cases we know more about the habits of the wealthy because they were more likely to document their relationships with their pets, as they did with all other aspects of their lives. The limited material wealth of people who are poor will result in fewer things to leave behind as evidence of their pets and how they cared for them. Following is an interesting and poignant example. At Ellis Island, a museum displays the history of the immigrants to America that passed through New York City. In one exhibit there was a small rabbit-fur muff, the sort used by girls to warm their hands while skating in the winter. It might provoke little interest or concern within the context of this discussion if it were not for a short note related to the history of that muff that was provided when it was donated to the museum. The woman who donated the muff recalls that when she and her sister were young girls they had a pet rabbit that they loved a great deal. Their family fell on hard times at one point, and her parents decided that the pet rabbit would need to be sacrificed to feed them. Knowing that this would be an emotional blow to the girls, the father carefully skinned the rabbit, prepared the pelt, and the mother sewed the beautiful muff so the children would be able to keep their memory of their pet (personal recollection). Coppinger and Coppinger (2001) tell stories of how people they met around the world related to the dogs that lived and worked with them. If anything, what seems to be the case is that the poor may have a more "straightforward" relationship with their pets. There are few extras available for either the pet or the people, and they survive and suffer the ordeals of life together.

COMPANION ANIMAL DEMOGRAPHICS

Given the apparent interest in pets in the United States, it is remarkable to find that there are few authoritative data available regarding the numbers of pets and their distribution in **households (HHs)**. Companion animals are not counted during the decennial census. Few, if any, communities make an effort to enumerate the number of dogs, cats, or other animals found within their bounds. Even dog license data are notorious for their inaccuracy. (As a demonstration, ask your friends and classmates if they have a dog and if it is licensed.) Most of what we know comes from a variety of surveys that are conducted by different organizations associated with pets. The two most significant studies that are conducted on a regular basis come from the American Veterinary Medical Association (AVMA) and the American Pet Products Manufacturers Association (APPMA). Various other industry or interest groups may also conduct more limited surveys of the public.

Pet Ownership Surveys

The AVMA conducts an extensive survey of pet owners about every five years. It conducts the survey with the assistance of a professional survey company, National Family Opinion (NFO). National Family Opinion maintains a panel of representative HHs from across the country that is periodically questioned on a variety of topics and issues. This is a reasonable and cost-effective way to collect representative data. Based on a recent survey, it was estimated that 58.9% of American HHs owned a pet sometime during 2001 (AVMA, 2002). Of these, 46.9% considered their pet a member of the family, 50.9% thought of them as pets and companions, and just 2.2% considered the animals in their home property that they owned and cared for. Dog owners were most likely to consider their pets as family members (51%), followed by cat owners (46.1%), bird owners (47.4%), and horse owners (39.1%). In 72.8% of pet-owning HHs, women were the primary caretakers. A majority of pet-owning HHs (60%) had more than one pet, with nearly one-fifth (19.3%) having five or more. Both dogs and cats were found in 14.7% of the pet-owning HHs (Figure 1-3).

Using data from the Bureau of the Census (<http://www.census.gov>), it was possible to estimate the number of pets in the country by multiplying the number of HHs by the fraction of HHs that have dogs by the average number of dogs (1.6) found in each dog-owning HH. This resulted in an estimate of 61.6 million dogs living in homes in 2001. Similarly, it was possible to estimate that there were 70.8 million cats living in homes. While fewer homes had

FIGURE 1-3

Contrary to popular thought, dogs and cats can share a home (or bed!) quite comfortably. *(Courtesy Dr. Stephen Zawistowski)*

cats than dogs (36.1% vs. 31.6%), cat-owning HHs averaged more cats per HH (2.1 cats compared with 1.6 dogs).

Similar data for other species include the following:

- Fish
  - 6.1% of HHs
  - 49.3 million fish
  - 7.7 fish/HH
- Birds
  - 4.6% of HHs
  - 10.1 million birds
  - 2.1 birds/HH
- Horses
  - 1.7% of HHs (does not count horses on ranches, farms, or other equine operations or industries)
  - 5.1 million horses (kept as companions)
  - 2.9 horses/HH
- Rabbits
  - 1.7% of HHs
  - 4.8 million rabbits
  - 2.7 rabbits/HH
- Ferrets
  - 0.5% of HHs
  - 1 million ferrets
  - 2.1 ferrets/HH

The survey of the APPMA (2005) provides very similar data in terms of ownership patterns in American HHs. The APPMA reported that 63% of HHs owned a pet and the breakdown by species is comparable to that reported by the AVMA:

- Dog—39% of HHs
- Cat—34% of HHs
- Freshwater fish—13% of HHs
- Bird—6% of HHs

FIGURE 1-4

Growth in the pet product field is being driven by the introduction of a wide range of premium and luxury items, in addition to the products that meet the everyday needs of pet owners. *(Kelly Cunningham)*

- Small animal—5% of HHs
- Reptile—4% of HHs
- Horse—4% of HHs
- Saltwater fish—0.7% of HHs

Tracking previous versions of various survey reports shows that there has been a steady increase in the rate of pet ownership and increasing numbers of animals kept as pets. This is helping to drive a pet care industry that reached close to $36 billion in 2005 and $38.5 billion in 2006 (Figure 1-4). This exceeds the $20 billion toy industry and the $24 billion candy business (Guthrie, 2005). In fact, spending on pets is increasing faster than the pet population, having doubled between 1994 and 2005. Much of this growth is driven by consumers with annual incomes over $70,000 who buy an expanding range of high-end and luxury products for their pets (D'Aquila & Dillon, 2007).

Estimating Cat and Dog Birth Rates and Death Rates

As part of a study sponsored by the National Council on Pet Population Study and Policy (NCPPSP), New et al. (2004) worked with the AVMA during its 1996 survey of pet owners to estimate the **birth rates** and **death rates** of owned cats and dogs in the United States (see Chapter 4 for more about the NCPPSP). They surveyed a subsample of 7,399 HHs that had responded to the AVMA survey. They collected additional data from these HHs on the numbers of cats or dogs added or lost during the previous year, and manner of the addition or loss. With results similar to the AVMA demographic studies, along with data from the Bureau of the Census, New et al. were able to generate national estimates for a variety of parameters.

They were able to project that 23% of HHs had a cat leave sometime during the year, and 21% of HHs had a dog leave the home. This would have resulted in 11.3 million cats and 7.1 million dogs leaving homes during the year. The most common reason for the loss of dog or cat from the home was death. They estimated that 9 million dogs and cats died during the year for all reasons combined. A total of 4.9 million cats died, yielding a crude estimated death rate of 8.3/100 cats and 4.16 million dogs died with a crude estimated death rate of 7.9/100 dogs. In 8% of HHs it was reported that cats just died or were killed (3.4 million cats) in some way and 7% of HHs reported dogs dying or being killed (2.4 million dogs). Cats were reported euthanized by 5% of cat-owning HHs and 6% of dog-owning HHs reported at least one dog being euthanized. The next most common reason for a cat to leave a HH was to *disappear*. This was reported by 5% of cat-owning HHs (1.9 million cats), as compared with 2% of dog-owning HHs (0.6 million dogs). The discrepancy between cat and dog disappearances may be accounted for by different concepts of cat and dog ownership and caretaking. Cats are more frequently allowed to roam freely, and less effort may be made to find them if they do disappear. Dogs and cats were given away to friends, relatives, and others by 4% of both dog- and cat-owning HHs (2 million dogs and 2.9 million cats). About 1% of dog- and cat-owning HHs reported relinquishing animals to an animal shelter.

The most common source of new dogs and cats in a home was offspring of animals already living in the home. They estimated that 5% of cat-owning HHs had at least one litter during the year, and 2.6% of dog-owning HHs had at least one litter during the year. Over three times as many cat litters were unplanned as planned (3.6% HHs vs. 1.1% HHs). Dog litters were as likely to be planned as unplanned (1.2% HHs vs. 1.4% HHs). The vast majority of kittens projected born during the year, 83%, would have been from unplanned litters, accounting for an estimated 5.5 million kittens. Less than half of the puppies born, 43% or an estimated 2.6 million, were unplanned. Combining kittens from planned and unplanned litters results in a total of 6.6 million kittens and an estimated crude birth rate of 11.2/100 cats. The estimate for the total number of puppies born was 6 million for a crude birth rate of 11.4/100 dogs. Nearly half of the kittens born were given away, 2.9 million, while 1.9 million puppies were given away. It was much more common for puppies to be sold, 2.3 million, than for kittens to be sold, 0.2 million. After births in the home, the next most common way HHs acquired a new cat was by taking in a stray (6%), followed by getting one from a friend or neighbor (3%). It is possible that many of these strays were among the cats that were reported to have disappeared by other HHs. While dog HHs were more likely to get a dog from a breeder (2.7% HHs) than any other source, it did not account for as many new dogs, since litters born in a home typically provided a greater number of individuals. After breeders, the next most common source of a new dog for a HH was friends and relatives (2.4% HHs).

The New et al. (2004) research provides several important points to consider. The first is that the estimated crude birth rate for dogs and cats in HHs exceeds the estimated crude death rate in HHs, and is consistent with the observed population growth of both dogs and cats in HHs. A second finding is that a large part of the acquisition and disposition of dogs and cats by HHs occurs outside of any organized exchange system. This is consistent with data collected by the APPMA (2005, p. xxiii) showing that friends and relatives are common sources for both dogs and cats, along with breeders for dogs and taking in strays for cats. Many puppies and kittens are the result of unplanned litters, and many of these are then passed on to friends and relatives, and still others are taken in as strays. As a result, it can be difficult for many of the associated industries and professions to impart their messages on proper medical care, humane treatment, or product availability at the time of acquisition. Finally, these data support what has become a truism in the pet care field, *cats ain't dogs*. These data are consistent with a growing body of evidence that people acquire and care for cats differently than they do for dogs. This has important implications for industries hoping to sell products to pet owners and organizations and professionals that advocate for the humane care of animals.

DISCUSSION QUESTIONS

1. Which organizations conduct surveys about U.S. pet ownership on a regular basis?
2. What role did Europeans play in the establishment of companion animals in American society?
3. How have the species that are kept as companion animals changed over the centuries?
4. How much money do Americans spend on their pets? How did the pet supply industry change during the twentieth century?
5. Describe the role of hobbyists in the development of the aquarium trade.
6. How do the research findings of New et al. (2004) demonstrate the differences in how Americans perceive ownership of cats versus dogs? Why do you think these differences exist?

REFERENCES

American Pet Products Manufacturers Association. (2005). *2005–2006 APPMA national pet owners survey*. Greenwich, CT: APPMA.

American Veterinary Medical Association. (2002). *U.S. pet ownership & demographic source book*. Schaumberg, IL: AVMA.

Axelrod, H. R., & Schultz, L. P. (1990). *The handbook of tropical aquarium fishes*. Neptune, NJ: T.F.H. Publications.

Axelrod, H. R., Burgess, W. E., Pronek, N., Axelrod, G. S., & Boruchowitz, D. E. (1998). *Aquarium fishes of the world*. Neptune City, NJ: T. F. H. Publications.

Caras, R. (1996). *A perfect harmony*. New York: Simon and Schuster.

Coppinger, R., & Coppinger, L. (2001). *Dogs: A startling new understanding of canine origins, behavior, and evolution*. New York: Scribner.

Coren, S. (2005, Summer). Confidants to kings, consolers to queens. *Modern Dog*, pp. 59–63.

D'Aquila, J., & Dillon, M. (2007, February). Are you barking up the right investment tree? *Mercanti Chronicle*.

Diamond, J. (1999). *Guns, germs, and steel*. New York: W. W. Norton.

Grier, K. C. (2006). *Pets in America: A history*. Chapel Hill, NC: University of North Carolina Press.

Guthrie, D. (2005, February 14). Pets become bigger part of family budget. *Detroit News*.

Horn, J. C., & Meer, J. (1984, August). The pleasure of their company: A report on Psychology Today's survey on pets and people. *Psychology Today*, 52–67.

Kirkwood, K. (2005, July). Hail to the dog. *Dog Fancy*, pp. 16–20.

Mugford, R. A. (1980). The social significance of pet ownership. In S. A. Corson and E. O'Leary Corson, *Ethology and non-verbal communication in mental health*, pp. 111–112. Oxford: Pergamon.

New, J. C. Jr., Kelch, W. J., Hutchinson, J. M., Salman, M. D., King, M., Scarlett, J. M., et al. (2004). Birth and death rate estimates of cats and dogs in U.S. households and related factors. *Journal of Applied Animal Welfare Science, 7*(4), 229–241.

Pennisi, E. (2006). Fishing for common ground. *Science, 311*, 766–767.

Saunders, M. (1893). *Beautiful Joe*. Philadelphia: Charles H. Banes.

Saunders, M. (1908). *My pets: Real happenings in my aviary*. Philadelphia: The Griffith and Rowland Press.

Shaw, M., & Fisher, J. (1939). *Animals as friends and how to keep them*. New York: Didier.

Zappler, G., & Villiard, P. (1981). *A pet of your own*. Garden City, NY: Doubleday.

CHAPTER 2

Development of Companion Animals

A cat is a lion in a jungle of small bushes.

—Indian Proverb

KEY TERMS

domestication	evolution	fancy
phenotypes	selective breeding	kennel club
socialization	genome	affiliative
traits	mitochondrial DNA (mtDNA)	livestock
feral		shelf pets
trap-neuter-release (TNR)	nuclear DNA	pocket pets
neoteny	purebred	selfs
breeds	tame	rare breeds

Companion animals are, in general, domesticated species. That is, they have been modified in various ways from some wild ancestor to make them more appropriate for living in close proximity with humans. **Domestication** is a process that takes place over a period of many generations of a species. It is important to recognize that species itself is a dynamic term that describes the relative state of a group of organisms in relation to other groups of organisms. Species are not static. A species is constantly responding to the environmental pressures and other species that surround it. It is on an evolutionary trajectory that will allow it to survive within its environment, or if the changes are too great or sudden, it will become extinct. Domestication occurs when humans alter that evolutionary trajectory for some part of a species' population, and instead of continuing to adapt to the natural environment where it has been found, it begins to adapt to an environment created by and associated with humans. This process is both biological and cultural (Clutton-Brock, 1992). As the species changes, or evolves, its place and role in relation to humans will change. It is an iterative process as changes in biology and behavior suit it to fill new niches with humans, and this in turn will suggest still other opportunities.

Domestication and its results are one of the most visible, and at the same time, poorly understood biological events to impact a species. All species have not been domesticated in the same way, and different species may be at different stages of the domestication process. Several key elements are generally common to all types of domestication. The first is that the population of animals (or plants for that matter) must show a range in **phenotypes**. This phenotypic variation must be correlated with underlying genetic variation. It is important that at least some members of the population are reasonably tolerant of humans being nearby, and through experience demonstrate a greater level of **socialization** to people. Most of the significant domesticated species were derived from wild ancestors that lived in social groups, whether herds, packs, or flocks. Dogs, horses, pigs, and chickens are just a few examples. Domestic cats are a rare exception. Other than lions, wild felines do not spend significant time in social groups. Contact with other members of the species is typically limited to mating, and mother-offspring interactions. Domestication has helped to make cats more tolerant of one another.

While individual members of a species may be tamed, allowing or tolerating a high degree of close human proximity or contact, it will retain instinctive behavior patterns, and will not directly pass its tameness on to it young. One of the first stages of domestication is the selection, whether or not intended, of individuals that are easiest to approach and socialize, and least likely to show instinctive patterns of attack or fleeing.

When young, many species will have a brief time frame when they will socialize to conspecifics. This sensitive period may be just several hours long, or extend for a number of weeks. Immediately after hatching, geese and ducks will *imprint* on the first moving object that they see (Immelman, 1972). They will then follow this object or individual, and use it as model for social bonding and orientation. This would usually be their parents and would ensure that their primary orientation is towards a conspecific. These birds are precocial, leaving the nest soon after hatching so it is critical that they develop a bond with their parents quickly after birth. Canines will generally give birth in some sort of den. The pups are helpless at birth and the mother provides extensive care to her young. They socialize gradually, first to their mother, then to siblings, and then to conspecifics that may be part of a pack or social group. Another aspect of domestication may be the lengthening of this socialization period. The critical period for dogs is quite long, lasting from about 3 to 19 weeks after birth. These expanded socialization windows make it easier to raise an animal that will be calm and accepting of humans, and other animals. Wild animals when taken as young may be tamed by socializing them with people. This is an uncertain process, however, and when the animal reaches sexual and physical maturity may revert to instinctive behaviors and present a danger to people and other animals.

A second change is the enhancement of the **reproductive potential** of the species. Domestication typically included an economic motivation that usually required the production of more individual animals. The reproduction of many wild species is modulated by seasons, social structure and status, the availability of prey or food, and other factors. Domestic species, when compared to wild counterparts, may begin breeding at a younger age, more frequently, have larger numbers of young, and be non-selective in the acceptance of mates, accepting the mate provided by humans.

Once a species is easy to handle and breed, it is possible to identify and select animals that exhibit desirable **traits**, and breed them for a specific purpose or appearance. The extent of variation that we may see in a domestic species will be a function of both the genetic structure and variation that was present in the original population of wild ancestors, and mutations that may be identified, preserved, and used in breeding programs. On the other hand, breeding animals within a small population can result in inbreeding and the loss of genetic variation and an increased probability of inherited diseases. Over time, accumulated differences between the wild and domestic forms may preclude hybridization, cross breeding, or the production of viable, fertile offspring.

There are times when domestic animals have escaped or otherwise returned to living on their own in a free-living or wild state. The mustangs of the American West are descendents of horses brought to the New World by explorers and settlers. Pigeons, common in large numbers in both urban and rural areas, are the prolific progeny of ancestors brought to America as a ready source of food in the early colonies (Figure 2-1). Over time, descendents of these domestic species living in such a fashion will begin to re-adapt to a free-living or wild existence. Descendents of a domestic species living in this way are called **feral**. Among the most common changes is a reacquired wariness toward people. This is likely a function of both limited or non-existent opportunities to socialize when young, and eventual genetic changes in the population. Unfortunately, one of the changes resulting from previous domestication that

FIGURE 2-1

When the first Europeans came to live in the New World they brought their domestic animals along with them. Pigeons were brought to serve as food, and were kept in dove cotes to facilitate access to the young pigeons, or squabs. Some were quite elaborate, such as this one at Colonial Williamsburg, Virginia. *(Courtesy Dr. Stephen Zawistowski)*

tends to persist is the increased reproductive capacity. As a result, a high rate of reproduction, often coupled with limited natural predators, may result in a population with unchecked growth that may threaten the local environment. Efforts to deal with feral cat populations had typically relied on lethal control methods. **Trap-neuter-release (TNR)** programs are now becoming a more common approach (Slater 2002). Feral cats are caught using live traps, sterilized, often vaccinated for rabies and other diseases, and then released back in the area where they were captured. Caretakers will then ensure that the cats are provided with food, and often some type of shelter in severe weather. This is a compromise approach. Feral cats are difficult to tame and it may not be possible to place them in new homes. At the same time there is limited public support for euthanizing cats that are generally healthy. In addition, lethal control has not been highly effective controlling populations of free-roaming cats. Sterilization helps to control population growth, and can result in a slow decline in numbers through attrition by natural causes.

Another common result of domestication is **neoteny**, or maintaining juvenile physical and behavioral characteristics into maturity. Many **breeds** of dogs maintain the turn down/flop ears, rounded face, and shortened muzzles of puppies. Adult dogs may also exhibit the solicitation behaviors usually seen in puppies. Both wolf and dog puppies will lick the muzzle of adult dogs in greeting them. It is not uncommon for adult dogs to offer their owners "kisses" when they greet them, or when they want to solicit food or attention. Domestic cats will retain the predatory play behavior observed in kittens of wild felines. In wild kittens, this play will develop into highly efficient predatory sequences. In domestic cats this play may or may not develop into the ability to catch and kill prey. It will be strongly dependent on early experiences as a kitten (Fitzgerald & Turner, 2000; Leyhausen, 1979).

In summary, domestication will generally have the following impact on a species:

- Selection, direct or indirect, for tameness, or tolerance for the presence of people
- Changes in reproduction
  - Earlier sexual maturity
  - Breed more frequently
  - Breed indiscriminately
  - Larger numbers of young
  - Neoteny of physical and behavioral traits
  - Overall changes in size—usually smaller domesticated versions, though "giants" may also be found

DOMESTICATION/DEVELOPMENT OF COMPANION ANIMALS

Dogs

Dogs are the most diverse single species on the planet. No other species shows the extraordinary range in size of the dogs. The size will range from the 0.5 kg Chihuahua to the 80 kg Great Dane. One could hardly imagine that the nearly

FIGURE 2-2

Dogs are the most widely variable species on earth. They can range in size from the (A) tiny 1–3 kg Chihuahua *(Courtesy ASPCA)* to the (B) enormous 45–60 kg Great Dane *(Artville)*. (C) The almost hairless Chinese crested *(www.fotolia.com)* looks to be a separate species from the (D) heavily corded puli *(Artville)*.

hairless hide of the Chinese crested and the long, corded coat of the puli could come from two individuals of the same species (Figure 2-2). There are probably several hundred different breeds and types of dogs worldwide, 152 of which are recognized by the **American Kennel Club** (Crowely & Adelman, 1998), and in one way or another they tend to fall between these different extremes. The variation in dog behaviors may be nearly as great as their physical differences. Some dogs point, others retrieve, still others herd and some guard. The only constant may be that whatever the look and behavior, someone, somewhere liked it, wanted it, and chose to breed dogs in a way to either develop or maintain the trait.

There has been a recent revolution in the study of dog **evolution** (Zawistowski & Patronek, 2004). New methods in molecular genetics have provided a vast new set of data revealing some unexpected surprises. When combined with traditional studies of fossils and archaeological evidence we have a very complicated story that we are only starting to understand. The dog has long been considered the first species domesticated by humans (Clutton-Brock, 1995). The wolf was thought to be the traditional ancestor of the dog. However, given the wide physical and behavioral variation seen in the dog, it was not uncommon to believe that other wild canids must have been involved in the development of the dog. Darwin studied a variety of domestic species and the role that **selective breeding** can play in the evolution of a species. He bred several varieties of domestic pigeons and frequently attended animal shows and talked to breeders. Based on his observations, Darwin did not believe that all dogs could have descended from a single ancestral species (Darwin, 1859/1979). Konrad Lorenz (1954) also believed that several different wild canids, including the jackal, gave rise to the many different breeds of dogs. Lorenz would later retract his statements regarding jackals and their role in dog evolution (Lorenz, 1975). He did believe that the northern, spitz breeds were those most closely related to wolves. In contrast to Darwin, who felt that the study of domestic animals could contribute to a better understanding of evolution and natural selection, Lorenz was among the modern scientists who disdained the value of studying domestic animals (Serpell, 1995). This attitude is changing, and the study of dog biology and behavior, combined with information on its **genome** may provide important insights into gene regulation and function (Ostrander & Kruglyak, 2000; Sutter & Ostrander, 2004).

Canine Archaeology. Archaeological evidence related to the origin of the dog can be difficult to interpret. Fossil bone fragments that came from a wolf or a dog may be very similar at the early stages of evolutionary separation. This is especially true in those areas where the wolf species of the time were small in size, closer to the size of the early dogs (Harrison, 1973; Olsen, 1985). There are also limited numbers of clearly identified dog fossils available for the earliest period when the separation from wolves may have been happening. At the same time, the animals giving rise to those fossils may have shown important behavioral differences. Since behavior does not fossilize, the context and locations in which remains are found are significant. Remains found in proximity to other human artifacts would suggest that the animal was associated with people. These data have in general placed the domestication of dogs in an era about 14,000 BP (Beneke, 1987; Clutton-Brock, 1995), or late Paleolithic period. An important find that also suggested a significant social relationship with dogs was of a puppy skeleton that had been buried with a human 12,000 BP (Davis & Valla, 1978). Found in the upper Jordan Valley of Israel, the canine skeleton was clearly that of dog, a 4- to 5-month-old puppy, and it had been buried with an elderly human. Other artifacts found at the site suggested that the location was inhabited by humans just making the transition from hunter-gathering to agriculture. Similar burial sites have been found in North America dating back over 8,000 years (Figure 2-3). Morey (2006) suggests that evaluation of the evidence associated with burial sites provides significant information about the timing of canine domestication and the special social relationship that early dogs had with humans. The early date for burial in Central North America (Illinois) suggests that dogs must have come with some of the earliest

FIGURE 2-3

This dog was carefully buried around 8,000 years ago in North America, suggesting that dogs came to the New World with the earliest humans to cross the land bridge from Asia. *(Courtesy of William S. Webb Museum of Anthropology, University of Kentucky)*

humans to settle North America, crossing the land bridge over the Bering Strait. Wherever and whenever dogs were domesticated, they quickly spread as humans moved throughout the world.

Canine Genetics and Evolution. Modern genetic analyses have provided a new window into the domestication of the dog and its derivation from wild canids. Taxonomists had long considered the dog to be a species closely related to the various wolf subspecies (*Canis lupus*), but clearly separate, and were classified as *Canis familiaris*. In 1993, based on the resemblance of genetic, morphological, and behavioral traits, the domestic dog was reclassified as a subspecies of wolf, *Canis lupus familiaris* (Wozencraft, 1993). At the same time that genetic evidence can be used to verify the relation of dogs to wolves, it has been used to examine the evolutionary history of when the dog lineage separated from that of wolves.

Using genetic data to trace relationships between different species and populations, and estimating the time frame for the separation of those groups into reproductively isolated populations requires two stages. The first requires the ability to identify and sequence homologous DNA sequences found in the genomes of each population. **Homologous sequences** are those that are similar in form and function and are assumed to have a common evolutionary history. As two populations separate, random changes or substitutions of nucleotide bases during replication will accumulate in the genomes of each population.

You can trace the lineage of a population by examining the number and order of these changes or substitutions. The fewer differences between two populations or samples, the more closely related they are. As the length of time that two populations are reproductively separated from one another increases, the number of nucleotide differences will tend to increase. If you are examining a large number of samples, this analysis rapidly grows in complexity. Genetic taxonomists have developed sophisticated statistical methods to compare substitution patterns between different samples, and determine probable relationships between the groups (for more information on molecular taxonomy see Page and Holmes, 2002).

Analysis of nucleotide substitutions can also permit estimation of time frames for the separation of two populations. It requires estimating a molecular clock. A simple description of the process requires making an estimate of how frequently nucleotide substitutions occur. If you assume that a substitution happens on average once every 5,000 years, then if two populations differ by five nucleotide substitutions in the DNA sequence being analyzed, you can estimate that the populations separated 25,000 years in the past (5,000 years multiplied by 5 substitutions). Use of this method requires a substantial number of assumptions, not the least of which is that the substitution rate has remained constant, and that your original estimate for the substitution or evolutionary rate was reasonably accurate. The DNA sequences used in an analysis may be from either nuclear-coded genes or **mitochondrial DNA (mtDNA)**. **Nuclear DNA** undergoes recombination during sexual reproduction. This can introduce additional complexity into the analysis. Mitochondria are cellular organelles that contain DNA independent of the DNA found in the nucleus. This DNA will code for proteins used in the metabolic functions that occur within the mitochondria. During meiotic cell division, the mitochondria, which are found in the cytoplasm, will be shared between the resulting daughter cells. Later, during fertilization of an egg cell, the nuclear DNA of the sperm is combined with the nuclear DNA of the egg. However, the sperm do not contribute cytoplasm or mitochondria to the resulting zygote. As a result, mtDNA can be traced along matrilineal lines, from mother to daughter.

Based on an analysis of mtDNA, Vilà et al. (1997) published a remarkable study adding a new twist to our thoughts on the evolution of dogs. They analyzed mtDNA sequences from 162 wolves from twenty-seven different locations around the world, 140 dogs representing 67 breeds and 5 crossbreeds, along with 5 coyotes, 2 golden jackals, 2 black-backed jackals, and 8 Simien jackals. Their resulting analysis confirmed the expectation that wolves were the ancestors of dogs. There were much greater similarities between wolf and dog mtDNA than between dog and either coyote or jackal mtDNA. There was extensive genetic diversity seen in the dogs and the authors indicate that this is likely due to domestication events from different female lineages of wolves and cross breeding between dogs and wolves multiple times during the early stages of divergence. More striking, however, was their suggestion that the wolf-dog separation may have happened as long as 135,000 years ago, which is substantially earlier than the fossil record has indicated. They postulate that the genetic divergence between dogs and wolves may have happened without substantial morphological change and that the changes seen in the fossil record are associated with a time when humans were transitioning from a hunter-gatherer existence to sedentary agrarian practices. Tsuda et al. (1997) also examined mtDNA of wolves and dogs and support the contention that dogs are descended from wolves. They also support the concept that domestication of the dog flowed from four or more female wolves, or that once domesticated, surviving dog lineages interbred multiple times with wolves.

Savolainen et al. also analyzed mtDNA sequences to evaluate the evolutionary origins of the domestic dog. They sampled 654 domestic dogs from Europe, Asia, Africa, and Arctic America along with 38 Eurasian wolves. Their results indicate that East Asian wolves were the most likely ancestors of dogs, about 15,000 years BP (2002). These data support an earlier proposition, based on fossil evidence, by Olsen and Olsen (1977), that the Chinese wolf was the ancestor of the dog. The evidence also suggests that domestication of dogs by humans in this region may have occurred multiple times.

While the archaeological evidence indicates that dogs were found in the New World by 8,000 to 9,000 years BP (Morey, 2006), the question is whether any dogs were domesticated in the New World, or were they brought over during human migration that occurred during the late Pleistocene era (Foster et al., 1996). Leonard et al. (2002) extracted mtDNA from the remains of 37 dog specimens found in the New World, deposited before the arrival of Columbus from sites in Mexico, Peru, and Bolivia. In addition, they examined remains from eleven dogs found in Alaska, but predating the arrival of European explorers in that region. They compared the nucleotide sequences they obtained to those of previously analyzed samples from wolves and dogs. Their analysis indicates that the these ancient New World dogs were not derived from New World wolves, but were in fact descended from Old World wolves. This would indicate that the humans who crossed the Bering Strait must have brought dogs with them. The level of genetic diversity observed in the ancient New World dog remains showed that these dogs must have represented a sample from a large, well-mixed, and diverse population of dogs in the Old World at that time. If the archaeological record for differentiation of the dog were accepted as being about 14,000 BP, this in turn would also suggest that dogs must have been common companions for humans as they migrated. Dogs would have moved from their origin in East Asia to the other side of the world in just a few thousand years.

While mtDNA studies have proved useful in tracing the dog's ancestry from wolves, it is less helpful for the study of different breeds of dogs. Most breeds of dogs in their current form have existed for only several hundred years (Crowley & Adelman, 1998; Wilcox & Walkowicz, 1995), and mtDNA does not evolve fast enough to generate the variation that would be required to distinguish between different breeds. Parker et al. (2004) analyzed nuclear microsatellites combined with phylogenetic analysis and genetic clustering techniques to elucidate the genetic relationships among 85 different breeds of dogs. They sampled from 414 **purebred** dogs and found that the variation among breeds accounts for 27% of the total genetic variation in the population studied. This is high compared to other domestic breeds and confirms the strong genetic isolation between breeds. Their cluster analysis detected four separate groupings (Figure 2-4). The first group was thought to be most ancient in origin since it also included a sample of wolves that were part of the analysis. This group may be the best

FIGURE 2-4

An analysis of canine microsatellite DNA results in four major groupings or clades of dog breeds. These clades are generally consistent with our knowledge of breed histories and similarities. *(From Fig. 3A, B from Parker et al., Science, 304, 1160–1164, 2004)*

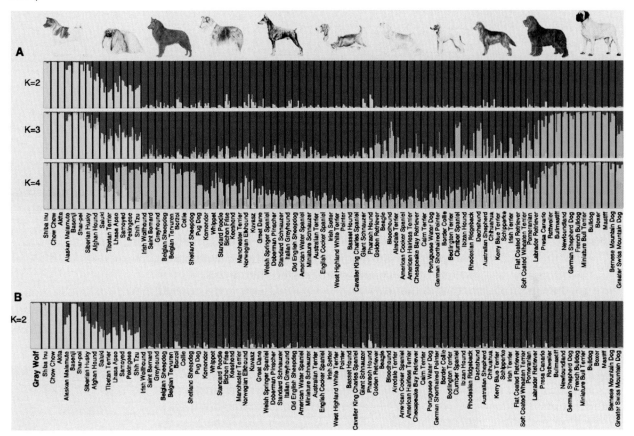

living representation of the ancient canine gene pool. It included the following breeds:

- Basenji (Africa)
- Saluki and Afghan (Middle East)
- Tibetan terrier and Lhasa apso (Tibet)
- Chow chow, Pekingese, shar-pei, and shih tzu (China)
- Akita and Shiba Inu (Japan)
- Alaskan malamute, Siberian husky, and Samoyed (Arctic)

The second group included the following:

- Mastiff, bulldog, boxer, and related breeds
- Rottweiler, Newfoundland, and Bernese mountain dog
- German shepherd dog

The third group included the following:
- Belgian sheepdog, Belgian Tervuren, Collie, Shetland sheepdog (herding)
- Irish wolfhound, greyhound, borzoi
- Saint Bernard

And a fourth group was composed of the following:
- Scent hounds
- Terriers
- Spaniels
- Pointers
- Retrievers

In many ways, the genetic clustering is consistent with our general understanding of canine breed history, and the morphological similarities observed. Wayne and Ostrander (1999) have suggested that some ancient breeds such as the dingo, New Guinea singing dog, greyhound, and mastiff may have developed in association with isolated human populations, where there was limited opportunity for them to interbreed with other types of dogs until travel become more common and easier (Figure 2-5).

An interesting result of the analysis is that several breeds generally considered among the oldest breeds of dogs, the Pharaoh hound and Ibizan hound, were not included in the first grouping of oldest breeds. While they may appear

FIGURE 2-5

Scientists are still investigating the unique qualities of the New Guinea singing dog. *(Courtesy Alice Moon-Fanelli, PhD, CAAB)*

to be quite similar in appearance to the dogs depicted in Egyptian tombs, their genomes indicate that the breed must have been re-created by breeders at some more recent time by combining other breeds of dogs. Previously, genetic analysis had shown that the Xoloitzcuintli, or Mexican hairless dog, long thought to be of ancient origin, and possibly derived from New World wolves, was in fact derived from dogs originating in the Old World (Vilà, Maldonado, & Wayne, 1999). Perhaps more surprising was that the "Xolo" was not related to the Chinese crested dog that it resembles physically. Continued analyses of this type will likely provide additional insights and other surprises in our understanding of dogs.

Canine Domestication. The question is how can we reconcile the archaeological findings with the molecular genetic data and come up with a coherent picture of how wolves became dogs. A traditional view was that Man the Hunter somehow joined up with Wolf the Hunter. It was quite a romantic narrative as the two great social predators stalked large game together. Wolf puppies would sometimes be taken into caves as "pets" and over time these wolves became more and more **tame**. Coppinger and Coppinger provide the following simplified version of this process:

- Capture a wolf
- Tame the wolf
- Train the wolf
- Breed the wolf to other tame, trained wolves
- And, presto! A domesticated dog (2001, p. 57)

The trouble with the story is that it did not make much sense. Well-educated experts, working with extensive facilities, have their hands full trying to tame wolf puppies born in captivity (Klinghammer & Goodman, 1987). The puppies must be taken from the litter before their eyes are opened, and then bottle-fed by humans. Even then, when they have matured, they are simply more tolerant of people. They are in no way "dogs." When stressed or challenged, they will revert to their instinctive response patterns with potentially devastating consequences for a human within reach. These patterns were only papered over with a thin veneer of tameness. It is difficult to conceive that our human ancestors, at a time when they were just starting to figure out projectile weapons, would have the wherewithal to capture not just one wolf puppy and raise it to sexual maturity, but many. Selection for tameness would require a number of individuals to choose from for breeding stock. They would then pair the ones they wanted to mate, take the puppies from that litter, and several others, and do it again (and again, and again). It is more likely that any wild wolf puppies that were caught stayed around until they could escape, or the first time one snarled and snapped and then ended up as that evening's dinner entrée.

The alternative process proposed by Coppinger and Coppinger (2001) goes as follows:

- People create a new niche, the village.
- Some wolves invade the new niche and gain access to a new food source.

- Those wolves that can use the new niche are genetically predisposed to show less "flight distance" than those that cannot.
- Those "tamer" wolves gain selective advantage in the new niche over the wilder ones.

This model is consistent with the archaeological evidence placing domestication of the dog at the about the same time that humans were transitioning to an agrarian lifestyle, with established villages. Villages would bring with them that unique human contribution to the environment—the dump. Leftover food, spoiled food, the bones, and the offal of animals killed and eaten would probably be tossed into a convenient location. Some wolves would come to feed at these locations. Wild wolves would flee every time someone approached the dump. Some would run a shorter distance than others; some would wait longer before they ran. These wolves would have first access to any new donations to the dump. All wild animals must acquire enough resources to reach sexual maturity, mate, and leave offspring if they are to be successful. Individuals able to exploit a resource more efficiently than others are at a distinct advantage. Those wolves most efficient at exploiting the food in the dump would not need to risk injury hunting. Over time, the flight distances for these wolves would become shorter and shorter, as individuals within this population become more and more tame. A significant part of domestication in this scenario would have taken place through natural selection.

An experiment that tested this basic scenario was carried out in Russia over a 40-year period with silver foxes, *Vulpes vulpes* (Trut, 1999). Dmitry Belyaev and colleagues began their work in 1959 by defining very clear selection criteria for the foxes that they would breed in succeeding generations. They started with 30 male foxes and 100 vixens. Since these foxes had already been bred in captivity for the fur trade they were already more tame than truly wild foxes, as they are adapted to being caged and fed by humans. When the first group of fox pups reached sexual maturity at seven to eight months of age they were tested for tameness. Class III foxes would flee at the approach of the experimenter, or attempt to bite when stroked or fed. Class II foxes let themselves be petted or handled, but did not show any friendly overtures to the experimenters. Class I foxes wagged their tails when petted or fed and whined for attention. The top 4 to 5 percent of male foxes and top 20 percent of vixens were chosen for breeding. By the time they reached the sixth selected generation the experimenters needed to add another class, IE, or "domesticated elite." These foxes were eager to solicit the attention of humans, and would run to greet them. While just 18 percent of the foxes were IE in the tenth generation, this had just about doubled to 35 percent in the twentieth generation, and by the time the research was reported in 1999, 70–80 percent of the population was classified as elite. These results are not surprising. There are numerous examples of successful selection for a wide variety of behaviors in many different species. In addition to the response to selection for behavioral tameness, they also observed a suite of correlated morphological changes. The foxes began to show piebald color patterns, floppy ears, and carried their tails curled up as adults, not down the way wild canid adults would normally carry their tails. They have also measured reduced adrenal cortex function, and changes in the shape of the skull. Finally, after 40 years, they have also seen some small changes in reproductive behavior. Wild foxes are strict seasonal breeders, reaching sexual maturity at about eight months of age. The selected vixens are reaching sexual maturity at about one month earlier, and are having litters that average about one pup larger. A few have even started to cycle twice per year.

What is remarkable about all of these physical changes is that these are the same general changes one would have predicted as part of a domestication scenario. However, the experimenters did not select for any physical traits, only for the operationally defined trait of tameness. If we go back to the wolves at the village garbage dump, would we see a similar set of correlated responses to "natural selection" for tameness among the wolf population feeding there? Some of the physical markers such as the tail curled up and floppy ears, would have given the humans a handy way to identify those wolves most likely to be friendly. It is a question that will deserve additional consideration in the years to come.

Once humans were able to approach and handle a domesticated wolf, it would have been possible to evaluate desirable characters, and begin selective breeding for a particular look or function. Fondon and Garner (2004) postulate that the canine genome is uniquely developed to provide high rates of genetic variation through gene-associated tandem repeat expansions and contractions, resulting in a major source of phenotypic variation. They examined a number of genes associated with developmental regulation and found that in most of the cases, allelic variations were a function of incremental differences in length, generally corresponding to two or three repeats of the length of the most common allele for that locus. These sorts of alterations at genes associated with the control of development events could have wide-ranging effects on the morphological development of dogs and could have accounted for the rapid evolution of highly diverse breeds shortly after the domestication of the dog from the wolf.

Another interesting view of dog evolution by McGhee (2002) brings us back full circle to the initial premise of Caras (1996) that the domestication process was a form of co-evolution of animals and humans. McGhee suggests that if the oldest dates of over 100,000 BP for separation of wolves and dogs were true, then those early "dogs" would have lived in concert with humans just barely more advanced socially than they were. These early human groups, only recently evolved from *Homo erectus*, probably lived in small, mobile family bands, organized in a way similar to modern chimpanzee bands. These wolf and human family groups may have coalesced among many of the same required resources, water, and food. These mobile bands of humans would not have left food or debris in any quantity in a single location as they hunted and gathered what they could. While these wolf-dogs may have accumulated the genetic signatures we can detect with modern analytic methods, they likely did not manifest anatomical changes that could be reflected in the fossil record. Once small groups of humans began to gather and stay in a particular location for extended periods of time, beginning the transition from hunter-gatherer to farmer-villager, this population of wolves would have been pre-adapted to begin the rapid changes in behavior and morphology that would allow them to exploit this new human-generated niche. McGhee raises the question of whether, in addition to the influence humans obviously had on the evolution of some wolves to become dogs, wolves influenced our own evolution. Did the human hunter bands that might be associated with wolf packs enjoy an advantage? Were these the groups that were most likely to thrive and go on to form the first villages, where wolves truly became dogs? He suggests that some elements of this story may be recorded in human folklore. There are numerous stories of wolves nurturing and caring for humans. Romulus and Remus, the founders of Rome were supposedly fostered by a she-wolf. He also sees that our frequent vilification and persecution of wolves by humans is part of this same close relationship. Our cultural depictions of the wolf are seldom neutral, being either villain or savior, and may be rooted in a history between humans and wolves that predates history and culture.

Social History of Dogs. If we accept that dogs were both genetically and physically differentiated from wolves by 12,000–14,000 year ago, it would be another 10,000 years until the next major steps in the development in dogs took place. Until this time dogs were pretty much a generic lot, probably very similar in size and appearance to the pariah dogs still found around the world today. In addition to scavenging around human villages they probably went along on hunting forays, helping to raise game, chase, and either bring down or locate wounded animals. They also would have proved useful helping to guard the next animals domesticated by humans, sheep and goats. Their natural territoriality and social and protective behaviors would have lent themselves to these tasks. In was not until 5,000–6,000 years ago we see the first evidence for development of dogs for specific purposes (Rogers, 2005). The simple village of humans has become far more sophisticated and different people might have different roles or tasks. Hierarchies and social structures had developed to the point that there was now a ruling class. This upper class no longer needed to raise or catch their own food. They depended on others for the basic needs. Hunting now became a form of recreation and the implements of the hunt were a sign of wealth and power. Among these signs were dogs, specially bred and raised for the hunt. The first of these were dogs very much like the salukis of today. Evidence of these dogs is found in the art and other artifacts of ancient Sumeria and Mesopotamia. These first *breeds* were long and lean with deep chests. They were sight hounds and beautifully adapted for chasing game on the open deserts. The hunters, mostly nobility, would ride out to the hunt in chariots with dogs kept on leashes until game was spotted. Once small antelope or other game was spotted the dogs would be let loose to chase and bring down the quarry.

The next type of dog to appear was the large, heavy-boned mastiff. There is evidence that mastiffs may have appeared independently in both northern India or Tibet and somewhere in the Mediterranean. Five thousand years ago the Island of Molossis, near Greece, was known for such dogs and for some time these large dogs were called Molossians. They were used as guards, hunting large dangerous game such as lions, and also for warfare. These war dogs might be outfitted with armor, and have a collar of blades. As they ran slashing an enemy's lines they would wound and cripple men and horses. At other times they were outfitted with torches and the flames would frighten horses and start brush and other materials on fire. By 4,000 years ago, the Egyptians had their own variety of sight hound, very similar to present day greyhounds. They also had mastiffs and here we now see the first small dogs. The Egyptians also memorialized their dogs in their art, and as part of their religion (Wilcox & Walkowicz, 1995). Anubis, the Egyptian god of the dead, is depicted with the body of a man and the head of a dog. It was Anubis who would gather the dead into his arms and transport them to the next world. The Egyptians were highly dependent on the annual flooding of the Nile River for continual renewal of their fertile croplands, the star that would presage this event each year was named Sirius, the Dog Star. The appearance of Sirius would forewarn the shepherds to move their flocks to higher ground.

The next type of dog to develop was the scent hound. These hounds were derived from the mastiffs. Some of the lighter, quicker forms were selected to trail game by scent. The loose skin and floppy ears of hounds reflect this heritage. The Greeks had highly developed hounds for hunting which they called Laconians. The Greeks respected the loyalty and honesty of dogs and Homer highlights this as part of the climax of his epic *Odyssey* (Rogers, 2005). When Odysseus finally makes it back to Ithaca after the exceptional ordeal that he and his men endured in the return from Troy, he finds that his home has been taken over by interlopers. Bedraggled and disheveled, he is not recognized by the members of his household. It is Argus, his loyal hound now old and discarded on a dung heap, that recognizes his master. Barely able to move, Argus shows his happiness in the return of Odysseus silently, before passing away. Odysseus, who would soon slaughter the interlopers who sought his kingdom, wealth, and hand of his wife, stifled a tear in gratitude for such love and loyalty. Unlike the Egyptians, the Greeks dreaded the dog days of summer signaled by the rise of Sirius. For them, these days were times of fever, sunstroke, and drought. They believed that dogs went mad, and a dog bite at this time would result in rabies.

The Romans, of course, traced the foundation of their culture to the twins Romulus and Remus, who were suckled by a she-wolf. They began to divide dogs into various groups depending on their use. They recognized the following common categories for dogs:

- House
- Shepherd
- Sporting
- Pugnacious or war
- Dogs that ran by scent
- Dogs that ran on sight (Wilcox & Walkowicz, 1995).

The Romans were particularly fond of their war dogs and used them not only in combat, but in the Coliseum as well. These dogs would battle one another, men, or wild animals such as lions and bears. When Julius Caesar invaded Britain, he brought along a great number of war dogs. Once there, he was impressed with the large, aggressive dogs that the local inhabitants possessed and also brought to battle. When the Romans found especially large and powerful dogs on an island in the Mediterranean Sea, they named the island after the dogs—the Canis or what became the Canary Islands. Later, the small songbirds found on the islands would be called canaries. The Romans were apparently fond of small house dogs as well, and kept miniature or toy dogs, among the first being the Maltese. The remains of many of different sizes were found in the ruins of Pompeii. Among these were the skeleton of an elderly person with their arms wrapped around the skeleton of what was likely a cherished pet.

During the Middle Ages hunting became very sophisticated as a pastime. It was closely linked to feudal lords and the ladies and knights of their courts. Many new breeds and types of dogs were developed for hunting specific types of game. Hunting might be done on horseback or afoot. Among the most popular animals to hunt were various deer and boar. Boar were a particular challenge since they might seek cover in thickets and were capable of doing significant damage to hound, horse, and huntsman with their tusks. The huntsman went afield with lance and spear, and hunting was pursued to develop the skills with arms and horseback needed for battle. Small game such as rabbits and hares might be taken by coursing with sight hounds. It was not uncommon to hunt with both dogs and falcons. The dogs would drive the game into the open where the falcon could stoop on it. Birds could also be taken in this way. By 1486, Juliana Barnes, the prioress of Sopwell Nunnery listed the following breeds in the *Boke of St. Albans*:

- Greyhound
- Bastard
- Mengrell
- Mastif
- Lemor
- Raches
- Kenettys
- Terroures
- Butcher's Hounds
- Dunghill Dogges
- Tyrndeltaylles
- Prycheryd Currys
- Small Ladyes Poppees That Bere Awayethe the Flees (Wilcox & Walkowicz, 1995).

These breeds included many dogs that would be familiar to us today: greyhounds, mastiffs, small and large scent hounds, terriers, bulldogs, and mongrels. The Small Ladyes Poppees were lap dogs, and in addition to companionship, and keeping someone warm in the drafty castles, were presumed to attract the fleas common to everyone to jump off the ladies and onto them. The next type of dog developed during this era was the spaniel. It was a bird specialist and would locate and then either hold the birds in place to be netted, or flushed for the falcons.

While the many hunting dogs, guard dogs, and lap dogs of the wealthy might be pampered, the dogs of the poor serfs shared their limited means and circumstances. Terriers were a particular help as they were small, aggressive dogs that needed little upkeep and were deadly efficient in ridding a field of badgers or barn of rats. A particularly onerous treatment was reserved for the dogs belonging to any peasants that lived near the protected estates and hunting lands of the lord. While it was accepted that dogs were needed to guard flocks and perform other tasks, the lords were so jealous of their hunting and game, steps were taken to ensure that peasant dogs would be unable to pursue or kill game. It was therefore not uncommon for the tendons in the hind legs to be cut, or the toes on the front feet. Thus hobbled, these dogs could limp about through their daily tasks, but were unable to threaten the royal deer. Peasants could not own hunting dogs. However, poachers developed their own dog, the lurcher. A cross between a greyhound and either a terrier or sheepdog, the lurcher was quick, silent, deadly, and efficient in its pursuit of small game that a poacher depended on. Lurchers are still bred and used for coursing hare and rabbit.

As the great era of European exploration of the world began at the end of the fifteenth century and on through the next several centuries, their dogs came with them. The Spanish brought fierce mastiffs with them to the New World and used them to intimidate the natives. On their return home, in addition to gold, spices, and other riches, they brought back some of the dogs they found on their travels. This resulted in additional new breeds being introduced to Europe, and their crosses with the existing breeds led to a proliferation of additional new breeds being developed.

The current culture of purebred dogs, the dog **fancy**, and the formal competitions for dogs developed during the 1800s. As the role of royalty and hereditary nobles began to decline in social influence, the growing wealth of merchants, financiers, and industrial magnates began to dominate. Among the many habits of the upper classes that they accepted and continued was an interest in dogs. Owning and breeding purebred dogs was a sign of refinement and a respected pastime for a Victorian gentleman. Informal shows, competitions, and exhibits were held to demonstrate the qualities of the top dogs. In 1859, the first formal show was held in Newcastle, England. It was sponsored by a gun maker, and was limited to bird dogs. The winning exhibitors were awarded guns as prizes by the sponsor. The Westminster Kennel Club held their first show at Madison Square Garden in New York in 1877. Henry Bergh, founder of **The American Society for the Prevention of Cruelty to Animals (ASPCA)**, was keynote speaker (Figure 2-6). He praised the owners for their dedication to dogs and the dogs for their devotion to humans: "They aim at the improvement of the race of animals which you are invited here tonight to inspect. . . . He is dedicated to his master, adopts his manners, defends his property, attaches himself to him until death, and all this springs not from necessity or restraint, but simple gratitude and true friendship." Bergh also called upon those in attendance to see to the welfare of all dogs, whether pedigreed or not. Accounts of the event suggest that few people may have heard Bergh's remarks or those of the other speakers as they were overwhelmed by the din of barking and howling canines. Nonetheless, the **kennel club** responded to Bergh by donating proceeds from the show to the ASPCA to provide care for stray dogs and those belonging to the poor and indigent (ASPCA Archives, 1877 press clips).

Much like the feudal lords who preceded them in the care of dogs, this developing class of dog fanciers could and would pamper their dogs. However, as in times past, dogs still suffered the lot of their masters. While dogs belonging

FIGURE 2-6

Entry badge for the first major dog show in America, held in New York in May 1877. Henry Bergh, founder of the ASPCA gave the keynote address, and proceeds from the show were donated for the care of homeless dogs. *(Courtesy ASPCA)*

to the wealthy might have special beds, and choice meals, dogs belonging to the poor needed to make do with what they might scavenge and scrounge. Instead of living in villages on the lands of a feudal lord, the poor were now often found crowded into cities. Their dogs were not needed to help with herding or guarding as many of these people worked in factories or manufacturing establishments. Others might be small vendors selling vegetables, odds and ends, or collecting rags. Those too poor to afford a horse to pull their carts would harness a dog to their carts. In many cases at the end of the day, the dog would be let loose to find his own food by scavenging through the streets and gutters. The wretched condition of these dogs was one concern of the early humane groups working to prevent cruelty to animals at this time (ASPCA Archives, Buffet notes; also see Chapter 2.4). There were small dogs called *turnspits* that trotted within wheels that turned the gears of rotisseries in shops and restaurants. Other dogs were used in the fighting pit against other dogs, wild animals, or rats. Still uncounted numbers wandered the streets where they harassed people and horses, and suffered. Every so often, when rabies might appear in a city's stray dogs, hundreds or thousands might be rounded up and killed by drowning, clubbing, or starvation.

During this same era dogs would be glorified for many of the same traits that endear them to people today: loyalty, courage, and honesty. One of the most famous stories from the early part of this era is that of Greyfriars Bobby. Bobby's master was a market policeman in Edinburgh. When his master died and was buried at Greyfriars Churchyard, Bobby did not go on to live with another master. Instead, for fourteen years he lived around the churchyard, surviving on the handouts of shopkeepers and townsfolk. Journalists and authors of the age wrote heartwarming stories about Bobby's love and loyalty to his master. He was eventually honored with a bronze statue in 1873 (Rogers, 2005). Loyalty has remained a popular theme in literature about dogs, and in 2004, two stories, one nonfiction and one fiction, were published for children about Hachiko (Newman, 2004; Turner, 2004). This dog belonged to a Japanese college professor who took a train to work each day. His dog, Hachiko, would follow him to the train station in the morning and then return home. In the afternoon, he returned to the station to wait for the professor. The professor died suddenly at work one day, and Hachiko loyally returned to the station each day to await the professor's return. This went on for almost ten years until Hachiko died. He is now remembered with a statue in the Shibuya train station.

The kennel clubs that formed sought to highlight the refined nature of dogs, those that were the very best representatives of their type. Purebred dogs were typically the passion of individuals who kept their own packs of hounds, or bird dogs, for hunting or some other type of dog that held a particular interest for them. They would breed and maintain their own records or studbooks. These records might be shared with other enthusiasts interested in the same type of dogs. Most of all, they bred for performance, doing the job that the breed was originally developed to perform. Each breeder would strive for some uniformity in size and appearance. This was particularly important in dogs that would work in packs, where the efficiency and aesthetics of the pack were dependent upon uniform size and speed, and even matching the color patterns and harmonizing the voices in the pack as it trailed game (Black, 1949). These breeders would outcross or breed their dogs to other lines or varieties of the same breed, or sometimes to other related breeds to add particular qualities or to correct faults. The formation of the American Kennel Club (AKC) in 1884 would bring written standards to the dog fancy. These standards would be developed and agreed upon by individual breed clubs, and these clubs together would form the AKC. It was no longer a single breeder looking to establish uniformity in the dogs that he or she bred. The goal would be to ensure that all registered dogs of that breed would have a uniform quality and appearance based on the written standards. Judging dogs at shows or field competitions would be based on the same standard. The development of these formal rules also resulted in the closure of the studbooks. Registration of a dog would now require that parents be registered.

In the years after World War II, a burgeoning middle class moved into the new suburbs outside the cities. There, with room and extra income, they completed their families by getting a dog. Walt Disney films like *Lady and the Tramp* and *Old Yeller* made it clear that dogs were a desirable and important part of families. These families now had televisions, and the post-war years brought them stories of Lassie, Rin Tin Tin, and the family fare of Walt Disney's weekly television shows, many of which featured dogs. In a marketing culture that now celebrated name brand products, purebred dogs became the name brand for dogs. The small group of fanciers and breeders working with the various breeds were unable to meet the new demand, and commercial breeding establishments were created to meet the need for pet dogs for retail sale. These *puppy mills* eventually became a significant concern for animal welfare groups, and there have been a number of campaigns conducted to regulate or eliminate them (see Controversies).

Feline Domestication

One story of the cat's origin has them coming forth from the sneeze of a lion on Noah's ark (Simpson, 1903). This explanation highlights two interesting aspects of the cat: that it was a late arrival in the family of animals (all of the other animals were already on the ark) and the remarkable similarity in behavior that the cat retains with its wild relatives. The domestication of cats is a more recent historical event than that of dogs. While there is some evidence that tamed wildcats may have been kept as "pets" as long as 9,500 years ago in Cyprus (Vigne et al., 2004) and 6,000 years ago elsewhere (Davis, 1987), their development as a domesticated species seems to be pretty well established in Egypt about 4,000 years ago (Mery, 1968). There has been some debate about the ancestral species that gave rise to the domestic cat, *Felis catus*. It has been difficult to track the evolution of the domestic cat since there is such a high degree of similarity in the morphology among the various small wildcat species, and there has been limited morphological differentiation between domestic cats and the most likely ancestral species. Several different lines of thought all converge on the African wildcat, *Felis sylvestris lybica*, as the most likely ancestor.

Cameron-Beaumont, Lowe, and Bradshaw (2002) performed a comparative behavioral survey of sixteen species and subspecies of small wildcats kept in zoos. They arranged for zookeepers to complete a survey form on the tameness and **affiliative** behaviors of cats in their care. They hoped to identify wild species that might have had a higher frequency of behaviors that would have lent themselves to the domestication process. They collected data on whether keepers were able to enter the cats' cages, and the extent to which the cats showed friendly behaviors similar to those showed by domestic cats such as rubbing or licking the keeper. Their results showed that the expression of affiliative behavior toward people was patchy, and widely spread among the species evaluated. This suggests that several of them could have been candidates for domestication. However, they conclude, "the tendency toward tameness among the small felids points for a localized human need as being the primary reason for the domestication of *F. s. lybica*, rather than any special features of its behavioral biology" (p. 365). Serpell (2000) indicates that the European wildcats, *Felis sylvestris sylvestris*, have a reputation for fierceness and are difficult to tame, even when reared as kittens. This is borne out of the data presented by Cameron-Beaumont et al.

Randi and Ragni (1991) analyzed populations of *F. sylvestris*, *lybica*, and *catus* populations from Sardinia, Sicily, and the Italian mainland and based on comparisons of morphometrics and allozyme variation concluded that *lybica* was the most likely ancestor of the domestic cat. More recently, Johnson et al. (2006) evaluated molecular genetic data related to cat evolution. Modern felid lineages appeared 11 million years ago in Asia. There was a substantial amount of movement, with at least ten intercontinental migrations associated with historic changes in sea level, allowing ancestral cat populations to move from one continent to another over land bridges. Cats were very successful predators, second only to humans. The lineage leading to the domestic cat appeared in Africa and Asia about 6.2 million years ago (Figure 2-7). It may have arisen from a family that had migrated to North America and then returned. A further branch appears in this lineage about 1 million years ago, leading to a family of small cats, and eventually to the ancestors of the domestic cat.

A further analysis of both domestic cat DNA samples (mtDNA), along with samples from five different subspecies of *Felis sylvestris* confirmed that cats were domesticated in the Near East (Driscoll et al., 2007). This likely happened in the Fertile Crescent region during the time when agricultural villages were developing about 9,000 years BP. There were at least five different founding females from across that region, and the resulting lineages were subsequently transported worldwide with human assistance.

In many ways, the archaeological evidence associating cat remains with human remains in ancient Egypt may be the most convincing evidence that *F. sylvestris lybica* gave rise to the modern domestic cat. A small tomb dated to about 2000 BCE yielded the bones of seventeen cats when it was excavated (Malek, 1993; Mery, 1968). Egypt was home to a rapidly expanding agricultural economy. Each year the Nile would flood and then recede to leave a rich layer of fertile silt to renew the fields. Grain was harvested in great quantities and stored against times of need. These stores attracted rats and mice, the natural prey of the wild cat, *F. s. lybica*. It is likely that the cat first came to live near people to prey upon the **rodents**. Over time, it is likely that they adapted to living in closer and closer proximity to humans. Ancient Egyptian religious practices assigned roles to a wide variety of animals and it would only be natural for cats that now protected their grain to be absorbed into these

FIGURE 2-7

Domestic cats are descended from wild cats that lived in North Africa. (From Fig. 2 from Johnson et al., SCIENCE 311: 73–77 (2006); Eustatic sea-level curve (on left) modified from B. U. Haq, J. Hardenbol, P. R. Vail, Science 235, 1156 (1987))

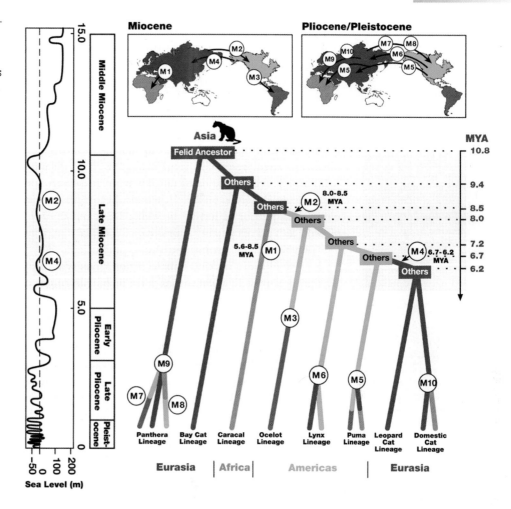

representations (Malek, 1993; Mellen, 1940). Mellen notes that the Egyptians had a history of taming a variety of creatures, including hawks, ibis, and others, and the cat would also attract their efforts.

Social Evolution of the Cat.   One manifestation of the Egyptian sun god Ra was as a male cat. Each night, in this form, Ra would battle the serpent of darkness (Malek, 1993). The Egyptian *Book of the Dead* mentions cats, and coffins from the Sixth Dynasty (2600 BCE) were inscribed with language from the book and painted with cat figures (Mellen, 1940). Eventually the cat would transition from primary representation as a male figure to an association with females. This would culminate during the Twenty-second Dynasty, 945–715 BCE (Serpell, 2000; p. 184), with a rise in the prominence of the goddess Bastet (Figure 2-8). Bastet had long been associated with the city of Bubastis. Egypt was a nation of clans and tribes, each with their own totemic animal (Mellen, 1940; Malek, 1993) and Bubastis was a center of cat worship, and their totem was Bastet (or Pasht). When a family from Bubastis came to prominence in ruling Egypt, they also brought their totem to prominence with them. Bastet was associated with fertility and plenty and the annual festival in her honor was popular and celebrated with vigor (Mellen, 1940). Serpell (2000) suggests that during this era, the cat held a place in Egyptian society equivalent

FIGURE 2-8

Bastet, the Egyptian goddess of fertility and plenty was depicted in cat form. *(University of Pennsylvania Museum, image #152022)*

to that held by the cow in present-day India. Many families owned cats and the death of the family cat would throw the family into mourning, shaving their eyebrows in a traditional sign of respect. Families that could afford the price would arrange to have their cat embalmed and buried in special cat cemeteries. Killing a cat was prohibited and might be punished by death. Mobs could become so enraged by the killing of a cat that they would fall upon the perpetrator and tear them to pieces (Mellen, 1940). In a strange twist, temple catteries were maintained for the purpose of providing cats for sacrifice and votive offerings (Armitage & Clutton-Brock, 1981). It was a great honor to care for these cats, and several species of catfish were raised in special ponds to feed these sacred cats (Mellen, 1940). However, before they reached the age of two they were strangled or killed by breaking their necks and then embalmed to be offered as votives during worship.

The Egyptians prohibited the export of cats (Mery, 1968), but Phoenician traders eventually carried cats to other areas where they traded (Mellen, 1940). By about 500 BCE, a cat appears on a marble block in Greece (Zeuner, 1963). Greeks were responsible for transporting cats to Rome. Neither the Greeks nor the Romans took the cat to heart as they did dogs. Mery (1968, p. 22) points out that while dog bones were found at Pompeii, no cat bones were recovered. Regardless, the Romans were likely responsible for introducing cats to the rest of Europe, Britain, and throughout their Empire (Zeuner, 1963). The cat appears again in a place of honor among the Norse in their mythology. Freyja, the goddess of love, like Bastet, was drawn in a cart by two cats. Norse maidens married if possible on Freyja's day (Friday), and if the sun shone during the ceremony it was a good omen and guests would note that the bride had "taken good care of cat" (Bulfinch, 1979; Mellen, 1940, p. 224).

The rise of Christianity in Europe during the Middle Ages brought hard times for cats. Extensive efforts were made to stamp out heretical and unorthodox teachings. Between the twelfth and fourteenth centuries, cat and devil worship were linked in attacks on various sects and individuals (Serpell, 2000). The lithe, sensuous behavior of the cat that lent itself to identification with female goddesses of fertility now brought the ire of a conservative, male, clerical hierarchy intent on erasing links to a past that were not consistent with their vision of orthodoxy (Russell, 1972). The next several centuries saw cats linked to the practice of witchcraft. Witches were variously accused of flying to their gatherings on the backs of demons in the form of black cats, using cats as their familiars or demonic companions and even taking on the shape of black cats themselves (Mellen, 1940; Mery, 1968; Russell, 1972). These many negative associations opened cats to a wide range of mistreatment and persecution throughout Europe in the Middle Ages (Lockwood, 2005). Cats were burned, strangled, and beaten to death in a wide range of traditional events and activities meant to "kill" or drive off the devil. This was in addition to the many casual cruelties visited upon the uncounted cats that roamed the streets of European cities. An interesting contradiction to this linkage of black cats and evil were the *matagots* in several parts of France. These were black cats that brought riches if they were brought into the house, loved, and fed (Mery, 1968).

Dislike and fear of cats was not restricted to the clerics and poorly educated peasants. Georges Buffon, the premier naturalist of the eighteenth and nineteenth centuries had a great appreciation for dogs (Rogers, 2005). Buffon was especially enamored with a dog's submissive nature, loyalty, and willingness to please its master. Apparently the independent nature of cats, combined with their "insatiable" sexuality, was not to Buffon's liking, and he went out of his way to revile them (Kete, 1994; Simpson, 1903).

While witches and black cats remain linked in the jovial celebration of Halloween in modern times, a whiff of superstition and concern remains, but this time to protect cats. Each year, at the end of October, it is not unusual for animal shelters to "hide" any black cats in their facilities until after Halloween, protecting them from adoption by people with evil intent. This happens in the absence of reasonable data to support this particular concern.

Development of Cat Fancy. In spite of the many challenges that confronted cats and the people who liked them, by the end of the nineteenth century an active cat fancy was in development. The first cat show was organized by Harrison Weir, and held at the Crystal Palace in London in 1871 (Simpson, 1903). Harrison went on to help organize and become president of the National Cat Club (NCC) in 1887. The NCC was organized for the following:

1. Promote honesty in the breeding of cats, in each breed and variety.
2. Determine the classification required for each breed or variety, and encourage the adoption of such classifications by breeders, exhibitors, judges, and the committees of all cat shows.
3. Maintain and keep the National Register of Cats.
4. Assist the showing and breeding of cats by holding cat shows under the best sanitary conditions, giving Championships and other prizes, and otherwise in all its power to protect and advance the interest of cats and their owners.

At the time, it cost 1 shilling to register a cat with the National Cat Club, and 5 shillings to enter it into the studbook (Simpson, 1903).

The first major American cat show was held in Madison Square Garden on May 8, 1895. It was organized by an Englishman, James T. Hyde, who had previously arranged horse shows at the Garden. He had visited the Crystal Palace cat show in England and imported the concept to the United States. This first show was won by a Maine coon cat named Cosie, owned by Mrs. Fred Brown (Figure 2-9). Among the prizes were a silver collar and a medal (Thomas, 2004). It was certainly fitting to have a Maine coon cat win that first American show. Maine coon cats were a breed developed in the United States and already had their own registry and competitions in Maine (Miller, 2004). A variety of cat clubs began to crop up in the United States during the years around the turn of the century. The first of these was the Beresford Cat Club of Chicago in 1899 (Thomas, 2004). The American Cat Association (ACA) formed in 1901 to provide a registry and pedigree service. There was a split in 1906 because of differences over registration policies, and the Cat Fanciers' Association was formed. The current form of the CFA came about in 1955 as an association of cat clubs.

The Early Cat Fancy. Simpson's book is a comprehensive look at the world of the cat fancy at the turn of the century (Simpson, 1903). She presents a variety of opinions and practices that remain common to this day. She begins by

FIGURE 2-9

Maine coon cats were developed in America, and Cosi won the first major cat show in the United States. *(Courtesy of the Cat Fanciers' Association Foundation)*

stating that "dogs are more essentially the friends of men, and cats may be considered as the chosen allies of womankind" (p. vii). This remains a common attitude and tends to be reflected in different ways that males and females interact with cats (Turner, 2000). Simpson goes on to point out that there was already a remarkable array of "catty" Christmas souvenirs, almanacs, calendars, and various other objects that would appeal to the cat lover. There was also a wide range of written materials for those involved with breeding and showing cats. *Fur and Feather* included information on cats, birds, rabbits, poultry, cavies, and mice, indicating that not just cats were popular as pets. *Our Cats* promoted itself as "the only newspaper in the world solely devoted to cats." In the United States, the *Cat Journal*, the *Pet Stock News*, and *Field and Fancy* all provided support for the growing interest in cats. The *Cat Courier*, except for some of the prices and trade names, is quite similar to the periodicals available to cat fanciers' today (Figure 2-10). It included ads for patent medicine for cats, Spratt's cat foods, the schedule for upcoming shows and the results of recent shows, and obituaries for well-known cat fanciers and highly successful and beloved cats. Studs were advertised at $10–$15, payable at the time of service, and pedigreed kittens could be purchased for $10–$25.

Simpson also provides a rare glimpse into the details of caring for pedigreed cats at the time (Simpson, 1903). This is of particular interest since commercial foods were not generally available, and veterinary care for cats was limited. She states that while a mix of table scraps would be perfectly fine for a couple of pet cats, caring for a number of cats in a cattery required some attention to a scheduled diet. She then gave an example of the diet that she followed for her own cattery:

- 2 days a week—fish with some rice or Freeman's Scientific Diet
- 2–3 days a week—raw meat

FIGURE 2-10

The *Cat Courier*, an early publication dedicated to the cat fancy featured articles and notices that would be familiar to anyone involved with cat shows and breeding today. *(Courtesy of the Cat Fanciers' Association Foundation)*

- 2 days a week—horsemeat
- 1 day—Spratt's biscuit soaked with stock
- 1 day—vegetables to purify the blood and keep the bowels in good condition. (pp. 37–57)

The cattery should have several four-inch deep metal tins filled with earth for the cats to use for elimination. Sand, ashes, paper, and sawdust were other common fillers used in cat boxes until the invention of cat litter in the 1950s.

She recommends that male cats that will not be used for shows and breeding be neutered at five to eight months of age, since keeping a male in the house after ten months of age would be unpleasant due to spraying to mark territories. Females should be neutered at a bit older age. Simpson points out that these altered cats will make great pets, and that they should receive greater attention at the various cat shows. Most interesting is that she also recommends that the active breeder install their own humane euthanasia box for humane killing of sick, old, or injured cats and kittens.

Cat breeds at this time were usually regional varieties with a unique appearance and known pedigree or history. As various cat populations were discovered in different parts of the world, they would eventually make their way to a serious cat breeder who would take on the task of standardizing the appearance of the "breed" and introducing it to other cat fanciers. The first documented, deliberate attempt to create a new breed did not happen until the 1920s, when the Himalayan was created by crossing stocky, longhaired Persian cats with the shorthaired, slender Siamese. After a number of generations, the popular Himalayan was accepted as a new breed. There are now three generally accepted ways for a new breed to be created:

1. Selective breeding to enhance a national or regional variety (these are often called *natural breeds*).

2. Hybridization of two or more different breeds. The Himalayan would be an example of this. There have been several breeds recently developed by crossing domestic cats with various small wild cats.

3. Mutation. The Scottish fold and munchkin cats are the result of isolated, spontaneous mutations that resulted in folded ears and shortened legs respectively that were preserved through breeding. (Thomas, 2004, pp. 3–13)

The Modern Cat Fancy. The Cat Fanciers' Association is a leading organization in the world of modern cat breeding, showing, and care. Its first two shows were held in 1906 in Detroit and Buffalo, and it now sanctions nearly 400 shows each year. The first cat registered by the Cat Fanciers' Association (CFA) was an orange and white longhaired male named Peter, born on May 2, 1906. By 1987 the CFA registered its *billionth* cat, Grand Champion Tosa Mahogany (Thomas, 2004). The CFA had undergone a variety of changes to reflect the needs of its constituency and currently adheres to the following objectives:

- The promotion of the welfare of cats and the improvement of their breed
- The registration of pedigrees of cats and kittens
- The promulgation of rules for the management of cat shows
- The licensing of cats shows held under the rules of this organization
- The promotion of interests of breeders and exhibitors of cats (Cat Fanciers' Association, 2003, p. 2)

Horses

The ancestors of what is the modern horse (*Equus caballus*) began its evolution 54 million years ago in North America. These primitive relatives were distinctly unlike the large, sleek animals that capture the affection and admiration of so many people and cultures of today. The *Eohippus* was about the size of a fox and had four toes on its front feet and three on its hind feet. By 35 million BP, *Mesohippus* was a larger animal and it now had three toes on the front feet, with an enlarged middle toe. As the forests of that era began to thin, and with more grasslands, the *Mesohippus* retained the teeth of a browser. As the grasslands became a more dominant part of the landscape, the *Merychippus* of 20 million BP had grown to 35 inches in height and its teeth were now adapted for grazing. The enlarged middle toe was now a weight-bearing hoof. The quick, social horse ancestor would eventually become a true *monodactyl*, or one-toed animal. That single toe had become a true hoof, providing excellent speed to evade predators on the wide-open grasslands. *Pliohippus* was highly successful and quickly multiplied and spread across a wide region. Some groups even crossed from North America into Asia over the Bering land bridge. This would prove significant for survival of the evolutionary line. As climates changed again, the original equines disappeared from the Western Hemisphere.

The ancestors that crossed into Asia rapidly spread through that continent and into Europe. They made quite an impression upon the early humans that lived at that time. Horses are among the most commonly depicted animals in early cave paintings. It was clear that humans valued horses as a food source, and the paintings frequently depicted hunting scenes with horses as the quarry. The presence of butchered bones in caves and at the base of cliffs, where herds may have been driven as part of the hunting strategy, provide strong ample evidence of this appreciation. When domestication of the horse began, probably in the region of the Asian steppes north of the Black Sea, it was as a source of food rather than transportation (Clutton-Brock, 1981).

Oxen were already being used for draft purposes in the Near East. As horses were introduced, the harnesses and yokes used with oxen were adapted for use with them. It took some time before horses found their true place in partnership with humans. Until wheeled carts and wagons were readily available, the stronger, slower oxen were better suited to hauling the sledges used to transport goods and materials. The rapid ascent of horses in value and appreciation came when they were put in front of light, wheeled, war chariots. Armies that employed horses in this way were dominant on the battlefield, and for centuries to come, horses were a fundamental element of every successful military force. Their value was so great that ownership was often limited to royalty and ruling classes. It was not until about 3,000 BP that intrepid souls made riding horses a common practice. By this time, kings and conquerors searched their lands and beyond for high-quality horses. The specialization of horses also began. Swift, spirited horses were selected and bred for war, and large, placid horses were bred for draft purposes. In nearly every place where horses were taken, they adapted to the conditions and needs of the people.

In 1519, the horse had its homecoming. Coronado brought 150 horses with him when he sailed to the New World to explore, conquer, and search for gold and other treasures. The natives they encountered were terrified of the massive, snorting beasts that Coronado and his men rode. Knowing their value, Coronado and the other conquistadors that followed prohibited the Indians from possessing horses. It was not long, however, before some horses were lost and others were stolen or otherwise ended up on their own. These horses found the land of their ancestors to their liking and once again wandered the grasslands of the Western Hemisphere. Along with horses that would come from settlers moving across the continent, strays from herds being moved, and those lost by military units, a new population of wild horses would again roam North America. Often called mustangs, these are not true wild horses, as they are descended from domestic stock. The only remaining true wild horses in the world are the Przewalski horses of Mongolia (*Equus przewalski*). In appearance, the Przewalski horse resembles the horses that appear in the ancient cave drawings. Short and stocky with brushy manes, they remain a link to an earlier era before horses were domesticated. These horses are now rare, and many zoos around the world have worked to breed them in captivity to preserve the species (Parker, 2003).

The role of the horse in modern times continues to evolve. First seen as a source of food, the horse eventually became the primary form of land transportation for humans from the time of the Pharaohs until the early part of the twentieth century. An aura developed around the breeding and care of horses unlike that for any other domestic animal. They were pampered and praised. Great racehorses were celebrities, as well known (or better) than any human athletes of their time. At the same time, their working brethren were beaten and driven until they died in their labors. It was no coincidence that it was the treatment of horses that gave rise to the humane movement in the United States. The role and place of horses in our modern society continues to evolve. While many horses still earn a living working on ranches and in other occupations, others are accepting the mantle of companions. They are fed and sheltered for the pleasure of their human companions. They now earn their living not by hauling loads, transporting people, racing, or performing. They are horses to ride for pleasure, not for work. This transition of horses from the role of **livestock** and work animals to companions is still evolving, and certainly not universal in our society. Debates on the use of horses for pulling carriages in the city, in rodeos, and other activities often generate intense confrontations. Perhaps nothing signifies the crux of this evolution as the debate over the slaughter of unwanted horses for consumption (see hot button issues). For people who see horses as an investment for work, racing, or other function, selling a horse for slaughter at the end of its useful life is a reasonable way to recoup some additional value. They will argue that horses have been on the menu for the entire known history of the association between humans and horses, which in fact predates other uses and functions for the horse. For others however, this is clearly beyond the pale. The horse is more than just another animal that we breed and use and extract final value.

Small Animals

Often called **shelf pets** or **pocket pets**, a variety of small animals have become common and popular in the home. While it is likely that people have been catching and keeping a variety of small animals as pets for a large part of human history, the domestication and adaptation of small animals as companions is a more recent development. In some cases these animals were originally used for another function. They might have been used as food at one time, or as subjects for laboratory research. What they generally share is small size, shorter life spans than dogs or cats, and less intense care requirements. Unfortunately, it is sometimes the case that people overlook the special needs that many of these species require. They may require very specific diets and environments.

FIGURE 2-11

Ferrets are derived from European stoats and have been used for hunting and to control vermin. *(Getty Images, Inc.)*

Ferrets. Ferrets currently kept as pets have descended from some form of polecat (*Musleta sp.*). More than likely it was the European polecat (*Mustela putorius*; Hemmer, 1990). The Romans used ferrets and probably domesticated them in the Mediterranean region about 3,000 years BP (Clutton-Brock, 1999). They were originally used to hunt rabbits and rats. Their slender bodies and quickness allowed them to drive rabbits out their burrows (Figure 2-11). Their skill and usefulness for this purpose are memorialized in the use of the phrase "ferret out" to describe digging into some topic to find and answer. Ferrets were brought to the New World in the seventeenth century to control rats. They did not become popular as pets until the 1970s. Ferrets are now bred in several different color forms. They are generally fastidious in their grooming, and can be litter-trained. Their size and cleanliness make them a desirable apartment pet. They are not welcome everywhere, however. New York City, the state of California, and other locales prohibit possession of ferrets. This apparently arises from concerns about their potential danger to children—they will bite sometimes—and their possible release and establishment as an introduced species that could threaten native birds and animals.

Rabbits. Rabbits (*Oryctolagus cuniculus*) originated in Europe and were probably domesticated during the Middle Ages (Gendron, 2000). The Romans, however, may have penned wild rabbits to fatten them before slaughter. Monks played an important role in breeding and domesticating rabbits. It is thought that one reason for the growth in the popularity in raising rabbits during the Middle Ages is that church teachings held that the unborn baby rabbits were not considered meat and could be eaten on days of abstinence when meat was to be avoided (Clutton-Brock, 1999, pp. 178–182). During this time, domestic rabbits became larger than their wild ancestors, with a change in skull shape that included a larger face and smaller cranium. During the Middle Ages, rabbits were not bred for coat color. However, as they spread through the world, many different coat colors, patterns, and fur types were selected in the formation of different breeds. The wide distribution of domestic rabbits resulted in many regional varieties or breeds. There are Dutch, Polish, Flemish, and many other breeds associated with their lands of origin. Some of these breeds were selected and maintained for their meat and others for their fur. The New Zealand white rabbit is hugely popular as a very versatile breed. It is hardy, provides good quality meat and white fur that can be used as is or easily dyed, and has also been widely used as a subject in biomedical research. A variety of dwarf breeds have been developed for pet keeping. Rabbits were first brought to the New World in the 1700s as an easy and convenient animal to raise for food. The number of breeds present in the United States increased dramatically in the 1950s, when keeping rabbits became more popular as a hobby (Gendron, 2000).

Rabbits were originally classified as a type of rodent until 1912. They are now correctly placed into the order Lagomorpha with pikas and hares. Lagomorphs have a unique digestive system. They feed primarily on grasses and other fibers. They are unable to retain the full nutritive value of these foods after the first pass through their digestive system. Their feces after the first pass through the system are soft and moist. Bacteria found in their colons further break down the nutrients in the food. The rabbits then eat these first feces to digest and absorb additional nutrients, especially B vitamins. This behavior is known as *coprophagy*.

In the past 15–20 years the concepts of how to keep and care for companion rabbits has changed dramatically. Rabbits were generally kept in outdoor hutches, even when being kept as pets. They were rarely considered a house pet. The term *house rabbit* is now common and is indicative of this changing philosophy. Rabbits can be litter-trained and allowed supervised liberty in a home. Care must be taken with any houseplants that may be present and electrical wires (to which rabbits seem to have an unnatural attraction). They may be confined in a cage or hutch indoors when it is not possible to supervise their activities. However, this move indoors with freedom to move about is

FIGURE 2-12

Domestic rabbits are a separate species from their wild counterparts. *(Alto Pets)*

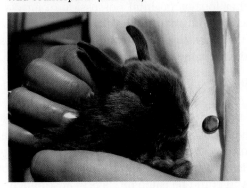

FIGURE 2-13

Gerbils will thump their hind legs as a warning and when excited. *(Istock photo)*

another step in the evolution of the rabbit (Figure 2-12). Rabbits have moved from the role of livestock as a source of food and fiber, to pets and objects of a hobby or fancy, to true companions for some people.

Gerbils. Gerbils (*Meriones unguiculatus*) were first brought to the United States for research by Dr. Victor Schwerker in 1954. They became a popular alternative to hamsters once they entered the pet trade. Their long tails are furred, blunting some of the squeamishness that people have stemming from experiences with wild rats and mice. They are also more active in the daytime than hamsters and other rodents, providing more opportunity for interaction with people. Clark and Galef (1980) studied and compared domestic and wild gerbils. They found that the domesticated gerbils were less aggressive, had smaller adrenal glands, and grew more rapidly than the gerbils from wild population. These are all changes one would have expected as a result of domestication. In addition to these changes, breeders and fanciers have been able to identify and select for a variety of coat colors. Gerbils can be found in colors ranging from white albinos, to cream, fawn, and completely black (Figure 2-13).

Rats and Mice. Keeping rats and mice as pets challenges many cultural and personal aversions to rodents. Wild rats and mice have long been a scourge of humans. As soon as humans were capable of gathering or producing enough food to store some part of it for use at a later time, it is likely that rats and mice were finding ways into the storehouses to eat their fill and spoil additional quantities with their feces and urine. As noted earlier, cats initially found a welcome in the world of humans when they became allies in the battle against rodents. The ferret was another species that was enlisted to assist in this effort. Several of these highly adaptable rodents tagged along on ships and other forms of transportation to follow humans and quickly spread from their Asian origins to worldwide distribution. Three rodent species have established themselves as commensal species of humans (Clutton-Brock, 1999). The black rat (*Rattus rattus*) is a native of Asia Minor and probably spread through Southern Europe with the Romans. It is believed to be a significant carrier of many diseases. The brown rat (*Rattus norvegicus*) spread across Europe in the mid-sixteenth century. The brown rat is highly adaptive and prolific. It has a wide-ranging omnivorous diet and can withstand significant climate variation. The house mouse (*Mus domesticus*) probably spread through the world before the rat

FIGURE 2-14

Mice come in a variety of colors other than the pink-eyed albinos most frequently seen in pet stores. *(Courtesy Dr. Stephen Zawistowski)*

species and may be the second most populous mammalian species on earth after humans.

Tamed mice were kept as pets for at least 300 years. In the mid-seventeenth century, white, grey, and dark varieties were noted. They became common subjects for biomedical research as the need grew for a small mammal that was easy to maintain and breed. Their short life cycle and prolific reproductive talents were well suited to the study of genetics. Gregor Mendel's research in genetics was rediscovered in the early 1900s, igniting a wave of research in the field that continues into this century. By 1909, Clarence Cook Little was producing inbred strains of mice, providing a significant advance and an important new tool for research. Genetically engineered strains of mice are now a significant resource for a wide range of research. The genetic diversity that made the mouse an ideal species for biomedical and genetics research also provided a rich substrate for fanciers interested in producing a variety of different types of mice. Pink-eyed, albino mice may be the form most familiar to most people. They are bred and sold in large numbers as pets, and as *feeders* for people who keep various reptiles as pets. Fancy mice that are bred and raised by hobbyists come in a wide range of colors, patterns, and several coat types (Figure 2-14). They are typically larger than the standard albino mouse. Coat colors may be creams, browns, blacks, in solids (or also called **selfs**), and various combinations in piebald and other patterns. Silky mice, with a fine, shiny coat, in gold or silver coat colors are particularly striking.

Brown rats or Norwegian rats have also been domesticated. They have also been bred for use in the laboratory and as pets. The albino rat became a standard subject for a great many studies in both biomedical research and psychology. Their ubiquitous presence in the mid-twentieth century in American psychology laboratories led Frank Beach to publish a classic paper, with the wonderful title, "The Snark Was a Boojum," on the weakness of comparative research when the work was so highly dependent on a single species for study (Beach, 1950). Rats are still used in biomedical research, and many different genetic lines have developed for this purpose. Many of the domestic rats kept as pets are colored varieties. Similar to fancy mice, various blacks, browns, and combinations are popular color forms for rats kept as pets (Figure 2-15). Rats are generally neat and tidy. When socialized, they will be

FIGURE 2-15

While people often think of rats of being dirty and living in and around trash, domestic rats are very clean by nature. *(Istock photo)*

FIGURE 2-16

Guinea pigs can be a delightful first pet for children. If they are handled gently, they seldom bite, and they are not so fast that they are likely to escape. *(Istock photo)*

quite tame, and will recognize familiar people and willingly perch on their shoulders and accept treats.

Guinea Pigs. The domestic guinea pig, or cavy (*Cavia porcellus*), is descended from the wild cavy of South America (Clutton-Brock, 1999). The guinea pig is one of the few New World species that was successfully domesticated. The Inca civilization in Peru raised guinea pigs for food. In villages there they may still be allowed to live within a home, and killed as needed for meals (Caras, 1996). Sailors returning from the New World would carry along a supply of cavies to provide fresh meat on the voyage home. Presumably, animals that made it back to the sailors' home ports alive might end up as intriguing pets for sweethearts and children. Their small size, chubby bodies, and habit of whistling would all seem to have predestined them to move from the pot to pet (Figure 2-16). The origin of their common name, guinea pig, is uncertain. One popular theory is that the ships returning to Europe frequently stopped in West African ports, including Guinea, and hence the name. Another theory is that when they first appeared in England they could be purchased for an English coin called a guinea. The "pig" part of the name is more than likely derived from their habit of squealing when aroused.

Hamsters. The origin of the common, or Syrian or golden hamster (*Mesocricetus auratus*), is well known (Clutton-Brock, 1999). In 1930, a single female hamster and her young were dug out of their burrow near Aleppo in Syria (Hemmer, 1990). Hamsters from this line were brought to England in 1931 and to America in 1938. They were initially used as laboratory subjects and later became popular as pets. A number of different varieties have been developed, including teddy bear with longer fur, and different coloring patterns (Figure 2-17). Several different, smaller, dwarf hamsters have entered the pet trade in the last fifteen years. The Russian dwarf (*Phodopus campbelli*), or Campbell's hamster, is among the most common of these. Others include the winter white (*P. sungous*), Chinese hamster (*Cricetulus griseus*), and Roborovski hamster (*P. roborovskii*), available less often in retail pet stores. It is striking that a nocturnal rodent with an occasionally pugnacious personality

FIGURE 2-17

Hamsters will pack their cheek pouches with food and carry it off to a hiding place somewhere in their habitat. *(Istock photo)*

would become such a popular pet for children. However, the hamster is small and generally inexpensive. It has the striking habit of packing its cheeks with food and can be comical in its behavior. Just as important perhaps has been the ingenuity of the pet product industry in the development of a vast array of imaginative cages and environments for the hamsters. Plastic tubes connecting a variety of modules for sleeping, eating, and exercise have continued to stimulate interest in the animals. The monetary value (or lack!) of the actual hamster to the retail industry is reflected in sales where the purchase of a complete habitat may come with a free hamster to live in it.

Livestock

A number of people raise and keep livestock as companions or as hobby. Many farm families hand-rear and raise orphans or *runts*, keeping them around as pets for the children. In the end, these pets may or may not end up going to market along with their brethren who had stayed with the flock or herd. The emotional dynamics of this process have stimulated quite a few children's books, including *Charlotte's Web* by E. B. White. The current growth in keeping livestock as a hobby is associated with an interest in unusual or **rare breeds**. Miniature horses are raised for show and will compete pulling small driving carriages and carts. They have even been bred and trained to fill the roles normally held by Seeing Eye or assistance dogs. Miniature cattle allow would-be ranchers to operate on small plots of land. They are also a curiosity that people enjoy seeing and petting (Figure 2-18).

The fanciers most serious about keeping livestock as a hobby may decide to concentrate on rare or endangered breeds. The development of modern production agricultural methods and a commodity-based economic structure has led to concentration on just a handful of breeds in each species. This has led to the loss of some older breeds that do not meet the needs of these high-volume production systems. The American Livestock Breeds Conservancy (ALBC) was formed in 1977 to conserve historic breeds and genetic diversity in livestock (see <http://albc-usa.org/>). These breeds include asses, cattle, goats, horses, sheep, pigs, rabbits, chickens, ducks, geese, and turkeys. They point out that these now-rare livestock breeds are an essential part of America's agricultural past and an important resource for the future. These breeds were developed to meet the needs of specific geographic regions and climates. Their genetic diversity may be a critical resource if agricultural changes require new adaptations in livestock.

FIGURE 2-18

Touchdown is a miniature longhorn, and is the mascot of the Harmony High School Longhorns in Osceola County, FL. *(Courtesy Dr. Stephen Zawistowski)*

The ALBC maintains a conservation priority list of over 150 different breeds based on genetic and numerical parameters. These parameters include (poultry have separate requirements):

1. The breed is from one of the seven traditional U.S. livestock species mentioned above.

2. The breed census satisfies numerical guidelines such as
   - Critical: Fewer than 200 annual registrations in the United States and estimated global population less than 2,000 (for rabbits less than 50 annual registrations and a global population under 500)
   - Threatened: Fewer than 1,000 annual registrations in the United States and estimated global population less than 5,000 (for rabbits fewer than 100 annual registrations and a global population under 1,000)
   - Watch: Fewer than 2,500 annual registrations in the United States and estimated global population less than 10,000 (for rabbits fewer than 200 annual registrations and a global population under 2,000); also included are breeds that present genetic or numerical concerns or have a limited geographic distribution
   - Recovering: Breeds that were once listed in another category and have exceeded Watch category numbers but are still in need of monitoring
   - Study: Breeds that are of genetic interest but either lack definition or lack genetic or historical documentation

3. The breed is a true genetic breed (when mated together, it produces the breed type).

4. The breed has an established and continuously breeding population in the United States since 1925. This requirement includes stipulations related to the number of breeding lines in the United States, the number of breeding females, and the presence of an association of breeders, among other specifics.

The Jacob sheep is one interesting and typical example of a breed listed as rare by the ALBC. They are a spotted sheep and descriptions date back to biblical accounts of Jacob, who bred spotted sheep. In the 1600s, spotted sheep were documented in England, and they became popular as an ornamental species. They are small and hardy, and have the interesting characteristic of being multihorned (Figure 2-19). The Jacob sheep were imported into the United States in the 1900s. They have grown in popularity as their fleece is favored by hand spinners and weavers practicing traditional crafts.

FIGURE 2-19

The multi-horned Jacob sheep is one of many rare breeds livestock that dedicated individuals are working to keep from going extinct. *(Courtesy Green Chimneys)*

In addition to hobby breeders, a number of historic villages and restorations have also begun to keep rare breeds of livestock on their grounds. The animals add to the pastoral setting, providing a glimpse of the intimate relationship that colonists had with the animals they kept. Maintaining the rare breeds is also consistent with their efforts to preserve and exhibit the culture of the era they are trying to represent.

DISCUSSION QUESTIONS

1. What is the relationship between domestication and tameness?
2. How does the reproductive potential of a domesticated species typically compare with that of its wild counterparts? Why does this occur? What are specific changes that occur?
3. What is neoteny? Besides those described in the text, give an example of physical neoteny and behavioral neoteny.
4. What are the limitations of using archaeological evidence to study the relation of dogs to wolves?
5. Compare and contrast the traditional view of canine domestication (i.e., wolf taming) with the alternative process involving the food source provided by human civilizations.
6. What morphological changes accompanied the increase in tameness of silver foxes observed in the classic study by Dmitry Belyaev and colleagues?
7. Describe the timing of and motivation behind the development of early dog breeds. What were the characteristics and purposes of the first breed types?
8. From the Middle Ages to the present, describe the spectrum of care received by dogs according to the socioeconomic status of their people.
9. Describe the development of kennel clubs and the dog fancy. What role has this process played in the establishment of today's puppy mills?
10. Describe the incorporation over time of cats into human civilizations. How did the cat's domestication differ from that of the dog?
11. How did the treatment and perception of cats during the Middle Ages differ from previous times? What factors were involved in this change?
12. Compare and contrast the development of the cat fancy and the dog fancy. Did cat breeds develop for the same reasons as dog breeds?
13. How has the role of the horse changed throughout history? What roles do horses play in modern American society?
14. Many of the small animals currently kept as pets were originally used for functions other than companionship. Select three of the small animals discussed in this chapter and describe their original functions. Are the animals still used for these purposes today?
15. How has domestication affected the coat color of small animals? Are similar effects observed in the cat and dog?
16. What is the goal of the American Livestock Breeds Conservancy (ALBC)? What led to its development? What types of animals are involved?

REFERENCES

Armitage, P. L., & Clutton-Brock, J. (1981). A radiological and histological investigation into the mummification of cats from ancient Egypt. *Journal of Archaeological Science, 8,* 185–196.

ASPCA Archives. (n.d.). Edward Buffet notes. Dog volume.

Beach, F. A. (1950). The snark was a boojum. *American Psychologist, 5,* 115–124.

Benecke, N. (1987). Studies of early dog remains from northern Europe. *Journal of Archaeological Science, 14,* 31–39.

Black, G. G. (1949). *American beagling.* New York: G. P. Putnam's Sons.

Bulfinch, T. (1979). *Bulfinch's mythology.* New York: Avenel Books.

Cameron-Beaumont, C., Lowe, S. E., & Bradshaw, J. W. S. (2002). Evidence suggesting preadaptation to domestication throughout the small Felidae. *Biological Journal of the Linnean Society, 75,* 361–366.

Caras, R. (1996). *A perfect harmony.* New York: Simon and Schuster.

The Cat Courier. (1917, July). No. 1.

Cat Fanciers' Association. (2003). *The Cat Fanciers' Association, Inc. 2004 yearbook.* Manasquan, NJ: Cat Fanciers' Association.

Clark, M. M., & Galef, B. G. (1980). Effects of rearing environment on adrenal weights, sexual development, and behavior in gerbils: An examination of Richter's Domestication Hypothesis. *Journal of Comparative and Physiological Psychology, 94*(5), 857–863.

Clutton-Brock, J. (1977). Man-made dogs. *Science, 197,* 1340–1342.

Clutton-Brock, J. (1981). *Domesticated animals from early times.* Austin, TX: University of Texas.

Clutton-Brock, J. (1992). The process of domestication. *Mammal Review, 22,* 79–85.

Clutton-Brock, J. (1995). Origins of the dog: Domestication and early history. In J. Serpell (Ed.), *The domestic dog: Its evolution, behavior and interactions with people.* Cambridge: Cambridge University Press.

Clutton-Brock, J. (1999). *A natural history of domesticated animals,* 2nd edition. Cambridge, MA: Cambridge University Press.

Coppinger, R., & Coppinger, L. (2001). *Dogs: A startling new understanding of canine origin, behavior and evolution.* New York: Scribner.

Coren, S. (2005, Summer). Confidants to kings, consolers to queens. *Modern Dog,* pp. 59–63.

Crowley, J., & Adelman, B. (Eds.). (1998). *The complete dog book: Official publication of the American Kennel Club,* 19th edition. New York: Howell Book House.

Darwin, C. (1979). *On the origin of species.* New York: Avenel Books (original publication 1859).

Davis, S. J. M. (1987). *Archaeology of animals.* London: Batsford.

Davis, S. J. M., & Valla, F. R. (1978). Evidence for the domestication of the dog 12,000 years ago in the Natufian of Israel. *Nature, 276,* 608–610.

Dent, T. (2004). In M. Siegal (Ed.), *The Cat Fanciers' Association complete cat book* (Foreword). New York: Harper Row.

Driscoll, C. A., Menotti-Raymond, M., Roca, A. L., Hupe, K., Johnson, W. E., Geffen, E., Harley, E., Delibes, M., Pontier, D., Kitchener, A. C., Yamaguchi, N., O'Brien, S. J., Macdonald, D. (2007). The Near Eastern origin of cat domestication. *Sceincexpress/* www.sciencexpress.org/28 June 2007/ Page 1/10.1126/science.1139518.

Fitzgerald, B. M., & Turner, D. (2000). Hunting behavior of domestic cats and their impact on prey populations. In D. Turner & P. Bateson (Eds.). *The domestic cat: The biology of its behaviour,* 2nd edition, (pp. 151–175). Cambridge: Cambridge University Press.

Fondon, J. W., & Garner, H. R. (2004). Molecular origins of rapid and continuous morphological evolution. *Proceedings of the National Academy of Sciences, 10*(52), 18058–18063.

Foster, P., Harding, R., Torroni, A., & Bardelt, H.-J. (1996). Origin and evolution of natural American mtDNA variation: A reappraisal. *American Journal of Human Genetics, 59,* 935–945.

Gendron, K. (2000). *The rabbit handbook: Purchase, care and breeding, understanding rabbit behavior.* New York: Barron's Educational Series.

Harrison, D. L. (1973). Some comparative features of the skulls of wolves (*Canis lupus* Linn.) and pariah dogs (*Canis familiaris,* Linn.) from the Arabian peninsula and neighboring lands. *Bonner Zoologische Beiträge, 24,* 185–191.

Hemmer, H. (1990). *Domestication: The decline of environmental appreciation.* Cambridge, MA: Cambridge University Press.

Immelman, K. (1972). Sexual and other long-term aspects of imprinting in birds and other species. *Advances in the Study of Behavior, 4,* 147–174.

Johnson, W. E., Eizirike, E., Pecon-Slattery, J. Murphy, W. J., Antunes, A., Teeling, E., et al. (2006). The late Miocene radiation of modern Felidae: A genetic assessment. *Science, 311,* 73–77.

Kete, K. (1994). *The Beast in the boudoir: Pet keeping in nineteenth-century Paris.* Berkely, CA: University of California Press.

Klinghammer, E., and Goodman, P. A. (1987). Socialization and management of wolves in captivity. In H. F. Dordrecht (Ed.), *Man and wolf.* The Netherlands: Dr. W. Junk Publishers.

Leonard, J. A., Wayne, R. K., Wheeler, J., Valadez, R., Guillén, S., & Vilà, C. (2002). Ancient DNA evidence for Old World origin of New World dogs. *Science, 298,* 1613–1616.

Leyhausen, P. (1979). *Cat behavior: The predatory and social behavior of domestic and wild cats.* New York: Garland STPM Press.

Lorenz, K. (1954). *Man meets dog.* London: Methuen.

Lorenz, K. (1975). In M. W. Fox (Ed.), *The wild canids: Their systematics, behavioural ecology and evolution* (Foreword). New York: Van Nostrand Reinhold.

Lockwood, R. (2005). Cruelty toward cats: Changing perspectives. In D. J. Salem & A. N. Rowan (Eds.), *State of the animals III* (pp. 15–26). Washington, DC: Humane Society Press.

Malek, J. (1993). *The cat in ancient Egypt.* London: British Museum Press.

McGhee, R. (2002). Co-evolution: New evidence suggests that to be truly human is to be partly wolf. *Alternatives Journal, 28*(12), 1–7.

Mellen, I. A. (1940). *The science and mysteries of the cat.* New York: Charles Scribner's Sons.

Mery, F. (1968). *The life, history and magic of the cat* (E. Street, Trans.). New York: Grosset B. Dunlap Publishers.

Miller, J. (2004). What is a pedigreed cat? In M. Siegal (Ed.), *The Cat Fanciers' Association complete cat book.* New York: Harper Row.

Morey, D. (2006). Burying evidence: The social bond between dogs and people. *Journal of Archaeological Science, 33,* 178–185.

Newman, L. (2004). *Hachiko waits.* New York: Henry Holt.

Olsen, S. J., & Olsen, J. W. (1977). The Chinese wolf, ancestor of New World dogs. *Science, 197,* 533–535.

Olsen, S. J. (1985). *Origins of the domestic dog.* Tucson, AZ: University of Arizona Press.

Ostrander, E. A., & Kurglyak, L. (2000). Unleashing the canine genome. *Genome Research, 10,* 1271–1274.

Page, R. D. M., & Holmes, E. C. (2002). *Molecular evolution: A phylogenetic approach.* Oxford: Blackwell.

Parker, R. (2003). *Equine science,* 2nd edition. Clifton Park, NY: Thomson Learning.

Parker, H. G., Kim, L. V., Sutter, N. B., Carlson, S., Lorentzen, T. B., Johnson, G. S., DeFrance, H. B., Ostrander, E. A., & Kruglyak, L. (2004). Genetic structure of the purebred domestic dog. *Science, 304,* 1160–1163.

Randi, E., & Ragni, B. (1991). Genetic variability and biochemical systematics of domestic and wild cat populations (*Felis sylvestris:* Felidae). *Journal of Mammology, 72,* 79–88.

Rogers, K. M. (2005). *First friend: A history of dogs and humans.* New York: St. Martin's Press.

Russell, J. B. (1972). *Witchcraft in the Middle Ages.* Ithaca, NY: Cornell University Press.

Savolainen, P., Zhang, Y., Luo, J., Lundberg, J., & Leitner, T. (2002). Genetic evidence for an East Asian origin of domestic dogs. *Science, 298,* 1610–1613.

Serpell, J. (1995). In J. Serpell (Ed.), *The domestic dog: Its evolution, behaviour and interactions with people* (Introduction). Cambridge, MA: Cambridge University Press.

Serpell, J. (2000). Domestication and history of the cat. In D. C. Turner & P. Bateson (Eds.), *The domestic cat: The biology and of its behaviour,* 2nd edition (pp. 179–192). Cambridge, MA: Cambridge University Press.

Simpson, F. (1903). *The book of the cat.* New York: Cassell.

Slater, M. (2002). *Community approaches to feral cats.* Washington, DC: Humane Society Press.

Sutter, N. B., & Ostrander, E. A. (2004). Dog star rising: The canine genetic system. *Nature Reviews, Genetics, 5,* 900–910.

Thomas, J. (2004). The way we were: A history of the Cat Fanciers' Association. In M. Siegel (Ed.), *The Cat Fanciers' Association complete cat book* (pp. 3–13). New York: Harper Row.

Trut, L. N. (1999). Early canid domestication: The farm fox experiment. *American Scientist, 87,* 160–169.

Tsuda, K., Kikkawa, Y., Yonekawa, K., & Tanabe, Y. (1997). Extensive interbreeding occurred among multiple matriarchal ancestors during the domestication of dogs: Evidence from inter- and intraspecies polymorphisms in the D-loop region of mitochondrial DNA between dogs and wolves. *Genes, Genetic Systems, 72,* 229–238.

Turner, D. C. (2000). The human-cat relationship. In D. C. Turner & P. Bateson (Eds.), *The domestic cat: The biology of its behaviour* (pp. 193–206). Cambridge, MA: Cambridge University Press.

Turner, P. (2004). *Hachiko: The true story of a loyal dog.* Boston: Houghton Mifflin.

Vigne, J. D., Guilaine, J., Debue, K. Haye, L., & Gérard, P. (2004). Early taming of the cat in Cyprus. *Science, 304,* 259.

Vilà, C., Savolainen, P., Maldonado, J. E., Amorim, I. R., Rice, J. E., Honeycutt, R. L., et al. (1997). Multiple and ancient origins of the domestic dog. *Science, 276,* 1687–1689.

Vilà, C., Maldonado, J. E., & Wayne, R. K. (1999). Phylogenetic relationships, evolution, and genetic diversity of the domestic dog. *Journal of Heredity, 90*(1), 71–77.

Wayne, R. K., & Ostrander, E. A. (1999). Origin and genetic diversity and genomic structure of the domestic dog. *Bio Essays, 21,* 247–257.

Wilcox, B., & Walkowicz, W. (1995). *Atlas of dog breeds of the world.* Neptune City, NJ: TFH Publications.

Wozencraft, W. C. (1993). Carnivora: Canidae. In D. E. Wilson & D. M. Reeder (Eds.), *Mammal species of the world—A taxonomic and geographic reference,* 2nd edition (pp. 279–288). Washington, DC: Smithsonian Institution Press.

Zawistowski, S., & Patronek, G. (2004). The dog in wolf's clothing. *Journal of Applied Animal Welfare Science, 7*(4): 277–278.

Zeuner, F. E. (1963). *A history of domesticated animals.* London: Hutchinson.

Development of Animal Protection

CHAPTER 3

> He who is cruel to animals becomes hard also in his dealings with men. We can judge the heart of a man by his treatment of animals.
>
> —Immanuel Kant

KEY TERMS

animal protection	animal sheltering	animal rights
Society for the Prevention of Cruelty to Animals (SPCA)	humane movement	new welfarism
	animal welfare	

Our current **animal protection** structures and practices can be clearly traced to the Victorian era in England. However, developments during that time were influenced by a variety of preceding elements. There is no clear linear narrative that connects these influences until they coalesce into a well-defined effort to provide animals with legal protection in the early 1820s. These elements included an expanding interest in nature, a prevailing sense of reformism, a desire to establish and promote *civilizing* influences on society, and a preoccupation with pain and its elimination in Victorian society (Turner, 1980). These early elements put as much emphasis on the damage that cruelty to animals did to the people as it did on the pain and suffering that it caused to animals.

EARLY EXAMPLES

The nineteenth century saw a steady change in the relationship between humans and the balance of the natural world. Dating back to Aristotle was the assumption that nature was organized in a clear hierarchy with lesser animals at the base progressing upward to the more sophisticated and complex animals. Humans held a place above the animals and just below deities. This *scala naturae*, or great chain of being, not only defined the natural order of things, it also provided the rationale for the treatment of animals. The position of humans as above and distinct from animals was alternately interpreted as giving license for the treatment and use of animals in whatever way desired, or imposing upon humans the obligation to be the stewards and caretakers of the world and other living things. It presumed that all living things were created in their current forms and were unchanging. A critical change occurred when Linnaeus introduced a new taxonomy for the organization of the natural world in the eighteenth century. Discarding previous assumptions about the subjective value or perfection of animals, Linnaeus based his structure on objective, observed similarities in anatomical structure. Most important was that he also placed humans into this structure based on the obvious shared characters with the other mammals and primates. This trend continued and in England was taken up by Erasmus Darwin, grandfather of Charles Darwin. Erasmus Darwin was a successful physician who took an interest in more general aspects of zoology. He wrote a lengthy poem, *Zoonomia*, in 1794 that hinted at some aspects of evolution. His ideas were radical for the time but anticipated further developments.

Interest in these ideas about animals and the natural world was strongly influenced in England by the many voyages of discovery taking place. As the British Empire stretched around the world, its sailors, merchants, soldiers, and scientists returned with specimens of plants and animals, alive and dead, from exotic and faraway places. Charles Darwin was part of one such expedition, circumnavigating the globe on the HMS *Beagle* from 1831 to 1836. Twenty years before he published his treatise on evolution, his *Voyage of the Beagle* in 1839 was a well-received contribution to this genre. When *Origin of Species* was published in 1859, it had an immediate impact. He had marshaled an extraordinary amount of evidence into a cogent and compelling argument that the biological world was not static, and that the observed similarities between living things was the result of a shared evolutionary history. Most important was the fact that Darwin's reasoning went beyond the physical similarities of anatomy and physiology. He included behavioral similarities in his analyses, an argument that he extended in *The Expression of the Emotions in Man and Animals* in 1872. This extension of the argument fit well with another preoccupation of Victorian England, pain and suffering.

The Victorian era saw a remarkable transition in society's tolerance and acceptance of pain and suffering. James Turner (1980) points out that the human desire to avoid pain is so pervasive that in our current time it may be too obvious to contemplate. However, to the Victorian mind it was something to avoid and prevent. It was during this era that the practice of medicine changed from a profession where the practitioner required the highest intolerance to the evidence or even the imposition of pain in a patient to one that sought to relieve pain. Many potential physicians chose other fields of endeavor due to their inability to tolerate the presence of pain found in the work. One of the reasons that Charles Darwin did not follow in the professional footsteps of his father and grandfather (both physicians) was because he was revolted by the sight of suffering patients and was unable to stomach the gruesome procedures common in medical practice at that time. This was when the physical strength and dexterity of a physician were highly prized. When surgical procedures, including amputation, were conducted without the benefit of anesthetics, speed was the only way to control the amount of pain inflicted by the cure! Introduction of morphine and other pain relieving medicines in the nineteenth century saw rapid acceptance. At the same time, it stimulated further the urge to eliminate pain and suffering where and when possible. If pain was anathema to the Victorian mind, then cruelty, the deliberate infliction of pain, must be stopped. The strength and influence of this conviction has been memorialized in the names of the first organizations formed to protect animals, Societies for the Prevention of Cruelty to Animals.

When the concern about pain was combined with the changing understanding of similarities between humans and animals, it became obvious that animals were also capable of suffering and feeling pain. Thinking on this had come full circle from the Cartesian assertion that animals were automatons incapable of feeling pain. Their squeals and thrashing were not just reflexive actions; this was evidence that they were suffering pain just as people would. This led to Jeremy Bentham's oft-repeated quote, "The question is not, Can they reason? Nor, Can they talk? But, Can they suffer?" Causing an animal unnecessary pain was the measure of animal cruelty. Overworking or beating a horse, bull and bear baiting, and many other practices came under scrutiny. However, vivisection, the experimentation on live, conscious animals, was considered the ultimate evil by many. The developing field of experimental physiology was practiced by educated men. Unlike members of the lower classes who might beat an animal in frustration or anger, or delight in the bloody display of two animals fighting, vivisectors performed acts of abject cruelty in a cold deliberate fashion. This would be the fundamental issue that would drive a wedge between animal protection advocates and scientists until our own time.

All of the above concerns happened within the context of a social era bent on reform. Abolition of slavery, temperance, and changing roles for women in society all reflected a time when significant attention was being placed on perfecting the social order. Among the issues considered were the activities and behaviors that resulted in unlawful or immoral behavior. There were numerous allusions to the observation that being cruel to animals when young would presage an adult life of crime and violence. One of the most dramatic depictions of this concept was a series of woodcut illustrations created by William Hogarth in 1751. His *Four Stages of Cruelty* depict the life of a young man that begins with the torture of animals as his entry into a life of moral depravity that included crimes that result in his eventual execution (Figure 3-1). This theme was reflected in the thoughts of numerous philosophers and social critics of the era. John Locke pointed out that children should be taught to not mistreat animals. The newly developing field of children's literature was strongly influenced by this thinking. Many of the earliest stories and books written for children included animals as part of the narrative and encouraged their proper treatment (Zawistowski, 1998). This literary tradition would reach its zenith in the publication of Anna Sewell's *Black Beauty* in 1877.

The first book published with animal protection as its explicit theme was *The Duty of Mercy and the Sin of Cruelty to Brute Animals* by Humphrey Primatt in 1776 (Linzey, 1998). Until this time, the question of animal treatment

FIGURE 3-1

William Hogarth's *Four Stages of Cruelty* depicts the life of Tom Nero. In the *First Stage* (A), the rampant cruelty inflicted upon animals in the streets of East London in the early 1700s. Tom Nero appears in the center of the frame, pushing an arrow in the anus of a dog. By the *Second Stage* (B) Tom Nero is an adult and is shown beating a cart horse that has collapsed. Hogarth makes his point on the connection between cruelty to animals and eventual violence to humans in the third picture in the series, *Perfection* (C). Tom Nero is now a grotesque appearing as a highwayman who has murdered a woman. The final scene, *Reward* (D), shows that after Tom Nero was convicted of his crime and hanged, his body was dissected, with a dog in the foreground eating his spilled entrails. *(Courtesy Dr. Randall Lockwood)*

(A)

(B)

(C)

(D)

was typically mentioned within a larger work dedicated to the discussion of human rights and responsibilities. Following the theme illustrated by William Hogarth, the primary concern was the fact that the cruel treatment of animals coarsened the nature of the man perpetrating the act and predisposed him to other still more heinous crimes. The brutal mistreatment of animals was wrong, but there was no consistent theory of how to deal with the issue.

Royal Society for the Prevention of Cruelty to Animals

All of these threads would finally begin to coalesce in 1800 with the first introduction of a bill to ban bull baiting in England. A common market day diversion, bull baiting was a boisterous and rowdy pastime. A bull would typically be tethered to a stake and *bull dogs* would be set upon it. The dogs were bred and trained to target the head of the bull, often grabbing the bull's nose in a tenacious bite. The bull suffered greatly and many dogs died gored or stomped by the bull. Local peasants would delight in the spectacle as an infrequent diversion in a life of hard work with limited reward and pleasure. Pressure to eliminate the practice hinged as much or more on the desire to deny the peasants their rowdy pleasure as it was to protect the animals. That bill failed, as did another bill presented in 1809 that would have prohibited cruelty to all domestic animals (Turner, 1980). In 1822, Richard Martin, a member of parliament representing Galway, secured passage of the Ill-Treatment of Cattle Act, the first codification of law to prohibit cruelty to animals (Unti, 1998c). The act's primary result was to eliminate the sport of bull baiting. The bill did not cover dogs or other non-livestock animals. An unintended consequence of the bill was that the owners of dogs looking for another pastime began to fight the dogs against one another, giving rise to the sport of dogfighting. Martin would succeed in passing a bill to prevent the cruel and improper treatment of dogs in 1826.

Once a law was in place, the effort was directed to the development of a structure to ensure enforcement of the law. In June 1824 Arthur Broome, an Anglican priest, called a meeting of like-minded individuals that resulted in the formation of the **Society for the Prevention of Cruelty to Animals (SPCA)** (Linzey, 1998). Richard Martin, along with a number of highly respectable members of society, was counted among the founders and early supporters of the SPCA. Among the early priorities was the investigation and enforcement of anticruelty laws championed by Martin in Parliament. The SPCA provided the financial resources needed to ensure that a cadre of constables was available to ensure that these laws were implemented. It rapidly became part of the respectable social order, and this was clearly demonstrated when Queen Victoria's patronage allowed for the addition of the prefix *Royal* in 1840.

ASPCA and Other U.S. Groups

The SPCA movement would be brought to America by a most unlikely candidate. Henry Bergh was the son of a wealthy New York City shipbuilder, Christian Bergh (Figure 3-2). Until midlife, it seemed that the younger Bergh had inherited none of his father's drive or ambition. When his father died and Henry inherited a substantial fortune, he and his wife idled away their time attending plays, parties, and traveling frequently. He was well connected

FIGURE 3-2

New York philanthropist and diplomat Henry Bergh brought the idea of a society to prevent cruelty to animals to America in 1866. *(Courtesy ASPCA)*

socially and politically, and eventually secured a diplomatic posting to the court of the Czar in St. Petersburg, Russia. While there in 1865 he experienced a transformation that would change his life and the lives of animals in the United States. While some of his early diaries suggested that Bergh was concerned about the mistreatment of animals, he had not yet made any move to act on their behalf. This all changed when he came across a Russian peasant beating his fallen horse along the road. Bergh had his driver stop his own carriage and he physically intervened to stop the peasant from beating the horse. This became a habit for him. When he left his post later that year, he stopped in England on his return to the United States. While there, he attended a meeting of the RSPCA and spoke at length with Lord Harrowsby, who was the head of the Society at that time. Bergh returned to the United States with the information and the will required to start a similar organization in New York City.

On February 8, 1866, Henry Bergh delivered a lecture to a packed assembly of politicians, publishers, business leaders, and social reform advocates. He detailed the many atrocities committed against animals and then described his plan to form an American Society for the Prevention of Cruelty to Animals. He quickly circulated a petition calling for the formation of such an organization. The signatories were a who's who of the politically powerful and social elite, including John Jacob Astor, Horace Greeley, and many others. He traveled to Albany, NY, with the petition in hand and on April 10, 1866, the New York State legislature granted a special charter for the formation of The American Society for the Prevention of Cruelty to Animals. Just nine days later Bergh secured the passage of a new anticruelty law in New York State that granted his fledgling organization the legal powers to enforce the law.

Henry Bergh was active immediately, stalking the streets of New York with a badge in one hand and a copy of the law in the other. The early cases prosecuted by Bergh and his agents were dominated by the mistreatment of the

FIGURE 3-3

The ASPCA influenced the formation of other societies throughout the country during the later half of the 1800s. Many of these societies adopted a version of the seal originally designed for the ASPCA by Frank Leslie. *(Courtesy ASPCA)*

many horses that worked to pull the carriages, carts, and trolleys in the city. One early case was a teamster caught beating his fallen horse with a spoke from one of the wheels. This event was memorialized by the publisher Frank Leslie, who designed the ASPCA seal depicting the scene with an avenging angel rising up, sword in hand, to stop the beating (Figure 3-3). In addition to the mistreatment of horses, the early cases of the Society included the transport of calves and other livestock to slaughter, sea turtles brought in for the restaurant trade, and companion animals. Dogfighting was common and Bergh engaged in an extended series of skirmishes with Kit Burns, an impresario of the fighting pit. Burns would arrange dogfights, ratting contests, and the occasional fight between bears and dogs, and other similar special events. Bergh enlisted the aid of the New York City police and once went so far as to drop through a skylight to surprise Burns and his patrons. Bergh and the ASPCA would prevail in this battle, eventually putting Kit Burns and his pit out of business.

The papers of the day provided extensive coverage of Bergh's exploits, along with frequent editorials lauding his efforts or criticizing him for exerting such an effort for animals and causing trouble for businesses and people looking to enjoy themselves. Bergh responded in kind and sent out a torrent of letters to the papers pointing out where animals were being mistreated and action was required, while at the same time defending himself and the ASPCA. Bergh engaged in a long-running contest with P. T. Barnum about the treatment of the animals in his menagerie and various shows and exhibitions. Barnum, the master showman, used these disputes to create even greater interest in his exhibitions. Bergh's earnest persistence would eventually win the grudging respect of Barnum, who would later serve as a pallbearer at Bergh's funeral and erect a statue in his honor in Bridgeport, CT, where he helped to start an SPCA.

The attention paid to Bergh's efforts stimulated the creation of additional SPCAs around the country in rapid order. The latent concern for animal treatment in this country would see its outlet in following the example of Henry Bergh and the ASPCA. The first to step forward was Marjorie Lord in 1867. She was the catalyst to gather the social elite of Buffalo, NY, including Millard Filmore, Henry Wells, and William Fargo to form a Buffalo branch of the ASPCA. This organization would eventually become the SPCA Serving Erie County. Bergh was already in communication with Mrs. William Appleton in Boston when a cross-country race between two horses in Massachusetts resulted in the death of both horses in February 1868. George Angell, a Boston attorney, read an account of the race and immediately responded with a letter to the paper calling for help starting a society to protect animals. Appleton and Angell soon combined their efforts, and with the information provided by Bergh formed the Massachusetts Society for the Prevention of Cruelty to Animals (MSPCA). Philadelphia was also among the first, largely because of the efforts of Carolyn Earle White (Unti, 1998b) with the formation of the Pennsylvania SPCA (Figure 3-4). Lord, Appleton, and White, despite their critical roles in the formation of these SPCAs, fell victim to the social conventions of the time and as women were unable to serve on the board of directors of the societies they helped to create. White would go on, however, to form a women's auxiliary that would open the first humane animal shelter in the United States.

FIGURE 3-4

Carolyn Earle White was one of many early leaders in the humane movement whose roles and influence were marginalized because they were women. *(Courtesy Dr. Stephen Zawistowski)*

The SPCA concept struck fertile ground throughout the country and by 1873 twenty-five states and territories had formed their own societies. Unlike the RSPCA in England, which was organized as a single national organization, each of these SPCAs in the United States was an independent organization with its own charter. While they were often explicit in their effort to follow the example of the ASPCA, using the original charter as a model and even making use of the image from the ASPCA seal, they did tend to focus their efforts and organizational structure to reflect local demands and resources.

The early work of these societies was typically focused on the protection of horses, cattle, and livestock. While dogs, cats, and other companion animals fell within their purview, they appear less frequently in their work and records for several reasons. Horses dominated the scene at the time. They were the primary source of transportation, and their mistreatment more often than not happened in public where it was easily observed. Companion animals were still uncommon and often limited to those people with enough money to engage in the extravagant luxury of feeding an animal that did not work or provide any other tangible benefit. Those companion animals that did exist may not have been readily visible if they were kept in the home, so their treatment would be more difficult to judge. The dogs and cats that were seen in public were often the strays that seemed to be everywhere in the urban landscape. Strays were rounded up occasionally and killed by a variety of horrific practices. One of the most emotional and explosive areas where dogs and cats turned up was in the laboratories of scientists. Vivisection at this time was practiced without the use of anesthetics or analgesics. Following the example of their colleagues in England, American animal protectionists mounted loud and vigorous campaigns against the use of animals in experiments. Bergh made frequent trips to

Albany where he would introduce legislation to outlaw or severely limit the practice of vivisection. While Bergh and his colleagues were never able to overcome the objections of the medical establishment in this area, it was part of a history that still frequently divides the fields of science and animal protection.

Other cats and dogs were kept and used for work of some sort or another. Cats and terriers were relied upon to kill rats and mice in stables and warehouses. Larger dogs might be used to pull carts and medium-size dogs were used to turn the rotisserie spits of eating establishments. When evidence of cruelty was observed in these various cases, SPCA representatives would attempt to intervene, often with mixed results. Bergh engaged in a long-term criticism of the New York City pound where strays were killed by drowning them in an iron cage plunged into the East River. In response to his criticism, City officials tried to have the ASPCA take over the management of the pound in New York. Bergh refused for years, believing that the City would not fulfill its promise to provide the financial resources required to do the job in a proper, humane fashion. It was not until six years after his death in 1894 that the ASPCA finally agreed to take on this task. At the same time, many of the other SPCAs around the country were also picking up the **animal sheltering** role. The introduction of electricity to power trolleys, and then the internal combustion engine resulted in the steady decline of the role horses played in the urban environment in the early twentieth century. By midcentury the animal sheltering role came to dominate the work of the SPCAs and **Humane Societies**.

American Humane Association

In 1877 John G. Shortall of the Illinois Humane Society called together leaders from the major SPCAs and humane societies to meet in Cleveland and consider issues related to the welfare of livestock being shipped by rail. This became the first meeting of what became the **American Humane Association (AHA)**. The group sought to overcome the difficulties that arose from the independent origin and activities of the many animal protection groups formed around the country. Its role would be to coordinate the work of the various regional organizations on issues that had a national significance. As its first priority the AHA concentrated on the care and treatment of livestock being transported by rail. In 1873, following on the advocacy of the American Humane Association and its member groups, combined with the literary efforts of Upton Sinclair and others, Congress passed a law that prohibited the shipment of animals for more than 28 hours without a rest stop.

Influenced by Henry Bergh's rescue of Mary Ellen in 1874, a number of the animal protection organizations began to add the protection of children to their work (see Mary Ellen sidebar). As a result, AHA would also assume a child protection role in its work. At this time, it was not unusual for groups to provide direct care services for both animals and children. Some facilities would have a single lobby with separate doors on either end for admission of animals or children. In the early 1880s, reflecting this dual role for children and animals, the AHA became an active advocate for humane education in the nation's schools.

The work of the AHA, and its materials, programs, and conventions, reflected a continued emphasis on the protection of animals being transported and slaughtered and the protection of children until the end of the 1800s. By the early 1900s, child labor law advocacy was added to the AHA's portfolio of activities. When World War I started, AHA joined with other American groups to provide aid and express concern about the treatment of horses and dogs that were used in combat and those that were victims of the fighting. In May 1915, the American Humane Association sponsored the first *Be Kind to Animals Week*, an event that continues to this day. The ensuing decades saw an extraordinary potpourri of topics appear in AHA publications and conferences, including ear cropping of dogs, lynching, cats being blamed for the decline of birds in the cities, giving children toy guns to play with, and the treatment of animals in the making of films. In 1941 the AHA came to agreement with the movie industry to be on the set and oversee the treatment of animals. This activity continues and the AHA will be acknowledged at the end of films indicating that no animals were harmed in the filming of the movie. Through its Red Star Animal Relief efforts, it also helped to rescue animals during natural disasters. At the start of World War II, Red Star Animal Relief issued recommendations for the care of animals during air raids. It recommended coordinating with Civil Defense to ensure that animals receive needed aid during wartime crises (Douglas, 1998).

Humane Society of the United States (HSUS)

In the 1950s a group within AHA saw that the organization was so caught up in the dog and cat animal sheltering issues that it no longer fulfilled its broader mission, especially when it came to opposition of vivisection and blood sports. In particular, they felt that AHA was ineffective in organizing national campaigns against rodeos, trapping, and vivisection, among other concerns. The period from World War I to World War II had seen a general decline in the vitality of the **humane movement**. By the early 1900s the original founders had passed away, along with their vision and energy. The early humane efforts had been part of reform attitude that had permeated the nation in the years following the Civil War. This synergy had also waned. A small group within AHA, led by Fred Myers, a former journalist, attempted to rekindle what they felt was a lost sense of urgency. The crux in this internal controversy was what they thought was the tepid response by AHA to the growing demand by the biomedical community for pound seizure—the release of animals in pounds and shelters for research purposes. When they were unable to stimulate the change they desired within AHA, they broke off to form a new group in 1954. The original name was the National Humane Society, but it was soon changed to the **Humane Society of the United States (HSUS)** when AHA objected to the similarity of the original name to their own. The HSUS deliberately chose Washington, DC, as its base of operations to exert the greatest possible influence on national issues.

The HSUS would begin with a strong connection to local humane organizations, helping them with their sheltering work, but also encouraging them to reclaim their role in other animal issues. Eventually the HSUS would eclipse the size and influence of the AHA, and add substantial staff to address the full range of **animal welfare** issues, including wildlife, vivisection, and agriculture in addition to companion animals (Unti, 2004).

Other Animal Groups

The formation of the HSUS presaged a flurry of activity in the animal protection field with many new organizations forming to address the concerns that led to the original development of the humane movement in Victorian England (Unti, 2004). While most of these groups place a substantial amount of their advocacy and campaign efforts on areas other than companion animals, they do continue to have an influence on companion animal issues.

- Animal Welfare Institute
  Founded in 1951 by Christine Stevens, the Animal Welfare Institute had a strong influence on early federal legislation related to the care and use of animals in laboratories. It was a key player in legislation related to preventing the theft of pets for use in experiments. This would later evolve into the Animal Welfare Act.

- Friends of Animals
  Founded in 1957, Friends of Animals is well known for its program promoting spaying and neutering of pets through certificates that subsidize the procedure.

- Fund for Animals
  Cleveland Amory, the author and commentator started the Fund for Animals in 1967. In 2005 it merged with the Humane Society of the United States.

- Animal Legal Defense Fund
  An organization of attorneys founded in 1979 by Joyce Tischler, the Animal Legal Defense Fund provides legal assistance to prosecutors handling cruelty cases, and can assist pet guardians fighting unfair laws targeting companion animals.

- People for the Ethical Treatment of Animals
  Perhaps the best known of the recent groups, People for the Ethical Treatment of Animals (PETA) was founded in 1980. It has launched a number of campaigns related to the use of companion animals used in research.

OTHER OPINIONS

While it is true that it is unlikely that you will find someone who supports the cruel treatment of animals, it is also true that there are other opinions and that the animal protection movement has faced significant opposition dating back to its origins. This opposition has come from a variety of different perspectives and will vary depending on the issue at hand. In some cases, animal protection advocates may differ with other people on whether a particular practice or treatment of animals is actually cruel, causing unnecessary suffering to animals. For example, animal protection advocates oppose the practice of docking tails and cropping the ears of dogs. They claim that this is a surgical procedure that is not needed for the welfare of the dogs and is painful. Dog breeders, on the other hand, indicate that the procedures are not painful and they are performed

FIGURE 3-5

Many humane groups are opposed to the cropping of dogs' ears as an unnecessary surgery (A)—natural ears; (B)—cropped ears). *(Courtesy ASPCA)*

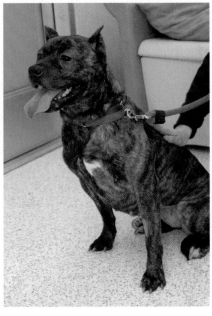

(A) (B)

by veterinarians under anesthesia and are essential to maintaining the integrity of the breeds that are typically docked and/or cropped (Figures 3-5). In other cases, the opposition may be based on economic arguments. While animal protection advocates are concerned about the treatment of animals raised for food in intensive confinement systems or *factory farming,* farmers and livestock managers who use confinement systems may counter with an argument that these systems allow for efficient animal husbandry and contribute to the low cost of food. They may also point out that these systems protect the animals from environmental extremes and disease. Protecting a traditional lifestyle can also be used to defend current practices opposed by animal protection advocates. This is a common argument used in response to opposition to hunting. These differences can be framed and debated within an animal welfare context. If the practice can be modified to limit or eliminate animal pain and suffering, it will be acceptable to some animal protection advocates. For example, agriculture practices that permit animals' greater freedom of movement, reduced population densities, and humane slaughter would be acceptable. In some cases, two groups that one would expect to disagree on nearly everything may end up in the same place on a specific issue. While animal protection groups typically oppose hunting, there is usually greater emphasis placed on efforts to eliminate "canned hunts" where wild animals are held in enclosures. In this case, traditional hunting groups that support a "fair chase" philosophy would also oppose the practice. Sportsmen's groups also joined with animal protection groups to support legislation that would prohibit hunting over the Internet, where a "hunter" can use a remote controlled gun to shoot animals.

National Animal Interest Alliance and Other Groups

The **National Animal Interest Alliance (NAIA)** represents a variety of groups that have come into conflict with animal protection groups at one time or another. The NAIA (<www.naiaonline.org>) opposes the abuse and cruel treatment of animals and supports animal welfare. NAIA draws a very distinct line separating **animal rights** from animal welfare. The NAIA defines animal welfare as the humane treatment and responsible use of animals. This use can include raising animals for food and fiber, research, circuses, and wildlife management by hunting. They describe animal rights as the opposition to most traditional uses of animals, including eating, owning pets, and hunting or for research. In addition to the NAIA, a number of other groups have been involved in the issues related to animal protection. The National Association for Biomedical Research (NABR; <http://www.nabr.org>) has supported the humane treatment and use of animals in medical and basic scientific research. The National Rifle Association (NRA; <http://www.nra.org>) is often at odds with animal groups over hunting issues and the American Kennel Club (AKC; <http://www.akc.org>) over dog-related issues. It is sometimes surprising to see that animal protection groups will on occasion be at odds with the American Veterinary Medical Association (AVMA) on some topics. These would include the docking and cropping of dogs, livestock management practices, and the use of animals in research or similar questions.

It is important to keep in mind, however, that there are many questions related to animals, and they exist within complex social and political environments. While animal protection groups may differ with the groups listed above on some topics, they may also agree on others. For example, animal protection groups, the NAIA, and the AKC are all opposed to laws that limit or restrict the ownership of particular breeds of dogs. It is probably true that most of the differences of opinion and position exist around a handful of issues and that there is general support across the board for a moderate animal welfare perspective.

 ## MODERN ANIMAL RIGHTS

On the surface, the modern animal rights movement would appear to have more in common with the early years of animal protection than the organizations and structures that developed over the past century and a half. Henry Stephens Salt anticipated many of the concepts that underlie animal rights philosophy in his work *Animal Rights* in 1892. His arguments and actions, forming the Humanitarian League in 1891 (Unti, 1998a), placed animal rights into the expanded context of human rights as expressed in concerns related to prisons, warfare, and social justice. However, while these early pioneers are frequently quoted and serve to inspire modern animal rights proponents, these heirs of Salt and others have expanded the philosophical reasoning they inherited as well as the tactics and methods they have adopted to promote their ideals. Just as the early animal protectionists were influenced by the prevailing social reform activities of their era, so too the modern animal rights proponents have adopted the tactics and practices honed by the social activists of the post–World War II time frame. Marches, protests, letter-writing campaigns, and numerous other strategies are now employed to put pressure on the government, corporations, and other entities identified as mistreating animals. While all of these had been used by activists of earlier time frames, they have now been adapted to make the best possible use of mass media opportunities that did not exist a century before.

A frequent question, and concern, is how does animal welfare differ from animal rights? Most frequently this distinction is delineated in the following fashion. Animal welfare is about limiting or preventing the cruel treatment of animals—that the human use of animals for food, research, and entertainment, among others is permissible if animal suffering is prevented, or kept to a minimum. Hence, animal welfare laws focus on humane slaughter methods, not the elimination of animal slaughter for food. A true animal rights orientation would argue that a humane slaughter law would be superfluous if animals were not killed for food at all. The animal rights position would be opposition to all use of animals by humans (Sunstein, 2004). In practice, however, it is not always that simple to make this distinction between animal rights and animal welfare positions. In many cases, organizations or individuals who describe themselves as animal rights advocates may support animal welfare efforts as an intermediate step on the path to a true animal rights solution. For example, while People for the Ethical Treatment of Animals (PETA) is a strong advocate for the adoption of a vegan diet, they also support the development of improved animal husbandry standards for animals raised for slaughter. There have been a number of other areas where animal welfare and animal rights positions have converged. It should be noted that Gary Francione (1998) has argued that this approach to animal rights should be called **new welfarism**. While this position will accept bigger cages as an intermediate step on the way to no cages, that is an improvement to reduce the suffering of animals until their use is completely abolished or otherwise eliminated; it is not a true animal rights position. In general, debate in this area typically centers on the best ways to reduce animal suffering, and how far this process should go.

Current animal rights philosophy is dominated by two primary lines of argument. The *Utilitarian* position, generally associated with Peter Singer (1975), expands upon Bentham's original proposition that every effort should be made to reduce or eliminate suffering and create opportunities for pleasure or well-being. Human use or interaction with animals could be endorsed *if* the overall benefit outweighed any pain or suffering that was caused. In most cases, however, the weight of the pain and suffering in the use of animals is presumed to greatly outweigh any expected benefit. The second line of argument arises from *Moral Rights*. The historical precedent to this reasoning arises from Immanuel Kant's supposition that strict limits be placed on individuals in the name of benefiting others (Regan, 1998). Tom Regan (1983) extended Kant's reasoning to animals. Animals have a basic moral right to be treated with respect, and this right is violated if we inflict harm on them to benefit humans.

The implications of these philosophical positions for companion animals are sometimes difficult to discern. In a very strict interpretation of animal rights, *owning pets* is questionable since it could involve the exploitation of animals for the benefit of humans. On the other hand, the proper care and guardianship of companion animals is beneficial to both sides of the human-animal bond. Complicating the picture is that the domestication of dogs, cats, and other companion animals, their extensive integration into society as pets, and their dependence on people for their care is such that denying this continued role would likely result in the extinction of these species. This would be contrary to the desires of many people who care greatly about animals and take great comfort in the relationships they share with their animal companions. In many cases animal rights and animal welfare groups utilize the bond that members of the public have with their companion animals as a bridge to appreciate the pain and suffering endured by animals used in experimentation or raised for food. At the same time it is difficult to overlook the pain and suffering that many companion animals suffer through irresponsible and negligent care, outright cruelty, and mistreatment and the problems that result from their commercial exploitation. There is broad agreement among both animal welfare and animal rights advocates that given the current circumstances, animals kept as companions should receive responsible care, efforts must be made to limit the production of unwanted progeny, and cruel treatment must be prohibited and punished. In addition to the specific question of whether animals should be kept as companions, there are a number of activities or uses of companion animals that come under criticism. Racing and other competitions, hunting with dogs, and many other activities go beyond simply keeping animals as companions. In these cases, the dogs, cats, and other animals may be exploited in ways that could bring harm to them or to other animals and raise significant animal rights concerns.

In what has been an influential book, Matthew Scully, a conservative political writer, has refashioned in *Dominion* (2002) many of the arguments originally brought forth by the early proponents of animal protection. The mistreatment of animals is wrong because it is a violation of the duty humans have to serve as stewards of animals and the natural world. His book is subtitled *The Power of Man, the Suffering of Animals, and the Call to Mercy*. It is not difficult to hear the echo of Henry Bergh's statement that *"Mercy to animals is essentially mercy to Mankind."* While much of his discussion is focused on animals raised for food, those used in research, or hunted, he does offer some insight into human relationships with pets. Regardless of philosophical underpinnings, pets *"look to us only for creaturely respect and whatever scraps of love we have to offer"* (p. 22).

Mary Ellen Wilson

In the late winter of 1873, Marietta "Etta" Angell Wheeler, a missionary and social worker in Hell's Kitchen, New York City, came to an elderly woman living alone in a tenement building. While the intent was to comfort and care for the woman, what she found instead was an appalling story of neglect and cruelty. The woman told her about a small child named **Mary Ellen Wilson,** who was kept confined in the apartment across the hall. She was haunted by the child's crying. She asked Wheeler if something might be done for the child. Wheeler went across the hall and was rebuffed by a woman there, but not before she caught a glimpse of a waif, disheveled and dressed in a tattered dress providing little protection against the cold December drafts. Wheeler went to the police and other authorities, but at the time no one was willing to take the bold step and intervene in the treatment of a child by her parents. While there were laws protecting children from mistreatment, there was no effective system in place to remove a child from an abusive home. In the midst of her despair, Wheeler's niece suggested that she ask **Henry Bergh** for help. Bergh was already well known for his work on behalf of animals. He was kind, but also bold, and willing to act when others would not.

Etta Wheeler did visit Bergh in the spring of 1874. She described little Mary Ellen and her condition to Bergh and asked if in some way he might be able to help. Bergh asked that she might provide him with a written account of her observations, and Wheeler readily assented. When Wheeler's documentation arrived at his offices, Bergh immediately sent them along to his attorney, Elbridge T. Gerry. Gerry was prepared. He requested a warrant under section 65 of the Habeas Corpus Act. This rare application of the Act relates to relieving someone of illegal confinement if there is reason to believe the individual may be carried out of the state or suffer significant injury. An ASPCA agent and two New York City detectives, writ in hand, seized Mary Ellen and brought her to the court on April 9, 1874. Bergh had already alerted the press, and they were on hand to see Mary Ellen arrive, still dressed in the simple, shabby garment she was wearing when first seen by Etta Wheeler. Since that time, Mary Ellen had "earned" a vicious cut across her face when struck with a pair of scissors by the woman who turned out to be her foster mother (Figure 3-6). The agents had wrapped her in a lap blanket from the carriage and carried her into the courthouse. The case worked its way through the court, and in the press. It became a bellwether for a pent-up movement to protect children from beatings, abuse, and hard and dangerous labor. When Mary Connolly, Mary Ellen's foster mother, was found guilty of assault and battery and sentenced to a year of hard labor, custody of the child fell to the courts. Eventually, Mary Ellen would live with Etta Wheeler's sister in upstate New York, and live to a comfortable old age.

As a result of Bergh's involvement in the case, a myth developed that Mary Ellen's rescue was the result of Bergh's declaring that if nothing else she would be protected under the same law used to outlaw cruelty to animals. Bergh was in fact so determined to keep the work of protecting animals and children separate that, along with Gerry and John D. Wright, he formed the Society for the Prevention of Cruelty to Children (Shelman & Lazoritz, 2005).

FIGURE 3-6

When Henry Bergh rescued Mary Ellen from an abusive home, he made it clear that he was not acting as president of the ASPCA. While he would eventually help to form the Society for the Prevention of Cruelty to Children, many animal protection groups would combine their efforts to protect both animals and children. *(Courtesy ASPCA)*

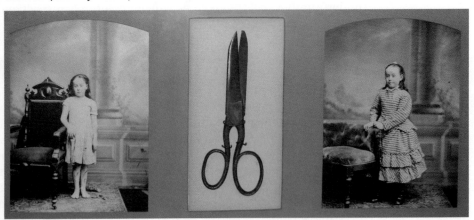

DISCUSSION QUESTIONS

1. Describe three elements of the Victorian era that contributed to our modern concepts of animal protection.
2. What is bull baiting? What impact did this activity have on the sport of dog fighting?
3. When was the Royal Society for the Prevention of Cruelty to Animals (RSPCA) established? When was the American Society for the Prevention of Cruelty to Animals (ASPCA) established? What was the relationship between the two societies?
4. How did Henry Bergh use his social and political connections to establish and grow the ASPCA?
5. What is the significance of the horse in the establishment and early years of the ASPCA?
6. How did the actions of Henry Bergh influence subsequent animal protection efforts in the United States?
7. On what types of animals did the early Societies for the Prevention of Cruelty to Animals (SPCA) focus? How is that different from modern SPCAs? What factors were/are involved in these differences?
8. As SPCAs developed throughout the United States, how were they connected to the ASPCA?
9. What factors influenced SPCAs to gradually take on a role in animal sheltering?
10. When and why was the American Humane Association (AHA) formed?
11. How did the work of the AHA influence the animal protection movement throughout the twentieth century?
12. When and why was the Humane Society of the United States (HSUS) formed?
13. What are two major issues that have brought opposition to the animal protection movement? Describe the opposing viewpoints. Provide at least one additional example not described in the text.
14. Explain the concept of *new welfarism* and its relevance to animal rights and animal welfare.

REFERENCES

Douglas, A. (1998). Archival notes on the American Humane Association.
Francione, G. L. (1998). Animal rights and new welfarism. In M. Bekoff (Ed.), *Encyclopedia of animal rights and animal welfare* (p. 45). Westport, CT: Greenwood Press.
Linzey, A. (1998). Primatt, Humphrey. In M. Bekoff (Ed.), *Encyclopedia of animal rights and animal welfare* (pp. 280–81). Westport, CT: Greenwood Press.
Linzey, A. (1998). Broome, Arthur. In M. Bekoff (Ed.), *Encyclopedia of animal rights and animal welfare* (pp. 95–96). Westport, CT: Greenwood Press.
Regan, T. (1983). *The case for animal rights*. Berkeley: University of California Press.
Regan, T. (1998). Animal rights. In M. Bekoff (Ed.), *Encyclopedia of animal rights and animal welfare* (pp. 42–43). Westport, CT: Greenwood Press.
Scully, M. (2002). *Dominion*. New York: St. Martin's Griffin.

Shelman, E., & Lazoritz, S. (2005). *The Mary Ellen Wilson child abuse case and the beginning of children's rights in 19th century America.* Jefferson, NC: McFarland.

Singer, P. (1975). *Animal liberation.* New York: New York Review of Books.

Sunstein, C. R. (2004). What are animal rights? In C. R Sunstein & M. C. Nussbaum (Eds.), *Animal rights: Current debates and new directions.* New York: Oxford University Press.

Turner, J. (1980). *Reckoning with the beast.* Baltimore, MD: Johns Hopkins University Press.

Unti., B. (1998a). Salt, Henry Stephens. In M. Bekoff (Ed.), *Encyclopedia of animal rights and animal welfare* (pp. 309–310). Westport, CT: Greenwood Press.

Unti, B. (1998b). White, Caroline Earle. In M. Bekoff (Ed.), *Encyclopedia of animal rights and animal welfare* (p. 362). Westport, CT: Greenwood Press.

Unti, B. (1998c). Martin, Richard. In M. Bekoff (Ed.), *Encyclopedia of animal rights and animal welfare* (p. 241). Westport, CT: Greenwood Press.

Unti, B. (2004). *Protecting all animals: A fifty year history of the Humane Society of the United States.* Washington, DC: Humane Society Press.

Zawistowski, S. (1998). Humane education. In M. Bekoff (Ed.), *Encyclopedia of animal rights and animal welfare* (pp. 189–191) Westport, CT: Greenwood Press.

Animal Shelters and Rescue

CHAPTER 4

No matter how much cats fight, there always seem to be plenty of kittens.

—Abraham Lincoln

KEY TERMS

poundmaster	disasters	Maddie's Fund
euthanasia	hurricanes	enrichment
zoonotic	shelter design	*Asilomar Accords*
animal control	Society of Animal Welfare Administrators	technology
decompression chambers		Internet
National Animal Control Association (NACA)	National Council on Pet Population Study and Policy (NCPPSP)	Petfinder
relinquishment		Radio Frequency Identification (RFID)
pound seizure	no-kill	microchips

Companion animal shelters arose from the impounds that towns and cities used to have for keeping stray and wandering livestock (Zawistowski & Morris, 2004). Households located within these towns typically needed to keep their own cows or goats for milk. They kept horses for transportation and may have kept chickens for eggs and a few pigs to fatten for slaughter or kitchen scraps. If these animals escaped from whatever confinement provided, the local **poundmaster** would capture the wandering animals and take them to the impound where they were held. If the owner wanted to reclaim the animal, they would need to pay the poundmaster a redemption fee. The poundmaster could either sell or slaughter animals that were not reclaimed. There was no salary for the poundmasters; they made their living based on the redemption fees paid, through the sales of animals not reclaimed, or selling or keeping the meat of the animals that were slaughtered. Occasionally, the poundmaster might need to capture and hold a dog or cat that was causing a nuisance. If a dog was valued for herding, hunting, or as a guard, the owner might make an effort to reclaim it, paying the redemption fee. Since cultural traditions did not accept dogs and cats as a food source, the poundmaster could not sell them for slaughter. As a result, excess dogs or cats would be killed by the poundmaster in the most convenient fashion. Clubbing, strangling, and drowning were among the most common methods employed. During the 1600s to early 1800s, when dogs and cats were uncommon as companion animals, they were not a significant part of the workload of the poundmaster.

However, as villages became towns and towns became cities, it became less common for the residents to maintain their own livestock. As the number of wandering cattle and other valuable animals for the poundmaster to catch declined, the numbers of dogs and cats that needed to be handled increased. While the numbers of dogs and cats grew, the system for their disposition did not change from that originally developed for livestock. Many were strays with no one to claim them. The ready availability of strays in the city for someone who wanted to acquire a cat or dog resulted in few dogs or cats being purchased from the pound. As a result, the numbers of excess dogs and cats that needed to be killed steadily increased. By the mid-1800s, the New York City pound had resorted to crowding dogs and cats into a large iron cage that was hoisted over and into the East River each day to drown the animals (Figure 4-1). There were limited efforts made to try to control the problem in a proactive fashion. Cleveland introduced dog licenses as a first try at control. In western parts of the United States, it was not uncommon for strays to be shot instead of being caught live. Dog license tags were introduced in Dodge City to help identify *owned strays*.

When Henry Bergh founded the American Society for the Prevention of Cruelty to Animals (ASPCA) in 1866, in addition to the plight of horses in New York City, he soon became enmeshed in the treatment of the dogs and cats at the pound (Figure 4-2). In addition to the issue of drowning the animals, the pound was criticized for corruption. Since the dogs and cats had limited or no *resale* value, the poundmaster and his helpers were dependent upon redemption fees for their compensation. It was not uncommon, then, for the pound workers

FIGURE 4-1

Unwanted stray animals from the pound in Manhattan were drowned daily in the East River in the 1800s. *(Courtesy ASPCA)*

Drowning strays at the New York City dog pound, 1877

FIGURE 4-2

Periodicals frequently made fun of Henry Bergh's concern about the treatment of animals, including the dogs and cats being taken to the pound. *(Courtesy ASPCA)*

FIGURE 4-3

The Women's Humane Society in Philadelphia built the first humane animal shelter to provide care for the city's homeless dogs and cats. *(Courtesy Women's Humane Society)*

to steal dogs from the yards of homes and then notify the owners that the dogs were in the pound and would be killed if they were not redeemed, for a fee. Essentially, the pound was supported by the kidnapping and ransom of dogs (Crossen, 2007). Some owners, after several trips to the pound to reclaim their dogs, eventually paid a type of protection fee to keep their dog from being taken again. While this practice occupied the efforts of the pound workers, they neglected the important job of collecting the many strays that roamed the streets of the city. These strays posed a danger when they spooked horses that were pulling carts or carriages, and a danger to harass and bite people. This concern was exacerbated by periodic outbreaks of rabies in the dog population. At these times, massive efforts would be made to round up all of the stray dogs in the city and take them to warehouses where they would eventually die or be killed. When Bergh's efforts on behalf of animals gained greater attention, city officials and the public called upon him and the ASPCA to take on the task of running the city pound. Henry Bergh refused to do this. He was quite familiar with the politics and political practices of New York's city hall at the time and did not want to risk the future of his fragile endeavor in what he anticipated would be a thankless effort that would drain the organization of energy and resources.

The first move toward a more humane method of animal sheltering was pioneered by **Carolyn Earle White** in Philadelphia. As a woman, she was unable to serve on the board of the organization that she helped to form, the Pennsylvania Society for the Prevention of Cruelty to Animals (PSPCA). She then organized the Women's Auxiliary of the PSPCA. It was White and this group that conceived, built, and operated the City Refuge for Lost and Suffering Animals (Figure 4-3). They took in the stray dogs and cats of Philadelphia and cared for them in the first *humane shelter* in the United States. They provided food and care for the strays, and made an effort to find new homes for as many of them as

What's in a Name?

If the first organizations created to protect animals were Societies for the Prevention of Cruelty to Animals, where did Humane Societies come from, and is there a difference? In Chapter 3, we saw that Henry Bergh not only founded the first organization in the United States to protect animals, the ASPCA, he also helped to found the first organization to protect children, the New York Society for the Prevention of Cruelty to Children (NYSPCC). The first humane society was formed in England by two doctors to rescue human drowning victims in 1774 and to promote the use of artificial resuscitation to save these people. Eventually *humane* became a common term used to describe the treatment and protection of people. As discussed in Chapter 3, when Bergh rescued Mary Ellen in 1874, this was clearly a humane action. Many of the SPCAs that he inspired to form and protect animals now took on the additional task of protecting children. However, unlike Bergh, they did not form new organizations but merged the child protection role into the animal protection role. In some cases, they assumed the ponderous title of *Society for the Prevention of Cruelty to Animals and Humane Society* and built a lobby with separate doors for accepting unwanted animals and children. Over time, a number of these organizations maintained their dual functions under the simpler title of Humane Society. This terminology was codified in 1885 when the American Humane Association passed a resolution that organizations providing services for both animals and children should be called Humane Societies. Still later, as the protection of children became a function of local government, the local humane society retained the animal care role and the new name. One of the most interesting examples of this transition traced the organization founded be Carolyn Earle White in Philadelphia. When she was unable to take on an active role in the management of the Pennsylvania Society for the Prevention of Cruelty to Animals that she helped to create, she founded the Women's Auxiliary of the Pennsylvania SPCA. This was eventually shortened to Women's SPCA. Later when the Women's SPCA took on a role protecting children the named was changed to Women's Humane Society, an organization that exists to this day as an animal protection group. It should be noted that in correspondence of the day Henry Bergh was vehement in his opposition to SPCAs taking on the additional task of protecting children. It was not that he thought children unworthy of protection; after all, he did rescue Mary Ellen and start the NYSPCC and served on its board until his death. He felt that if the two tasks were joined one would invariably suffer from lack of attention and support within the organization. In the end, there really are no differences between societies for the prevention of cruelty to animals and humane societies. They are largely interchangeable terms. The names do not denote differences in function or philosophy, nor do they reflect membership as chapters of centrally managed national organizations.

possible. White even commissioned the development of a more humane method of killing the excess animals that could not be placed into new homes. Their humane **euthanasia** chamber used gas to asphyxiate the animals.

In 1894, six years after the death of Henry Bergh, the ASPCA agreed to manage the New York City pound. Dog licenses were introduced to the city at that time. The fees collected from the licenses were then used to provide salaries for the men who worked at the animal shelter. Since the workers were no longer dependent upon redemption fees for their compensation, stealing of owned dogs was eliminated and the work concentrated on the capture of strays. They also made it a point to include cats as part of their efforts, the first time that cats became more than an incidental part of the pound/animal shelter effort. Strays were no longer killed by drowning. Following the example of Philadelphia, a chamber was developed that used the illuminating gas used to light homes to asphyxiate the animals (Figure 4-4).

FIGURE 4-4

The first improvement in the euthanasia of unwanted dogs and cats was the development of gas chambers. (A) When opened cages of dogs and cats could slide into the chamber. (B) The closed chamber would be infiltrated with a gas that would asphyxiate the animals. *(Courtesy ASPCA)*

(A)

(B)

Over the next several decades, many communities would solve their animal control needs by having the job taken over by the local SPCA or Humane Society. This would have a dramatic effect on the continued evolution of the American animal welfare movement. As Bergh, White, Angell, and the other founders passed away in the late 1800s and early 1900s, their organizations became more and more occupied with running animal shelters and the care of dogs and cats. As a result, a number of the other issues that were at the forefront in the early days—vivisection, hunting and blood sports, livestock transport and slaughter—became secondary concerns. It would not be until the years after World War II that these issues would again stimulate renewed interest (see Chapter 3).

Many new communities were being created at the turn of century. They needed to have animal control services, and as a result they met the need with variety of methods. In some cases, SPCAs and humane societies were among the first organizations to take root in these communities, and they assumed the job of running the animal shelters. In other cases, the city or town government might need to organize and run the shelter if there were no humane groups, or if it did not have the resources to run the shelter and animal control program. Combined with the fact that the humane groups were still founded and operated as independent organizations, not as branches of a single national organization, this resulted in the hodge-podge nature of the animal-sheltering system that currently exists in the United States.

ANIMAL SHELTER ORGANIZATIONS

It is important to recognize that animal shelters serve two important roles in the community. One function is to ensure that lost, stray, and homeless animals are provided with humane care and treatment. The second role is to protect public health from the potential dangers of animal bites or attacks, **zoonotic** diseases, and nuisances caused by animals. Historically, humane groups generally took on the former task while government fulfilled the latter. The second role has most often been called *animal control*, with its primary focus on the public interest. A recent trend has been to rename these programs *animal care and control*, recognizing that the public health function is consistent with providing humane treatment for the animals. The essential elements of a well-organized animal care and control program include the following (Handy, 2001):

- Uniformly enforce laws related to public health and safety
- Respond to nuisance complaints in a timely manner
- Investigate complaints of abuse and neglect
- Rescue mistreated and injured animals
- Shelter stray and homeless animals
- Work to reunite lost pets with their families
- Place healthy, behaviorally sound animals in responsible homes
- Euthanize suffering animals as well as those that are neither reclaimed nor adopted
- Promote mandatory identification of both dogs and cats
- Create incentives for the public to have pets sterilized
- Deter future problems through education programs

Three basic models have become the most common ways to organize and run the animal sheltering/control needs of a community. The local government may build, staff, and run the animal shelter. Funding for operations will most often come from the sale of dog licenses, redemption/reclaim fees for lost animals, fees for adoption of animals, and fines that might be collected for violation of various animal control regulations and ordinances. This funding may be supplemented with additional monies from the government budget. Operational control for the shelter will frequently be placed with the local health department or law enforcement. In these communities, a local humane group may also operate an independent animal shelter. A second option is when the local humane group enters into an agreement with the local government to provide the animal shelter/control needs for the community. Funding for operations may come through contract payments from the government, various fees collected by the humane group or some combination. The third option is really a broad range of intermediate arrangements. For example, the humane group may own and run a shelter. The government might provide field services such as picking up strays and enforcing regulations and have a contract with the humane group for housing any animals that are picked up or seized. In another example, the government may own the shelter, but the humane group may agree to run the shelter and provide the staffing. An arrangement found in some smaller communities will find the local government providing field services either through one or two animal control officers, or as a part of the law enforcement function and then contracting for boarding services through a veterinary clinic or pet boarding facility.

Overall, it is a very complicated and often difficult set of circumstances to understand. As a result, the funding for animal-sheltering services can vary

TABLE 4-1

Animal control funding for a sample of U.S. cities. *(Courtesy ASPCA)*

AREA	POPULATION in millions		BUDGET in millions	PER CAPITA	TOTAL INTAKE
	PEOPLE	ANIMALS		SPENT PER PERSON	
	2000 CENSUS	2004 ESTIMATED NUMBER	COMBINED ANIMAL SHELTER BUDGETS	COMBINED BUDGETS	TOTAL ANIMALS RECEIVED AT SHELTERS
New York	8.01	3.48	16.88	2.11	53,458
Los Angeles	3.69	1.61	19.60	5.31	67,204
Chicago	2.90	1.26	8.62	2.97	40,715
San Diego County	2.81	1.22	17.99	6.40	33,536
Houston	1.95	0.85	6.85	3.51	69,722
Broward County (Ft. Lauderdale)	1.62	0.60	7.40	4.57	54,816
Dallas	1.19	0.52	11.17	9.39	57,625
Erie County (Buffalo)	0.98	0.43	2.90	3.69	19,431
San Francisco	0.78	0.34	11.01	14.12	15,477
Cleveland	0.48	0.21	3.45	7.19	19,890

greatly from one community to another, as well as the range of services (Table 4-1). It is generally accepted that running an adequate animal control program for a community requires approximately $4 for each citizen (Handy, 2001). Sources for these funds would typically include the following:

- Income from licenses, registrations, and permits
- Impoundment fees charged when someone's animal is picked up or held at the shelter
- Redemption fees charged when someone claims a lost dog or cat at the shelter
- Fines collected for citations related to animal control regulations or ordinances
- Adoption fees
- Private donations, grants, or other funds provided specifically for the community's animal-sheltering program

FIGURE 4-5

Dogs were frequently housed in large colony kennels, with mixed sizes and sexes, until the years after World War II. *(Courtesy ASPCA)*

ANIMAL SHELTER DEVELOPMENTS

Animal shelters have seen several trends in their development. In the beginning, the primary goal in the advancement of humane animal sheltering was to improve the conditions of the shelter and find more humane ways to euthanize dogs and cats that were not claimed by owners or placed in new homes. These early shelters tended to keep dogs and cats in group pens or cages (Figure 4-5). It was common for a shelter to have seven large kennels or pens for dogs, one for each day of the week. All of the dogs that came in on Monday went into the first pen; the dogs that arrived on Tuesday went into the second pen, and so on through the week. People who came in to look for a lost dog or find a pet to adopt would be invited to look in each of the kennels. Each morning, all of the remaining dogs in the cage, kennel, or pen for that day would be removed and killed, the kennel would be cleaned, and the cycle would start over. In many locations, the local shelter or pound did not take cats at all. As a result, we have the familiar term *dog pound*. Even though most shelters may not have taken cats, they were apparently a common element of urban life. Statistics from New York City, where the ASPCA did accept and count cats as part of its work, showed that cats outnumbered dogs.

At first, the change from drowning, clubbing, or strangling unwanted animals took some time to change. The various gas chamber configurations were the first innovations. By the turn of the twentieth century, the availability of electricity began to encourage the development of various methods of electrocution. The next major development was the introduction of the decompression chamber following World War II. It is important to note that during this time limited attention was paid to the question of how many animals were killed in animal shelters; the focus was on how, not how many. In fact, the pet care booklets prepared and distributed by various humane groups made limited mention of sterilizing dogs and cats as a way to prevent overpopulation of unwanted animals. Spaying and neutering were more often recommended to manage problem behaviors. Male cats were neutered to control spraying; female dogs were spayed to keep male dogs from gathering around the house. It was not until the 1960s that concern for how many animals were being killed became an important question in the animal-sheltering community. At that time, Phyllis Wright, who worked at the Humane Society of the United States (HSUS), began to advocate a three-pronged program to reduce the numbers of unwanted animals in the United States (Unti, 2004). She promoted sterilization, education, and legislation as the way to attack the problem at its roots. Laws that required the licensing of dogs and keeping them from roaming free would help to ensure that strays were kept to a minimum and the use of a license for identification would help families find their lost dogs. Members of the public would need to be educated about the proper care for their companion animals and this would now include the importance of sterilization as a way to prevent the birth of additional unwanted dogs and cats.

By the 1970s, efforts were made to phase out the use of the **decompression chambers** that were now seen as an inhumane form of euthanasia. This period saw the introduction of an overdose of sodium pentobarbital as the preferred method of euthanizing animals in shelters. The American Veterinary Medical Association (AVMA) has recently reaffirmed their position that euthanasia by injection performed by trained personnel is the most acceptable method for dogs and cats in animals shelters (Table 4-2). A variety of organizations including American Humane Association (AHA), Humane Society of the United States (HSUS), and **National Animal Control Association (NACA)**, as well as some state health departments, provide training and certification for euthanasia technicians to ensure that they are prepared to perform the function in an effective, efficient, and humane fashion and that they understand the regulatory requirements for the use of the euthanasia drugs. Continued efforts will be required to ensure that all shelters move toward this method. A variety of reasons, including concerns about cost, availability of trained staff, the ability of shelters to purchase the required drug (it is a controlled substance), and old-fashioned fear of change, have resulted in some facilities continuing to use carbon monoxide from vehicle exhaust, gunshot, and other unacceptable methods.

ANIMAL SHELTER PROGRAMS

Animals come to shelters through a variety of routes. The most common forms of intake are through **relinquishment** by an owner who is no longer able or interested in keeping the pet, or those brought into the shelter by animal control field officers. Other animals might be seized as part of an animal cruelty investigation or held if the owner is sick, arrested, or otherwise not able to keep the animal. Disposition of animals is generally through reclaim of lost animals by owners (Return to Owner, RTO), placement of the animal in a new home (adoption), or euthanasia for humane reasons if the animal is sick or injured (Euthanized for Humane Reasons, EHR) or euthanized to make space for more animals (Table 4-3). An increasingly frequent category of disposition is when animals are transferred from one shelter to another shelter or rescue/adoption group. Many of these groups make a special effort to take dogs and cats from

TABLE 4-2

AVMA recommended euthanasia methods *(adapted from Appendix I, AVMA, 2000, Report of the AVMA panel on euthanasia)*

SPECIES	ACCEPTABLE*	CONDITIONALLY ACCEPTABLE†
Amphibians	Barbiturates, inhalant anesthetics (in appropriate species), CO_2, CO, tricaine methane sulfonate (TMS, MS 222), benzocaine hydrochloride, double pithing.	Penetrating captive bolt, gunshot, stunning and decapitation, decapitation and pithing.
Birds	Barbiturates, inhalant anesthetics, CO_2, CO. Gunshot–free-ranging only.	N_2, Ar, cervical dislocation, decapitation, thoracic compression (small, free-ranging only).
Cats	Barbiturates, inhalant anesthetics, CO_2, CO, potassium chloride in conjunction with a general anesthetic.	N_2, Ar.
Dogs	Barbiturates, inhalant anesthetics, CO_2, CO, potassium chloride in conjunction with a general anesthetic.	N_2, Ar, penetrating captive bolt, electrocution.
Fish	Barbiturates, inhalant anesthetics, CO_2, tricaine methane sulfonate (TMS, MS 222), benzocaine, hydrochloride, 2-phenoxyethanol.	Decapitation and pithing, stunning and decapitation/pithing.
Horses	Barbiturates, potassium chloride in conjunction with general anesthesia, penetrating captive bolt.	Chloral hydrate (IV, after sedation), gunshot, electrocution.
Rabbits	Barbiturates, inhalant anesthetics, CO_2, CO, potassium chloride in conjunction with a general anesthetic.	N_2, Ar, cervical dislocation (<1 kg), decapitation, penetrating captive bolt.
Reptiles	Barbiturates, inhalant anesthetics (in appropriate species), CO_2 (in appropriate species).	Penetrating captive bolt, gunshot, decapitation and pithing, stunning and decapitation.
Rodents and other small mammals	Barbiturates, inhalant anesthetics, CO_2, CO, potassium chloride in conjunction with a general anesthetic, microwave irradiation.	Methoxyflurane, ether, N_2, Ar, cervical dislocation (rats <200 g), decapitation.

*Acceptable methods are those that consistently produce a humane death when used as the sole means of euthanasia.
†Conditional acceptable methods are those that by the nature of the technique or because of greater potential for operator error or safety hazards, may not consistently produce humane death or are methods not well documented in the scientific literature.

TABLE 4-3

Shelter Intake and Disposition Data *(From Zawistowski et al., 1998)*

DISPOSITION OF DOGS AND CATS ENTERING SHELTERS REPORTING TO 1994 AND 1995 SURVEYS												
	ADOPTED (%)		RECLAIMED BY OWNER (%)		EUTHANIZED (%)		OTHER (%)		UNKNOWN (%)		TOTAL NUMBER	
TYPE	1994	1995	1994	1995	1994	1995	1994	1995	1994	1995	1994	1995
Dogs	25.4	25.6	15.6	16.1	56.0	55.0	3.0	2.3	NA	0.9	2,031,909	1,863,727
Cats	22.6	23.4	2.1	2.2	71.9	71.2	3.5	2.4	NA	0.6	1,576,087	1,424,830
Either*	20.4	13.6	9.1	8.3	67.8	74.4	2.7	3.3	NA	0.2	523,836	424,017
Overall	23.6	23.4	9.6	9.9	63.6	63.5	3.2	2.4	NA	0.7	4,131,831	3,712,375

*Some shelters did not separate statistics for dogs and cats.

FIGURE 4-6

PetSmart Charities provides funds to transport dogs from places with excess numbers to areas where they will have a better chance of being adopted. *(Courtesy PetSmart Charities)*

overburdened animal control shelters where they might have a limited chance for placement in a new home. Some of these groups will focus on a particular breed of dog, providing the special care and knowledge that might be needed. In some cases, dogs may be shipped in from another shelter in another region of the country. At this time, it is most frequently dogs from the Midwest or the South being shipped to shelters in the Northeast or West Coast.

A very effective version of this activity has been the *Rescue Waggin'* funded by PetSmart Charities (Figure 4-6). In addition to transporting dogs from areas where shelters are overburdened to shelters where there is a demand for more adoptable pets, it also works to subsidize spay/neuter programs in the source communities to reduce the numbers of unwanted dogs. Nearly 15,000 dogs are transported a total of 100,000 miles each year (PetSmart Charities, 2005).

The most controversial way for a dog or cat to leave a shelter is through **pound seizure**. This is when the animals are sent to a laboratory where they will be used as research subjects. This practice became common in the years following World War II when biomedical research underwent a substantial upsurge and animal shelters had a substantial surplus of dogs and cats. The practice outraged animal protection advocates and once again drove a wedge of distrust between them and scientists. During the intervening years some jurisdictions have passed laws that prohibit the practice, while in a few others, there is a specific requirement for the shelter to provide animals for research purposes.

Disasters provide a special challenge to shelters. Earthquakes, floods, **hurricanes**, and fires all wreak havoc on a community's infrastructure. Homes and buildings may be destroyed and people are forced to flee the danger or may be trapped in life threatening circumstances. Animals are also endangered at this time. Red Cross emergency shelters do not accept animals, so people who are evacuating with their pets need to find an alternative shelter for themselves or find somewhere for their pets. In some cases, animal shelters are able to accept the pets and hold them until they can be reclaimed by their owners. However, when the shelter itself is in the path of danger it may be necessary to set up a temporary shelter outside the danger zone. In many cases, people will leave their pets in their homes, hoping for the best. On the heels of a disaster, it is not uncommon for animal shelter staff to be among the first to enter the area and begin efforts to capture or rescue animals that are trapped, injured, or running free. Every effort is then made to post lists of recovered animals so people can reclaim their pets. In the days following the terrorist attacks on September 11, 2001, ASPCA agents rescued over 200 pets from buildings that were evacuated around ground zero immediately after the attack on the World Trade Center. In August 2005, Hurricane Katrina struck New Orleans, LA and the Gulf Coast of the United States with ferocious winds and a storm surge that devastated vast areas of the region. Thousands of experienced animal rescue professionals and volunteers from around the country were needed to assist with the rescue of animals from the affected communities. The Louisiana SPCA in New Orleans was destroyed by the floods and animals that had been in the shelter were evacuated to the Houston SPCA before the storm.

A Perfect Storm

On August 29, 2005, Hurricane Katrina rushed across the Gulf of Mexico slamming into shore along Texas, Mississippi, Louisiana, and Florida. Winds over 100 mph uprooted trees, sucked the roofs off building, and drove torrential rains and a powerful storm surge of water from the Gulf deep inland. In the wake of the storm, rescue teams swung into action to rescue people and animals stranded by the storm. In the midst of these efforts, New Orleans was dealt a further blow—storm-weakened levees gave way and a tidal wave poured out from Lake Ponchartrain to inundate much of the city. What had been a natural disaster was now a catastrophe of unprecedented proportions.

Even with early warnings before the storm, thousands of people could not, or would not, evacuate their homes. Many did not have the assistance or transportation needed to evacuate. Storm evacuation shelters would not accept animals and follow up interviews revealed that many would not leave their pets behind (Baker, 2006). Rescue efforts were further complicated when the Coast Guard, National Guard, or other organizations refused to evacuate pets along with their owners.

Animal groups across the country mobilized thousands of staff and volunteers to rescue and care for the animals that were left on their own. Many of these responders were highly trained in technical animal rescue and brought their own equipment to support their efforts, including boats, protective clothing, and animal capture tools (Figure 4-7). Still others came with not much more than the desire to help in some way. Temporary shelters were established in several locations in the region, and an infrastructure evolved to rescue and transport animals to these locations. In Gonzales, LA, located between New Orleans and Baton Rouge, a horse exposition center was adapted for use. Kennel crates were used to house individual dogs, cats, and other animals, with several crates in each horse stall. Once there, animals were given medical care, bathed, and fed. A tracking system was established to identify animals and then they were often shipped out of the area for foster care until they could be claimed by their original owners. Corporations and private individuals offered the use of vans, trucks, trailers, and even private jets to transport animals to safety.

Mother Nature had one more trick up her sleeve. On September 21, 2005, Hurricane Rita rocked the Gulf Coast region. Officials in Houston quickly put some new lessons to work by encouraging people evacuating homes in their city to take their pets with them. Officials also arranged for the pets to be sheltered along with evacuees. Meanwhile, the rescue teams still on the ground in New Orleans settled in to weather the storm. The temporary shelter locations were beehives of activity as

FIGURE 4-7

Rescue workers in New Orleans in the days after Hurricane Katrina dealt with water, waste, and wind as they sought lost and abandoned animals. *(Courtesy ASPCA)*

rescuers readied as many animals as possible for transport to safer locations. Once most of the animals were shipped off, the tent city where the rescue workers lived was broken down and people were evacuated, leaving a skeleton crew of the most experienced to weather the storm. As Rita neared, those remaining literally circled the wagons. Trucks, vans, RVs, and even pallets of pet food and cat litter were used to surround the open-sided barns where the animals were housed. As Rita howled through that night, people took turns walking the barns to calm dogs, cats, and horses. The next morning found people bleary eyed, but everyone was unharmed. Throughout the morning and the balance of the day, the evacuated workers streamed back to the site to help clean up the mess and get things running again.

Thousands of dogs, cats, birds, reptiles, and other pets were rescued following Katrina. Most of these animals ended up being moved out of the region, similar to the human population that required temporary housing. Many of these rescued animals were eventually reunited with their original families, sometimes far way from their homes in New Orleans. Others found new homes with families in the communities where they were transported for foster care. In some cases, these adopted animals were reclaimed by their original owners. Legal battles have erupted over ownership and custody of these pets, and those cases may take several years to work their way through the courts.

Among the lessons learned from Hurricanes Katrina and Rita is that people will risk, and indeed lose, their lives to protect their companion animals. In the months following the storms federal legislation was passed, the Pet Evacuation and Transportation Standards Act (PETS Act). This legislation requires that communities seeking funds from the Federal Emergency Management Agency (FEMA) for disaster preparedness must include family pets and service animals in its plans for evacuation and sheltering. It was a hard way to learn a lesson, but there is hope that the next disaster will find all of us better prepared.

SUPPORT AND PROFESSIONAL ORGANIZATIONS

A number of groups have developed over the years to provide a range of support services for animal shelter administrators and staff. These groups provide training, consultation, resource materials, and financial assistance. In the absence of a formal structure to organize animal shelters these groups have become de facto umbrellas to encourage communication and improved practices.

The American Humane Association (AHA)

Founded in 1877, the AHA has sponsored an annual conference for most of its existence. Dating to its earliest days it has brought together leaders in the field to share ideas and information and support joint action on a range of issues important to the animal welfare field. It has offered training in technical animal rescue, euthanasia certification, equine abuse, and humane education among numerous other topics. One of its more interesting roles has been the oversight of the treatment of animals that appear in films and television.

The Humane Society of the United States (HSUS)

The HSUS is the sponsor of Animal Care Expo (Figure 4-8), the largest annual conference related to animal shelter work. It brings together a wide range of individuals including animal shelter professionals, veterinarians, experts in animal behavior, fundraising, corporations, and vendors. The HSUS also provides a consultation/evaluation service for animal shelters. Agencies can arrange to have an HSUS team evaluate their facility programs and services, providing a series of recommendations for change or improvement. *Animal*

FIGURE 4-8

Shelter and humane workers from around the country attend the annual HSUS Animal Care Expo for continuing education and to network with vendors and colleagues. *(Courtesy Humane Society of the United States, http://www.hsus.org)*

Sheltering is a monthly publication for animal shelters that provides in-depth articles on important topics and developments for animal shelters. Its *Pets for Life* program emphasizes problem-solving skills to keep pets with their families, especially the management and prevention of pet behavioral problems (www.petsforlife.org).

The American Society for the Prevention of Cruelty to Animals (ASPCA)

When the ASPCA did not renew its animal control contract with New York City in 1994 it created a National Shelter Outreach department to provide training and assistance for animal shelters around the country. Among its offerings are training in animal cruelty investigation, animal shelter medicine, shelter management, and strategic planning. It emphasizes a community-based approach to animal issues and provides an innovation bank of effective programs that been developed and used in different parts of the country.

The Society of Animal Welfare Administrators (SAWA)

The leaders of several prominent animal welfare groups formed SAWA in 1970 to encourage professional development and training in the field. They have recently developed a certification program for animal welfare administrators. Certification is based on passing a 100-question multiple-choice test that reflects the knowledge and skills required. The breadth of topics required acknowledges the increasing complexity of the work in the field and the sophistication needed to be an effective manager of a nonprofit organization. The areas and topics covered in the test include the following:

- Administration and management, including strategic planning, accounting, budgeting and financial policies, contract negotiation, and rules related to nonprofit tax status
- Personnel supervision and leadership, including recruitment, selection, training and performance evaluation, labor relations, compensation, and benefits
- Public relations and fundraising, including media and presentation skills, customer service policies, fundraising, and development
- Animal care and treatment, including humane animal treatment, animal care and control laws, animal health and welfare, and **shelter design**
- Reasoning related to problem solving, information analysis and synthesis, and discretion

The **Society of Animal Welfare Administrators** holds an annual conference that emphasizes new developments in the field, especially those that influence successful organizational management. The organization also sponsors an annual meeting for directors of operations where the focus is on the day-to-day management requirements for an animal shelter.

National Animal Control Association (NACA)

The National Animal Control Association (NACA) was formed to support the movement toward more professional animal control management. It sponsors an annual training conference, a bimonthly newsletter (*NACA News*) and the NACA 100 Training Academy. The academy, combined with the NACA Animal Control Training Manual, provides an important step in formalizing the skills required by animal control personnel to protect public safety and ensure the humane treatment of animals. The National Animal Control Association also offers training certification in defensive driving, chemical immobilization, euthanasia, and the use of a bite stick (used for protection against an aggressive dog).

Other National and Regional Federations and Conferences

In addition to the national groups and organizations noted in this chapter, there are a number of other national, local, and regional federations that sponsor their own training conferences and activities. The New England Federation of Humane Societies includes a number of groups from the Northeast and sponsors an annual conference. Prairie States is an annual meeting for groups in the Midwest, and a number of states have their own animal control associations or humane federations to sponsor training and coordinated activities. The no-kill movement as noted on the following pages has resulted in national conferences that promote and support activities and programs consistent with the philosophy that unwanted dogs and cats should not be killed to manage their populations.

In 2002, the Alliance for the Contraception of Cats and Dogs (ACCD) held its first meeting to address developments in nonsurgical methods of pet sterilization. A second meeting was held in 2004. The ACCD was a loose group of scientists, animal welfare experts, veterinarians, and others with an interest in this line of research and its application to the control of dog and cat populations. The ACCD was registered as nonprofit, tax-exempt organizations in 2006 to pursue the research, development, testing, and approval of drugs and chemicals for this purpose (http://www.acc-d.org).

National Council on Pet Population Study and Policy (NCPPSP)

As more and more attention was being paid to the issue of how many unwanted dogs and cats were being killed in America's animal shelters each year, anger, distrust, and blame often led to bitter arguments between animal welfare groups, veterinarians, and the dog and cat breeders and fanciers. An animal welfare group might blame "backyard breeders" as the source of the excess, while breed clubs would accuse humane groups of inefficient and ineffective shelter management resulting in limited adoptions and high rates of euthanasia. A common thread through these arguments was the frequent absence of empirical information. Little was actually known about how many dogs and cats entered shelters each year, why or how they ended up there, and how many were actually euthanized. Several conferences were held to address these questions, and there was general agreement that good statistical information would be needed to understand the nature of the problem and develop interventions to address these issues (Anchel, 1990). Unfortunately, the decentralized organization of the nation's animal shelters made this a difficult, if not impossible, task. In 1992, Mo Salman of Colorado State University and Patricia Olson of the University of Minnesota co-organized a meeting in Minnesota. They brought together representatives from humane groups, breeders, and veterinary associations. The results of that meeting set the stage for a follow-up meeting that stimulated the formation of the **National Council on Pet Population Study and Policy (NCPPSP)**. The original ten groups that formed the NCPPSP included the following:

- American Animal Hospital Association
- American Humane Association
- American Kennel Club
- American Society for the Prevention of Cruelty to Animals
- American Veterinary Medical Association
- Association for Veterinary Epidemiology and Preventive Medicine
- Cat Fancier's Association
- Humane Society of the United States
- Massachusetts Society for the Prevention of Cruelty to Animals
- National Animal Control Association

The formation of the Council required patience and sensitivity to each organization's mission and constituents. One demonstration of this is that the name of the council did not include the term *pet overpopulation*, since not all groups agreed that this was indeed the issue. The mission stated for the NCPPSP is "to gather and analyze reliable data that further characterize the number, origin, and disposition of pets (dogs and cats) in the United States; to promote responsible stewardship of these companion animals; and based on data gathered, to recommend programs to reduce the number of unwanted pets in the United States." To accomplish its mission, the NCPPSP solicited funding from a variety of foundations and sources and then commissioned a bold research agenda. Three initiatives were authorized:

- A shelter statistics survey to identify all the animal shelters in the country and gather statistics on animal intake and disposition
- A shelter relinquishment study collecting information on why people relinquish their pets to animal shelters
- A household survey to evaluate the movement of pets in and out of households

The results of these studies were discussed in Chapter 1. The Society of Animal Welfare Administrators would later join the NCPPSP, followed by the American Pet Product Manufacturers Association (APPMA), while the American Kennel Club (AKC) would eventually resign.

NO-KILL MOVEMENT

The next step in the evolution of animal shelters was in the late 1980s when the question of euthanasia moved from how and how many to why? Duvin (1989) published an influential essay that questioned the role animal shelters played in the killing of healthy dogs and cats. This served as one of the sparks that ignited an ongoing discussion on **No-Kill**, the premise that no healthy, behaviorally sound dog or cat should be killed simply for the convenience of eliminating an unwanted animal. This concept stood in contrast to the Victorian premise that helped to stimulate the early development of the humane movement—pain was to be avoided by preventing cruelty, or if needed, killing an animal to prevent further suffering. Animals in shelters had been killed for decades in the belief that doing so would prevent further suffering from the elements and starvation as a stray, or mistreatment by people. The no-kill philosophy challenges this assumption, and it stimulated heated and at times vicious debate in the 1990s. Many new sheltering organizations and rescue groups were formed with the understanding that they would not kill healthy animals that they accepted. To do so, however, they needed to limit the number of animals they could accept, turning away dogs and cats if the shelter was full. Traditional sheltering facilities took to calling themselves *Open Admission*, noting that they would take any animal that was in need, in contrast to what they called *Limited Admission* shelters. They noted with regret that if a new home could not be found, the dog or cat might be euthanized. This was better than suffering as a stray or living the balance of its life in a cage in a shelter that would not euthanize. Even the use of the term *euthanize* has come under scrutiny. Strictly defined, euthanasia is killing an individual in immediate distress from pain and discomfort, and critics contend that killing healthy animals to prevent *possible* suffering does not meet this criterion.

A second event in 1989 played a further role in this controversy. The San Francisco SPCA, one of the oldest in the country and now under the leadership of Richard Avanzino, announced that it would not renew its contract to provide animal control services for the city of San Francisco. Following a transition period to allow the city government time to develop its own animal control program, the SFSPCA would become a no-kill sheltering organization. During the following decade the relationship between the SFSPCA and what became San Francisco Animal Care and Control (SF AC & C) would wax and wane. Eventually, the SPCA, then led by Ed Sayres, and AC & C, led by Carl Friedman, would sign a precedent-setting pact that would require the two organizations to cooperate to ensure that no adoptable animal would be euthanized in the city of San Francisco. At this point, no healthy, behaviorally sound dog or cat is being euthanized at either the SPCA or AC & C and San Francisco has the lowest overall per capita euthanasia rate of dogs and cats for any major city in America (Table 4-4). The events in San Francisco inspired similar changes in cities across the country as many of the oldest SPCAs and humane societies gave up the arrangements they had with local governments to work toward a no-kill community. The ASPCA would end its 100-year agreement with New York City in 1994, along with Richmond SPCA and others.

TABLE 4-4

Per capita pet euthanasia table *(Based on Animal People, July/August 2006, p. 18)*

CITY	ANIMALS EUTHANIZED PER 1000 HUMAN POPULATION	YEAR
Ithaca, NY	2.2	2003
New York, NY	2.6	2005
Pittsburgh, PA	8.6	2003
Baltimore, MD	9.2	2003
Philadelphia, PA	19.7	2002
Richmond, VA	7.7	2004
Augusta, GA	45.3	2004
Nashville, TN	18.9	2004
Dallas/Ft. Worth, TX	23.0	2002
Denver, CO	5.8	2002
Phoenix/Maricopa, AZ	16.0	2005
Chicago, IL	6.9	2005
Indianapolis, IN	18.5	2005
San Francisco, CA	2.5	2004
Visalia, CA	81.1	2002
Amarillo, TX	50.9	2002

Maddie's Fund

Inspired by the success in San Francisco, the Duffield family endowed the **Maddie's Fund** in 1994 to promote and fund similar cooperative community-based efforts in other cities across the country with the ultimate goal of having these no-kill cities become the foundation for a no-kill nation. From 1999 to 2005, $33 million has been distributed to support projects ranging from the development of shelter veterinary programs at University of California, Davis, Iowa State University, and Auburn University to community collaborations in Utah and Austin, TX. The Maddie's Fund approach is two-pronged seeking to increase the numbers of animals adopted from animal shelters and rescue groups through more efficient marketing and reducing the inflow of animals to the shelter by promoting the sterilization of dogs and cats to prevent unwanted litters.

Asilomar Accords

In 2004, a number of leaders in the animal-sheltering community met to discuss differences of opinion and misunderstandings related to the no-kill movement. The results of this meeting that was held in California are known as the **Asilomar Accords**. The *Asilomar Accords* set out a framework to calculate the *Annual Live Release Rate* for individual shelters and for a community—that is, to calculate the fraction of animals that enter into the shelter/rescue systems in a community that are placed in new homes, returned to their owners, or transferred to other shelters or rescue groups. Owner-requested euthanasia for dogs and cats that are very sick, badly injured, or have severe behavioral problems that would make them a risk to return to the community are excluded. The *Asilomar Accords* provided three important contributions to the effort to reduce significantly the number of healthy and treatable companion animals euthanized at animal shelters each year:

- *Guiding Principles* to govern the interpretation of the statistics and the relationships between individuals and organizations involved with animal shelters
- Consensus definitions that use nonjudgmental terminology to define various classes of dogs and cats that enter the shelter, and their dispositions
- A standard set of formulas for calculating the *Annual Live Release Rate*

It will be useful to consider these contributions since they are likely to have a significant impact on animal sheltering for the next couple of decades. These materials are provided in a separate section at the end of this chapter.

The Guiding Principles start by making it clear that there is a tangible goal—saving the lives of animals. That goal can best be achieved by having groups in a community work together in a positive fashion that is free from blame setting, in an environment of mutual respect. Various stakeholders are encouraged to eliminate insulting and "problem" language and terminology. Progress toward the goal of saving animal lives can be measured, and formulas are provided to facilitate calculations, along with standard definitions to assure uniform data collection and analysis. This may seem to be a small contribution to the problem at hand, but previous sections of this chapter have already shown the lack of standard structure and procedures in the animal-sheltering field. Within that context, the *Asilomar Accords* are a remarkable achievement.

SHELTER DESIGN

In addition to the philosophical shifts in the animal-sheltering community, there have been a number of operational and structural changes in the past 30 years. In the years following World War II, as our knowledge of medicine and disease transmission improved, shelters moved away from the group housing systems that they had been using to individual kennels or cages for each animal. Vaccinations for dogs and cats in shelters became more common as well. At this time, kennel runs were generally built with concrete block with wire fencing for the doors and perhaps the upper half of the kennel run. The run may or may not have had a door that opened to an individual outdoor run that provided the dog with a chance to get some sun and air and facilitated cleaning of the indoor half while the animal was outside. The concrete block and floor would either be painted or preferably glazed to keep dirt from collecting in the pores of the concrete. Ventilation for the facility typically depended on windows, skylights, and the occasional fan. As with all animal facilities, odor and general appearance relied heavily upon the skill and dedication of the staff and management. However, given the high animal population levels in the shelters and frequently undermanned staffs, shelters were often dank smelly places that merely exacerbated the low opinion the public might have based on the fact that dogs and cats were killed there on a daily basis. Overall, it was not a situation that lent itself to high levels of public interest and support.

The 1990s saw a change in shelter design and program function. Many of the newer facilities were built with sophisticated heating, ventilation, and air conditioning systems (HVAC) that helped to provide greater levels of comfort for the animals. It also reduced odors and the transmission of airborne diseases. It is now more common to expect that an animal shelter will have 12–15 air exchanges per hour, similar to those used for an animal laboratory facility (for comparison, most office facilities maintain 6–8 air exchanges per hour). Floors and walls are now coated with epoxy, which does not deteriorate like paint and is easier to sanitize. Shelters are better lighted than before, with larger open, airy areas that are more pleasant for visitors and the public (Figure 4-9).

It is important to recognize that proper shelter design and construction is not just an aesthetic luxury, it is an integral part of shelter function and performance. Proper material and quality workmanship that eliminates gaps with and between surfaces facilitates sanitation and cleaning. It keeps animals and staff safe by preventing escapes and prevents harm to animals from loose materials and sharp edges. Well-designed traffic patterns within the shelter

FIGURE 4-9

Modern shelter design emphasizes natural light, easy-to-clean surfaces, efficient air systems, and open, inviting spaces to provide better care for the animals, and to be more inviting to the public. *(Courtesy Julie Morris)*

help to prevent the spread of disease from one section to another. The heating and ventilation systems can keep animals comfortable and less likely to contract disease and help to prevent pathogens from spreading through the shelter. Stress is reduced for animals (and staff) by providing sound attenuation and appropriate lighting that approximates natural daylight. In the end, a well-conceived shelter provides animals with a humane place to live, for as long as they are there, a comfortable place for staff to work, and an inviting place for the public to visit.

SHELTER MEDICINE

Along with the changes in the physical structure of animal shelters, there have also been important changes in the operational procedures of shelters. A key advance has been the development of the specialty of *Animal Shelter Medicine* (Miller & Zawistowski, 2004). This discipline combines elements of general companion animal care with aspects of herd management. Medical care is provided for individual animals as needed, but special attention is paid to preventive care through the use of quarantine, isolation, and sanitation protocols. Veterinarians working in this arena have made significant advances in perfecting highly efficient, high-volume spay/neuter procedures for dogs and cats to stem the production of additional unwanted animals. Depending on the size and breadth of the shelter's functions, other aspects of the work might include veterinary forensics to support cruelty investigations and preparation for disasters. The primary function of a shelter medicine program is to support the basic functions of a high-quality animal shelter program:

1. A clean, comfortable, and sanitary environment in which stress is minimized
2. A husbandry program that focuses on proper diet, exercise, behavioral **enrichment**, and maintenance of comfortable environmental conditions
3. Foster care for sick and debilitated animals that are adoptable
4. A health care program that addresses the needs of the animals while making optimal use of resources
5. Ongoing staff and volunteer training and development
6. Humane euthanasia when appropriate according to shelter policy and the needs of the community and animals (Miller, 2004)

These efforts are not the sole domain of the veterinary professional. They must include all members of a shelter staff. However, they certainly cannot be achieved without the input and expertise of an experienced veterinarian.

A shelter medicine course has been offered at Cornell University College of Veterinary Medicine since 1999. There has been a shelter medicine track offered at the American Humane Association Annual Conference with continuing education credit available since 1992. Specialized tracks in shelter medicine have also been available at several of the major veterinary conferences, including North American Veterinary Conference and Western States. Continuing education credit is also available though courses taught on-line through Veterinary Information Network since 2004.

BEHAVIOR PROGRAMS

Beginning in the late 1980s and early 1990s, shelters also began to implement animal behavior programs (Reid, Goldman, & Zawistowski, 2004). These programs served several basic functions:

- Evaluate animals in the shelter to determine if they were appropriate to place into adoption programs.
- Provide a behavioral profile of dogs, and cats when possible, to make the best possible match for placement into a new home.
- Provide enrichment, training, and rehabilitation for animals in the shelter to reduce levels of stress, correct some behavior problems, and prepare pets for placement into new homes.

Behavior Evaluations

Behavior problems are one of the most common reasons for the relinquishment of dogs and cats to animal shelters (Salman et al., 2000). These problems include hyperactivity, household destruction, and aggression. As a result, shelters have begun to evaluate the animals that come in to determine if they are appropriate for placement in a new home. In nearly all of these cases, the evaluations are performed on dogs. There are two reasons for the evaluations. The first is to identify potential behavior problems that would result in the dog failing in a new home and coming back to the shelter. The second is to protect the safety of the new family and the community at large, as well as limiting liability that the shelter might face if the dog harmed someone after placement in a new home (Lacroix, 2004).

The procedures available to evaluate the behaviors of dogs in an animal shelter can vary greatly from one shelter to another and will depend a great deal on the training and experience of the staff conducting evaluations (Reid, Goldman, & Zawistowski, 2004). One example of a simple, commonly employed evaluation is *SAFER* (*Safety Assessment for Evaluating Rehoming*; American Humane Association, 2002), developed by certified applied animal behaviorist Emily Weiss, Ph.D., and the Kansas Humane Society. The evaluation includes six parts that focus on determining whether the dog shows a potential for dangerous aggression. The test is meant to be performed on dogs six months of age or older by staff members trained to recognize the various forms of canine communication including body postures, ear and tail positions, and vocalizations. The basic parts of the test are as follows:

- Stare test. The handler sits in a chair and holds both sides of the dog's head and looks directly into the dog's eyes. The dog's response may vary from tail wagging and attempts to lick the handler to growls and attempts to bite.
- Sensitivity test. The handler sits in a chair and, holding the dog's collar, kneads and pinches large handfuls of skin working from the ears and head back to the rump. The dog may lean into the handler, accepting the touch; remain neutral, aloof, and tolerant; or attempt to growl or bite.

- Tag test. An effort is made to engage the dog in play by calling in a high-pitched voice and reaching and giving a light, quick touch to one of the dog's hind legs, and then quickly moving back. The dog accepts this invitation to play, tries to avoid the handler, ignores the handler, or threatens the handler.
- Pinch test. The test is conducted twice. When the dog is calm, the seated handler will say "pinch" and then run one hand down the dog's front leg and pinch between two toes. The test should be repeated after one minute. Dogs may gently pull their paw from the handler's hand, lightly mouth the handler's hand, or respond in an aggressive fashion.
- Food Aggression test. The dog is presented with a bowl of dry kibble mixed with highly palatable canned food. When the dog has been eating for a short time, the handler uses an artificial hand (Assess-a-Hand) to pull the bowl away, saying, "Give me your food." The dog may allow the handler to remove the food bowl or will follow the bowl with his head. If this happens, the handler strokes the dog's head, face, and body with the hand (Figure 4-10).
- Dog-Dog Aggression test. A "helper" dog is attached to a secure wall anchor with a collar and leash. The helper should be a well-socialized and calm dog. The dog to be tested is also on a leash and brought up to the helper dog by the handler. The behavior of the dog being tested may range from attempting to play with the helper dog to growling and attempts to attack.

FIGURE 4-10

During shelter behavior evaluations, dogs may be tested for food aggression by touching them with a plastic hand on a stick—the Assess-a-Hand. *(Courtesy Dr. Emily Weiss)*

The *SAFER* test is meant to help shelters with their decision-making process on whether dogs should be placed for adoption. If the dog being tested growls or tries to bite during any stage of the evaluation, the evaluation is terminated. The dog may be retested later, or the shelter staff may decide to euthanize the dog as an inappropriate adoption risk. Any effort to place such a dog in a new home would require very careful thought, extensive behavior modification, and restrictive placement to an experienced home, typically without children. Dogs with minor issues may receive some behavior modification and training or be placed for adoption in carefully selected experienced homes.

These evaluations are controversial. If a dog *fails* the evaluation, its most likely fate will be euthanasia. Unlike medical decisions for euthanasia that are usually supported by specific, detailed diagnostic tests, euthanasia decisions for behavioral reasons may not receive the same level of acceptance among staff, volunteers, or the even the public. At the same time, the tragic consequences of new owners or members of the public being seriously harmed or killed by adopted dogs clearly demonstrate that some sort of evaluation is necessary to a well-structured animal placement program.

The *Meet Your Match Canineality Adoption Program*, also developed by Emily Weiss, is designed to help increase the likelihood that recently adopted dogs stay in their new homes. It was developed to be used after some sort of aggression evaluation has already been performed. It also includes five parts that are meant to determine friendliness, playfulness, energy level, and motivation or drive. Based on the dog's behavior during each segment of the evaluation, it is placed into one of three different categories: *Easy, Average,* or *High Maintenance*. Each of these groupings has three subgroups, resulting in nine different categories for the dogs that are available for adoption. Each of the nine categories has its own standard description of the dog's basic personality/canineality:

- High Maintenance
  - *Life of the Party*—I think everything is fun, interesting, and meant for play, especially you. Anything you do, I'll want to do too. With my own brand of surprises, life with me will keep you constantly on your toes, and the fun is guaranteed.
  - *Go-Getter*—Want to get more exercise? Action is my middle name. My "Let's Go!" lifestyle will keep you motivated to get outside and move. I have tons of energy; and just like the sun, I'm burning and working 24 hours a day, seven days a week. I'll run for miles, chase a ball for hours, and still want to play at the end of the day.
  - *Free Spirit*—Intelligent, independent, confident, and clever, I prefer making my own decisions but will listen to you if you make a good case. We're partners in this adventure. Treat me like one and we'll both live happily ever after.

- Average Maintenance
  - *Wallflower*—Shy yet charming canine searching for patient owner with relaxed lifestyle. Looking for gentle guidance to help me come out of my shell. Treat me sweet and kind and I'll blossom.

- *Busy Bee*—I'm a naturally playful, curious, and trusting canine. Take me for a big walk every day; give me something to do. After my job's done, I'll curl up in front of the fire with you in the evenings. I'm a dog on a mission to please you and myself.
- *Goofball*—I'm a fun-loving, happy-all-the-time, glass-is-half-full kind of dog looking for someone who loves to laugh and play around. Must have a great sense of humor and a bunch of tennis balls.

• Easy Maintenance

- *Couch Potato*—Like the easy life? I think I'm the perfect match for you. I'm a relaxed, laid-back kind of dog that enjoys long naps, watching movies, curling up on laps, and walking very short distances from the couch to the food bowl and back.
- *Constant Companion*—Looking for an emotionally secure, mutually satisfying, low maintenance relationship? I am all you need. Let me sit at your feet, walk by your side, and I'll be your devoted companion forever.
- *Teacher's Pet*—I've got the whole package—smart, fuzzy, four legs, love to learn, and live to please. Go ahead, teach me anything. Sit, stay, balance your checkbook, I can do it all. Keep me entertained and I'll be yours forever.

Note that all of the descriptions emphasize the positive aspects of the dog's behavior, while at the same time pointing out what the owner will need to do in terms of training or exercise. The kennel cards for the dogs will reflect these categories (Figure 4-11A–C). People coming in to adopt will then answer a simple survey and, based on their experience and desires in a new pet's behavior, will be directed to one of the three categories as their best possible match (Table 4-5). The survey form provides a standard format for adoption counselors to talk to adopters about the time and cost investment required, and their expectations regarding what they want to do with their new pet. The number of answers in columns M, L, and K are totaled and these designate the

FIGURE 4-11

The Meet Your Match™ adoption program tries to match potential adopters with dogs based on the expectations of the people and the behavior profiles of the dogs. Bright Kennel Cards provide behavior descriptions that correspond to evaluations of the dogs, and can be matched with surveys completed by the interested adopter. *(Courtesy ASPCA)*

(A)

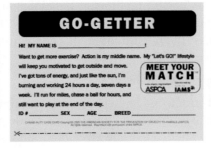
(B)

(C)

TABLE 4-5
Dog Adopter Survey *(ASPCA)*

DOG ADOPTER SURVEY

1	I have owned a dog before.	YES	NO		Currently own dog(s)
2	The last time I had a dog was...	2–10 years ago	More than 10 years ago		Not currently, but within the last year
3	My dog needs to get along with my other dogs.	NO			YES ← List their names, ages, genders, and breeds
4	My dog needs to be good with *(circle all that apply)*	Children over 8 years old	Children under 8 years old / Elderly People		Cats / Animals other than dogs or cats
5	My dog will primarily be an...	Inside dog			Outside dog
6	How many hours will your dog spend outside per day?				_____ hours
7	My dog needs to be able to be alone...	4 hours or less per day	8-10 hours per day	2 hours or less per day	12 hours per day
8	When I'm at home, I want my dog to be by my side...		All of the time	Some of the time	Little of the time
9	When I'm not at home, my dog will spend its time.	In the garage / In a crate in the house	In the yard		Loose in the house / Confined to one room in the house
10	I want a guard dog.	NO			YES
11	I want my dog to hunt or herd with me.	NO			YES
12	I want my dog to be the type that is very enthusiastic in the way s/he shows s/he loves people:		Not at all	Somewhat	Very
13	I want my dog to be playful:		Not at all	Somewhat	Very
14	I want my dog to be laid back:		Very	Somewhat	Not at all
15	I am comfortable doing some training with my dog to improve manners such as jumping, stealing food, and pulling on the leash:		No training	Some training	A lot of training
16	I (or my children) want to participate in Agility, Flyball, or Obedience with our dog.		NO	YES	
17	I am interested in a dog with "special needs" (medical or behavioral).		NO	YES	
18	How much do you think you'll spend yearly for the care of your dog? (Food, medical care, boarding, toys, etc.)				$_____

FOR OFFICE USE ONLY: N: M: L: K: D: 1-2-3-4-5-6-7-9-10-11-18

Side 1 of 2

DATE: _____ FIRST NAME _____ LAST NAME _____
ADDRESS _____ ST _____ ZIP _____ CITY _____
HOME PHONE: _____ EMAIL _____ WORK PHONE _____

Copyright © 2005 THE ASPCA®. All rights reserved. Reprinted with permission of the ASPCA.

most compatible match between adopter and dog. The column with the highest total corresponds to one of the three general categories for the dogs:

- Column M = Easy Maintenance dogs
- Column L = Average Maintenance dogs
- Column K = High Maintenance dogs

These matches are based on research conducted by Emily Weiss that included testing dogs being boarded at a kennel and collecting information from the dogs' owners regarding the behavior of their dogs and how they interacted with them. The system is meant to replicate the matches that were observed under those conditions (*Meet Your Match* information and materials provided courtesy of Dr. Emily Weiss and the American Society for the Prevention of Cruelty to Animals).

Enrichment

The National Research Council's *Guide for the Care and Use of Laboratory Animals* indicates, "Animals should be housed with the goal of maximizing species-specific behaviors and minimizing stress-induced behavior. For social species this normally requires housing in compatible pairs or groups" (National Research Council, 1996, p. 22). Meeting these standards in an animal shelter environment can be a significant challenge due to shortcomings in staff and facility design. This is especially true given the restrictions on group housing to manage disease conditions, and the difficulty managing behaviors of poorly socialized animals. Progress is being made, however, as newer shelters are being designed with group housing options for appropriately behaved animals, especially cats. More effort is also being made to provide dogs with opportunities for exercise through regular walks, temporary playgroups in dog runs, and obedience training classes. Active physical periods should be combined with options for object play and variety in feeding procedures.

It is important to keep in mind that companion animals have evolved to prosper in human company. Ensuring regular and positive human contact for dogs and cats is both beneficial and indeed necessary to ensure that individual animals do not deteriorate psychologically while in confinement in the shelter, and to either maintain or improve their social skills with people in anticipation of placement into a new home. Obedience training classes for dogs in the shelter will provide them with both physical and psychological stimulation. At the same time, the skills developed during obedience classes will make the dog a more likely candidate for placement with a new family. Object play and feeding activities can be used to occupy the dog when volunteers or staff are not available. Food or other treats can be placed into various puzzle boxes or other objects so the dog can *work* for its meals. It is important to build the amount of food provided in this fashion, or as treats during training into the overall feeding plan to ensure the animals get the proper nutritional balance and not become obese.

If the shelter has the appropriate staff on hand, or experienced volunteers available, targeted rehabilitation can be attempted to correct specific problems. These problems may be identified during the evaluation process, while observing the dog or cat's behavior in the shelter, or based on information provided by the previous owner during relinquishment to the shelter. Rehabilitation will typically take the form of some type of behavior modification. Fears and anxieties may be reduced through habituation or desensitization to the evoking stimuli, and more appropriate responses developed by positive reinforcement. A number of shelters have had success when they place dogs and cats with behavior problems into foster homes with specially trained volunteers to provide intensive daily interaction and treatment.

TECHNOLOGY

Shelters were slow to implement new forms of **technology** to assist in their work. That is changing and many sheltering and rescue groups are now using state-of-the-art technology. It may be hard to believe, but until the early 1990s, most animal shelters still kept their records in logbooks or on handwritten kennel cards. Looking for information on an animal in the shelter, trying to connect a family with a lost pet or even compiling a simple monthly statistical report required laborious hand sorting of these records. There are now a number of readily available software packages to support animal shelter management. These systems are making it easier for shelters to keep track of intake and disposition of animals, and maintain accurate statistical information that can be used to target resources and efforts where needed.

One of the most significant technology developments has been the use of the **Internet** to promote and support pet adoptions. The leading Web site here is **Petfinder** (<www.Petfinder.com>). Started in suburban New Jersey by Betsy and Jared Saul as a New Year's Eve resolution in 1995, it grew rapidly. It now includes over 9,000 shelters and rescue groups that post pets available for adoption. Shelters are able to provide information on the dogs, cats, and other animals that are available, along with a picture and brief description. Members of the public looking for a pet can search by species, breed, geographic area, and other criteria. Using the Internet has been especially helpful for rescue groups that do not have a facility that people can come to visit. These groups often foster the animals that they have available in the homes of volunteers, making it difficult for potential adopters to see the animals. It has also been helpful for shelters that are located in out-of-the-way, difficult-to-find locations. People can scan their listings and call ahead if they see a pet they would be interested in adopting listed, saving a trip if the pet has already been placed, or if no pet that would meet their needs is available.

One area of technology that has not fulfilled its potential is microchips, or **Radio Frequency Identification (RFID)**. **Microchips** are small (about the size of a grain of rice). The glass capsule holds a small radio transmitter that is injected under the skin of an animal. When the animal is scanned with a device that emits a radio wave, the microchip echoes back a signal with an identification number similar to systems that scan bar codes at cash registers (Figure 4-12). The ID number can then be used to trace the owner of the pet. One of the important advantages of this system is that it provides permanent identification for an animal that cannot be lost or fall off. About 5 percent of the pets in the country are currently microchipped. Several problems have hindered wide-scale implementation of microchip identification systems. First is that there are

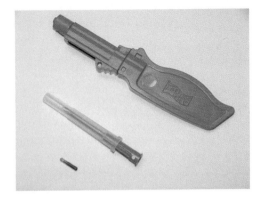

FIGURE 4-12

Radio Frequency Identification (RFID) or microchips are a permanent method of providing pets with ID. It depends on an updated database with the microchip's number and contact information for the owner, along with a system of readers for the microchips at animal shelters and veterinary clinics. *(Courtesy Allflex)*

several competing commercial interests and there have been problems with compatibility of their products. Microchips distributed by one company may not always be read by the scanners manufactured by another company. Second is the fact that the while the microchips provide a form of digital identification, that ID must then be matched with information on the name, address, and other contact information of the owner. This requires a robust and well-maintained database with owner information that must be updated regularly to keep it current. The American Kennel Club (AKC) has maintained the Companion Animal Recovery system (CAR) as a registry for microchipped pets. However, there are other registries. A number of animal care agencies have banded together to form a coalition (Coalition for Reuniting Pets and Families) to promote a uniform system of microchips and scanners that support full capacity to read all microchips, and a network of coordinated databases to simplify matching a microchips ID number with owner contact information.

ASILOMAR ACCORDS
Guiding Principles

1. The mission of those involved in creating the *Asilomar Accords* is to work together to save the lives of all healthy and treatable companion animals.

2. We recognize that all stakeholders in the animal welfare community have a passion for and are dedicated to the mutual goal of saving animals' lives.

3. We acknowledge that the euthanasia of healthy and treatable animals is the sad responsibility of some animal welfare organizations that neither desired nor sought this task. We believe that the euthanasia of healthy and treatable animals is a community-wide problem requiring community-based solutions. We also recognize that animal welfare organizations can be leaders in bringing about a change in social order and other factors that result in the euthanasia of healthy and treatable animals, including the compounding problems of some pet owners'/guardians' failure to spay and neuter; properly socialize and train; be tolerant of; provide veterinary care to; or take responsibility for companion animals.

4. We, as animal welfare stakeholders, agree to foster a mutual respect for one another. When discussing differences of policy and opinion, either

publicly or within and among our own agencies, we agree to refrain from denigrating or speaking ill of one another. We will also encourage those other individuals and organizations in our sphere of influence to do the same.

5. We encourage all communities to embrace the vision and spirit of these Accords, while acknowledging that differences exist between various communities and geographic regions of the country.

6. We encourage the creation of local "community coalitions" consisting of a variety of organizations (e.g., governmental animal control agencies, nonprofit shelters, grassroots foster care providers, feral cat groups, funders and veterinary associations) for the purpose of saving the lives of healthy and treatable animals. We are committed to the belief that no one organization or type of organization can achiever this goal alone, and that we need one another, and that the only true solution is to work together. We need to find common ground, put aside our difference and work collaboratively to reach the ultimate goal of ending the euthanasia of healthy and treatable companion animals.

7. While we understand that other types of programs and efforts (including adoption, spay and neuter programs, education, cruelty investigations, enforcement of animal control laws and regulation, behavior and training assistance and feral cat management) play a critical role in impacting euthanasia figures, for purposes of this nationwide initiative we have elected to leave these programs in the hands of local organizations and encourage them to continue offering, and expanding upon, these critical services.

8. In order to achieve harmony and forward progress, we encourage each community coalition to discuss language and terminology which has been historically viewed as hurtful or divisive by some animal welfare stakeholders (whether intentional or inadvertent), identify "problem" language, and reach a consensus to modify or phase out language and terminology accordingly.

9. We believe in the importance of transparency and the open sharing of accurate, complete animal-sheltering data and statistics in a manner which that is clear to both the animal welfare community and the public.

10. We believe it is essential to utilize a uniform method of collecting and reporting shelter data in order to promote transparency and better assess the euthanasia rate of healthy and treatable animals. We determined that uniform method of reporting needs to include the collection and analysis of animal-sheltering data as set forth in the "Animal Statistics Table." These statistics need to be collected for each individual organization and for the community as a whole and need to be reported to the public annually (e.g., Web sites, newsletters, annual reports). In addition, we determined that each community's "Live Release Rate" needs to be calculated, shared, and reported annually to the public—individually by each organization and jointly by each community coalition. Both individual organizations and community coalitions should strive for

continuous improvement of these numbers. The "Animal Statistics Table" and formulas for calculating the "Live Release Rate" are forth in Section IV of these Accords.

11. We developed several standard "definitions" to enable uniform and accurate collection, analysis, and reporting of animal-sheltering data and statistics. We encourage all communities to adopt the definitions which are set forth in Section III, and implement the principles of these Accords.

12. While we recognize that many animal welfare organizations provide services to companion animals other than dogs and cats, for purposes of this nationwide initiative we have elected to collect and share data solely as it related to dogs and cats.

13. We are committed to continuing dialogue, analysis, and potential modification of this vision as needs change and as progress is made toward achieving our mission.

14. Those involved in the development of the *Asilomar Accords* have agreed to make a personal commitment to ensure the furtherance of these accords, and to use their professional influence to bring about nationwide adoption of this vision.

Definitions

In order to facilitate the data collection process and assure consistent reporting across agencies, the following definitions have been developed. The Asilomar participants hope that these definitions are applied as a standard for categorizing dogs and cats in each organization. The definitions, however, are not meant to define the outcome for each animal entrusted to our care.

- Healthy. The term *healthy* means and includes all dogs and cats eight weeks of age or older that, at or subsequent to the time the animal is taken into possession, have manifested no sign of a behavioral or temperamental characteristic that could pose a health or safety risk or otherwise make the animal unsuitable for placement as a pet, and that have manifested no sign of disease, injury, a congenital or hereditary condition that adversely affects the health of the animal or that is likely to aversely affect the animal's health in the future.

- Treatable. The term *treatable* means and includes all dogs and cats that are rehabilitatable and all dogs and cats that are manageable.

- Rehabilitatable: The term *rehabilitatable* means and includes all dogs and cats that are not healthy, but are likely to become healthy, if given medical, foster, behavioral, or other care equivalent to the care typically provided to pets by reasonable and caring pet owners/guardians in the community.

- Manageable: The term *manageable* means and includes all dogs and cats that are not healthy and that are not likely to become healthy, regardless of the care provided. These animals would likely maintain a satisfactory quality of life if given medical, foster, behavioral, or other care (including long-term care) equivalent to the care typically provided to pets by reasonable and

IV. Annual Animal Statistics & Live Release Rate Formulas

1. ANNUAL ANIMAL STATISTICS TABLE		DOG	CAT	TOTAL
A	BEGINNING SHELTER COUNT (date)			
	INTAKE (Live Dogs & Cats Only)			
B	From the Public			
C	Incoming Transfers from Organizations within Community/Coalition			
D	Incoming Transfers from Organizations outside Community/Coalition			
E	From Owners/Guardians Requesting Euthanasia			
F	Total Intake [B + C + D + E]			
G	Owner/Guardian Requested Euthanasia (Unhealthy & Untreatable Only)			
H	ADJUSTED TOTAL INTAKE [F minus G]			
I	ADOPTIONS			
J	OUTGOING TRANSFERS to Organizations within Community/Coalition			
K	OUTGOING TRANSFERS to Organizations outside Community/Coalition			
L	RETURN TO OWNER/GUARDIAN			
	DOGS & CATS EUTHANIZED			
M	Healthy (Includes Owner/Guardian Requested Euthanasia)			
N	Treatable—Rehabilitatable (Includes Owner/Guardian Requested Euthanasia)			
O	Treatable—Manageable (Includes Owner/Guardian Requested Euthanasia)			

Continued

1. ANNUAL ANIMAL STATISTICS TABLE			DOG	CAT	TOTAL
P		Unhealthy & Untreatable (Includes Owner/Guardian Requested Euthanasia)			
Q		Total Euthanasia [M + N + O + P]			
R		Owner/Guardian Requested Euthanasia (Unhealthy & Untreatable Only)			
S		ADJUSTED TOTAL EUTHANASIA [Q minus R]			
T		SUBTOTAL OUTCOMES [I + J + K + L + S] (Unhealthy & Untreatable Only) Excludes Owner/Guardian Requested Euthanasia			
U		DIED OR LOST IN SHELTER/CARE			
V		TOTAL OUTCOMES [T + U] Excludes Owner/Guardian Requested Euthanasia (Unhealthy & Untreatable Only)			
W		ENDING SHELTER COUNT (date)			

To check the accuracy of the shelter data you've compiled, the Beginning Shelter Count (A) plus the Adjusted Total Intake (H) should equal the Total Outcomes (V) plus the Ending Shelter Count (W): A + H = V + W.

caring owners/guardians in the community. The term *manageable* does not include any dog or cat that is determined to pose a significant risk to human health or safety or to the health or safety of other animals.

- Unhealthy and Untreatable. The term *unhealthy and untreatable* means and includes all dogs and cats that, at or subsequent to the time they are taken into possession (1) have a behavioral or temperamental characteristic that poses a health or safety risk or otherwise makes the animal unsuitable for placement as a pet, and are not likely to become healthy or treatable even if provided the care typically provided to pets by reasonable and caring pet owner/guardians in the community; (2) are suffering from a disease, injury, or congenital or hereditary condition that adversely affects the animal's health or is likely to adversely affect the animal's health in the future, and are not likely to become healthy or treatable even if provided the care typically provided to pets by reasonable and caring pet/guardians in the community; or (3) are under the age of eight weeks and are not likely to become healthy or treatable, even if provided the care typically provided to pets by reasonable and care pet owners/guardians in the community.

ANNUAL LIVE RELEASE RATE FORMULAS

The Annual Live Release Rate is calculated by dividing total live outcomes (adoptions, outgoing transfers, and return to owner/guardian) by total outcomes (total live outcomes plus euthanasia not including owner/guardian requested euthanasia or died/lost in shelter/care).

Calculation for an individual agency:

Adoptions + All Outgoing Transfers + Return to Owner/Guardian divided by Total Outcomes excluding owner/guardian-requested euthanasia (unhealthy and untreatable) and dogs and cats that died or were lost in the shelter/care.

Annual Live Release Rate = $(I + J + K + L)/(T) * 100 = $ _____ %

NOTE: The Annual Live Release Rate Formula is different for an individual agency and a coalition or community due to transfers between agencies.

When reporting the Annual Live Release Rate for an individual agency, you should include the following statement: *The Annual Live Release Rate does not include* _____ *owner/guardian requested euthanasia that were unhealthy and untreatable [see Line R] and* _____ *dogs and cats that died or were lost in the shelter/care [see Line U].*

Calculation for community or coalition:

Adoptions + Return to Owner/Guardian divided by Total Outcomes excluding all outgoing transfers, owner/guardian requested euthanasia (unhealthy and untreatable), and dogs and cats that died or were lost in the shelter/care.

Annual Live Release Rate = $(I + K + L)/(T - J) * 100 = $ _____ %

When reporting the Annual Live Release Rate for the community or a coalition, you should include the following statement: *The Annual Live Release Rate does not include* _____ *owner/guardian requested euthanasia that were unhealthy and untreatable [see Line R] and* _____ *dogs and cats that died or were lost in the shelter/care [see Line U].*

Glossary of Terms for Calculation of Annual Animal Statistics

(A) **Beginning Shelter Count (Date):** The number of dogs and cats in your shelter or in your care including fosters at the beginning of the reporting period. The reporting period is annual—either a calendar year or a fiscal year. (Date) refers to the first day of the reporting period written in the following format: month/day/year.

Intake (Live Dogs and Cats Only): This table only deals with live dogs and cats for which your shelter or animal group assumed responsibility. Dogs and cats categorized as "dead on arrival," or DOA, are not included in these statistics. For intake animals, status is determined at the time paperwork is initiated.

(B) **From the Public:** The number of live dogs and cats your shelter or animal group received from the public. This includes dogs and cats turned in or surrendered by their owners/guardians; stray dogs and cats turned in by the public; stray dogs and cats picked up in the field; and dogs and cats impounded for cruelty investigation, custody care, and statutory/ordinance impoundment.

(C) **Incoming Transfers from Organizations within Community/Coalition:** The number of dogs and cats your shelter or animal group received from other animal organizations participating in your collaborative group. (This only applies if the reporting organization is working collaboratively with other shelters/groups in their area.) NOTE: On the community or coalition level, C (*Incoming Transfers from Organizations within Community/Coalition*) should equal J (*Outgoing Transfers to Organizations within Community/Coalition*).

(D) **Incoming Transfers from Organizations outside Community/Coalition:** The number of dogs and cats your shelter or animal group received from animal organizations that are not participating in your collaborative group.

NOTE: If you are not part of a collaboration that is compiling statistics, then all your incoming transfers would be listed here.

(E) **From Owners/Guardians Requesting Euthanasia:** The number of dogs and cats turned in or surrendered to your shelter or animal group by their owners/guardians for the purpose of euthanasia. This includes all categories of dogs and cats (healthy, treatable-rehabilitatable, treatable-manageable, unhealthy and untreatable). [See M, N, O, and P for definitions of healthy, treatable-rehabilitatable, treatable-manageable, and unhealthy and untreatable.]

(F) **Total Intake:** The sum of lines B through E. This includes all live dogs and cats for which your shelter or animal group assumed responsibility.

(G) **Owner/Guardian Requested Euthanasia (Unhealthy and Untreatable Only):** The number of unhealthy and untreatable dogs and cats your shelter or animal group euthanized at the request of their owners/guardians and the number of dogs and cats ordered to be euthanized by legislative, judicial or administrative action. Do not include any dogs and cats your shelter or animal group euthanized at the request of their owners/guardians and that were considered to be healthy, treatable-rehabilitatable, or treatable-manageable at the time of death. [See M, N, O, P for definitions of healthy, treatable-rehabilitatable, treatable-manageable, and unhealthy and untreatable.]

(H) Adjusted Total Intake: Lines F minus G. Total Intake minus the number of unhealthy and untreatable dogs and cats your shelter or animal group euthanized at the request of their owners/guardians. [See P for definition of unhealthy and untreatable category.]

(I) Adoptions: The number of dogs and cats your shelter or animal group placed with members of the public. Do not include dogs and cats in foster homes or dogs and cats transferred to other animal welfare organizations.

(J) Outgoing Transfers to Organizations within Community/Coalition: The number of dogs and cats your shelter or animal group turned over to other animal organizations within your collaborative group. (This only applies if the reporting organization is working collaboratively with other shelters/groups in their area.) NOTE: On the community or coalition level, J (*Outgoing Transfers to Organizations within Community/Coalition*) *should be equal to C (Incoming Transfers from Organizations within Community/Coalition).*

NOTE: If you are not part of a collaboration that is compiling statistics, then all your outgoing transfers would be listed here.

(K) Outgoing Transfers to Organizations outside Community/Coalition: The number of dogs and cats your shelter or animal group turned over to animal organizations that are not part of your collaborative group.

(L) Return to Owner/Guardian: The number of stray dogs and cats your shelter or animal group reunited with their owners/guardians and the number of dogs and cats reclaimed by their owners/guardians. Dogs and Cats Euthanized: The number of dogs and cats your shelter or animal group euthanized, broken down into the following categories: healthy; treatable-rehabilitatable; treatable-manageable; and unhealthy and untreatable. Dogs and cats are categorized at the time of euthanasia. [See M, N, O, and P for definitions of healthy, treatable-rehabilitatable, treatable-manageable, and unhealthy and untreatable.]

(M) Healthy (Includes Owner/Guardian Requested Euthanasia): The number of healthy dogs and cats that your shelter or animal group euthanized including the number of healthy dogs and cats your shelter or animal group euthanized at the request of their owners/guardians.

The term *healthy* means and includes *all dogs and cats eight weeks of age or older that, at or subsequent to the time the animal is taken into possession, have manifested no sign of a behavioral or temperamental characteristic that could pose a health or safety risk or otherwise make the animal unsuitable for placement as a pet, and have manifested no sign of disease, injury, or congenital or hereditary condition that adversely affects the health of the animal or that is likely to adversely affect the animal's health in the future.*

(N) Treatable-Rehabilitatable (Includes Owner/Guardian Requested Euthanasia): The number of treatable-rehabilitatable dogs and cats that your shelter or animal group euthanized including the number of treatable-rehabilitatable dogs and cats your shelter or animal group euthanized at the request of their owners/guardians. (These conditions are generally considered to be curable.)

The term treatable means and includes all dogs and cats that are rehabilitatable and all dogs and cats that are manageable. The term rehabilitatable means and includes all dogs and cats that are not healthy, but that are likely to become healthy, if given medical, foster, behavioral, or other care equivalent to the care typically provided to pets by reasonable and caring pet owners/guardians in the community.

(O) Treatable-Manageable (Includes Owner/Guardian Requested Euthanasia): The number of treatable-manageable dogs and cats that your shelter or animal group euthanized including the number of treatable-manageable dogs and cats your shelter or animal group euthanized at the request of their owners/guardians. (These conditions are generally considered to be chronic.)

The term treatable means and includes all dogs and cats that are rehabilitatable and all dogs and cats that are manageable. The term manageable means and includes all dogs and cats that are not healthy and that are not likely to become healthy, regardless of the care provided. These animals would likely maintain a satisfactory quality of life, if given medical, foster, behavioral, or other care (including long-term care) equivalent to the care typically provided to pets by reasonable and caring pet owners/guardians in the community. The term manageable does not include any dog or cat that is determined to pose a significant risk to human health or safety or to the health or safety of other animals.

(P) Unhealthy and Untreatable (Includes Owner/Guardian Requested Euthanasia): The number of unhealthy and untreatable dogs and cats that your shelter or animal group euthanized including the number of unhealthy and untreatable dogs and cats your shelter or animal group euthanized at the request of their owners/guardians and the number of dogs and cats ordered to be euthanized by legislative, judicial, or administrative action.

The term unhealthy and untreatable means and includes all dogs and cats that, at or subsequent to the time they are taken into possession, (1) have a behavioral or temperamental characteristic that poses a health or safety risk or otherwise makes the animal unsuitable for placement as a pet, and are not likely to become healthy or treatable even if provided the care typically provided to pets by reasonable and caring pet owners/ guardians in the community; (2) are suffering from a disease, injury, or congenital or hereditary condition that adversely affects the animal's health or is likely to adversely affect the animal's health in the future, and are not likely to become "healthy" or "treatable" even if provided the care typically provided to pets by reasonable and caring pet owners/guardians in the community; or (3) are under the age of eight weeks and are not likely to become "healthy" or "treatable," even if provided the care typically provided to pets by reasonable and caring pet owners/guardians in the community.

(Q) Total Euthanasia: Sum of lines M through P. This includes all dogs and cats your shelter or animal group euthanized (Healthy, Treatable-Rehabilitatable, Treatable-Manageable, and Unhealthy and Untreatable). [See M, N, O, and P for definitions of healthy, treatable-rehabilitatable, treatable-manageable, and unhealthy and untreatable.]

(R) Owner/Guardian Requested Euthanasia (Unhealthy and Untreatable Only): The number of unhealthy and untreatable dogs and cats that your shelter or animal group euthanized at the request of their owners/guardians and the number of dogs and cats ordered to be euthanized by legislative, judicial, or administrative action. Do not include any dogs and cats your shelter or animal group euthanized at the request of their owners/guardians and that were considered to be healthy, treatable-rehabilitatable, or treatable-manageable at the time of death. [See M, N, O, and P for definitions of healthy, treatable-rehabilitatable, treatable-manageable, and unhealthy and untreatable.]

(S) Adjusted Total Euthanasia: Total Euthanasia minus Owner/Guardian Request Euthanasia (Unhealthy and Untreatable Only). [See P for definition of unhealthy and untreatable category.]

(T) Subtotal Outcomes: Sum of lines I through L plus S. This includes the number of dogs and cats that your shelter or animal group adopted, transferred, or returned to owner/guardian. Do not include the number of dogs and cats that died or were lost while in your shelter or in your care or the number of unhealthy and untreatable dogs and cats that your shelter or animal group euthanized at the request of their owners/guardians or the number of dogs and cats ordered to be euthanized by legislative, judicial, or administrative action. [See P for definition of unhealthy and untreatable category.]

(U) Died or Lost in Shelter/Care: The number of dogs and cats for which your shelter or animal group assumed responsibility and that died or could not be accounted for. This includes the number of dogs and cats that died of medical complications (and were not euthanized), died in foster care or in transit, or were lost or stolen from the shelter.

(V) Total Outcomes: Sum of lines T and U. This is the total number of dog and cat outcomes that includes the number of dogs and cats your shelter or animal group adopted, transferred, returned to owner/guardian plus the number of dogs and cats for which your shelter or animal group assumed responsibility and that died of medical complications (and were not euthanized) or were lost or stolen (from the shelter or foster care). Total outcomes do not include the number of unhealthy and untreatable dogs and cats that your shelter or animal group euthanized at the request of their owners/guardians or the number of dogs and cats ordered to be euthanized by legislative, judicial, or administrative action. [See P for definition of unhealthy and untreatable category.]

(W) Ending Shelter Count (Date): The number of dogs and cats in your shelter or in your care including fosters at the end of the reporting period. The reporting period is annual—either a calendar year or a fiscal year. (Date) refers to the last day of the reporting period written in the following format: month/day/year.

Special Note: *To calculate the Annual Live Release Rate for your community or coalition, each participating shelter or animal group will need to fill out the Annual Animal Statistics Table for its individual organization. The reporting organization for the community/coalition will then compile this information for all the groups in one table and follow the instructions for calculating the community/coalition rate.*

DISCUSSION QUESTIONS

1. Describe the workload of the poundmaster in the mid-1800s. How had this changed over the previous two centuries?
2. What role did Carolyn Earle White play in the history of animal sheltering? Describe the relevant events that immediately preceded her actions.
3. What major changes occurred when the ASPCA took over management of the New York City pound in 1894?
4. Do communities in the United States address their animal-sheltering needs consistently? What determines how a community addresses these needs?
5. What was Phyllis Wright's three-prong approach to reducing the homeless animal population?
6. How have the methods of killing unwanted/homeless animals changed from the early 1800s to the present?
7. Describe the multiple challenges presented by the decentralized organization of the nation's animal shelters.
8. What were the three initiatives of the research agenda of the National Council on Pet Population Study and Policy (NCPPSP)?
9. Choose three of the national or regional support organizations and describe the major services they provide to animal sheltering administrators and staff. What is the underlying need for these organizations?
10. What is the difference between "open admission" and "limited admission" animal shelters?
11. What are the *Asilomar Accords*?
12. What are the major considerations of proper animal shelter design?
13. What are special considerations of a shelter medicine program that distinguish it from a clinical veterinary practice?
14. What is the purpose of the *SAFER* test? Briefly describe its components.
15. What is the purpose of the *Meet Your Match Canine-ality Adoption Program*? Briefly describe its components.
16. What are the benefits and challenges of providing group housing in an animal shelter?
17. What technological advances have been made in animal sheltering in recent decades?
18. Why is the Annual Live Release Rate formula different for individual agencies and those in a community and/or coalition?

REFERENCES

American Humane Association. (2002). *SAFER: The Safety Assessment for Evaluating Rehoming*. Denver, CO.

Anchel, M. (Ed). (1990). *Overpopulation of cats and dogs: Causes, effects, and prevention*. New York: Fordham University Press.

AVMA Panel on Euthanasia. (2001). Report of the AVMA panel on euthanasia. *Journal of the American Veterinary Medical Association, 218*(5), 669–696.

Baker, B. (2006). No friend left behind. *AARP Bulletin, 47*(5), 22–23.

Crossen, C. (2007, February 5). Dogs' role in society evolved: Their catcher never won our hearts. *The Wall Street Journal*, B1.

Duvin, E. (1989). *In the name of mercy*. Retrieved August 9, 2007, from www.BestFriends.org/nomorehomelesspets/pdf/mercy.pdf.

Handy, G. (2001). *Animal control management: A guide for local governments*. Washington, DC: International City/County Management Association.

LaCroix, C. (2004). Legal concerns for shelters and shelter veterinarians. In L. Miller & S. Zawistowski (Eds.), *Animal shelter medicine for veterinarians and staff* (pp. 35–45). Ames, IA: Blackwell Publishing.

Maddie's Fund. (2004). *Annual report*. Alameda, CA: Author.

Miller, L. (2004). Dog and cat care in the animal shelter. In L. Miller & S. Zawistowski (Eds.), *Animal shelter medicine for veterinarians and staff* (pp. 95–119). Ames, IA: Blackwell Publishing.

Miller, L. & Zawistowski, S. (Eds.). (2004). *Animal shelter medicine for veterinarians and staff*. Ames, IA: Blackwell Publishing.

National Research Council. (1996). *Guide for the care and use of laboratory animals* (7th ed.). Washington, DC: National Academy Press.

PetSmart Charities, Inc. (2004). *Annual report 2004: A new road home*. Phoenix, AZ: PetSmart Charities.

Reid, P., Goldman, J. & Zawistowski, S. (2004). Animal shelter behavior programs. In L. Miller & S. Zawistowski (Eds.), *Animal shelter medicine for veterinarians and staff* (pp. 317–331). Ames, IA: Blackwell Publishing.

Salman, M. D., Hutchinson, J., Ruch-Gallie, R., Kogan, L., New, J. C., Kass, et al. (2000). Behavioral reasons for relinquishment of dogs and cats to 12 shelters. *Journal of Applied Animal Welfare Science, 3*(2), 93–106.

Unti, B. (2004). *Protecting all animals: A fifty-year history of the humane society of the United States*. Washington, D.C.: Humane Society Press.

Zawistowski, S. & Morris, J. (2004). The evolving animal shelter. In L. Miller & S. Zawistowski (Eds.), *Animal shelter medicine for veterinarians and staff* (pp. 3–9) Ames, IA: Blackwell Publishing.

Zawistowski, S., Morris, J., Salman, M.D. & Ruch-Gallie, R. (1998). Population dynamics, overpopulation, and the welfare of companion animals: New insights on old and new data. *Journal of Applied Animal Welfare Science, 1*(3), 193–206.

CHAPTER 5

Companion Animals, Law, and Animal Cruelty

No man shall exercise any Tirrany or Crueltie towards any bruite Creature which are usuallie kept for mans use.

—The Body of Liberties, Massachusetts Bay Colonie, 1641
(as cited in Ascione & Arkow, 1999, p. 271)

KEY TERMS

law	exotic animals	abuse
wildlife	Convention on International Trade in Endangered Species of Wild Fauna and Flora (CITES)	neglect
state law		hoarding
federal laws		Hoarding of Animals Research Consortium
local laws		
Animal Welfare Act (AWA)	trade	bestiality
regulations	liability	zoophilia
dealers	insurance	The Link
animal fighting	breed-specific bans	violence
licensing	cruelty	legislation

Animals are covered by a complex mix of federal, state, and local laws. These laws sometimes overlap and sometimes are contradicting. Even the definition of *animal* will vary from one place to another. This is a rapidly changing field and *Animal Law* is an accepted specialty in the legal profession. Courses are taught at major **law** schools, there are specialty journals devoted to the field, and many bar associations are forming animal law subcommittees. The first part of this chapter will provide a brief introduction to the range of laws that impact companion animals, owners, and others associated with their care. The second part of this chapter will focus on animal cruelty laws and the connection between animal cruelty and other forms of violent and abusive behavior.

Rules related to the treatment of animals in America date back to the earliest days of European colonization. These typically reflected concerns related to the value of animals as property or concerns about the moral impact that mistreatment of animals had on humans (Favre, 2003). The ensuing centuries have seen the development of a complicated web of legal authority that sometimes overlap in jurisdiction and sometimes omit or overlook issues related to animals. The earliest laws were most often directed toward the treatment of livestock and horses. These animals were frequently among the most valuable possessions that an individual might have, and were critical for transportation and work, food, and fiber. As such, these laws were typically concerned with the protection and treatment of animals as *property*. Indeed, the very definition of *animal* as covered under these laws reflected the extent to which the animal could be owned, controlled, and have value. At one end of the scale was **wildlife**, which could not be owned, controlled, or possessed and therefore had no legal protection. At the other end was livestock, which could be owned, controlled, and had value. Dogs, cats, caged birds, and other animals that we now think of as pets or companions existed in a nether region of the law (Favre, 2003).

A New York **state law** passed in 1866, due to the influence of Henry Bergh, founder of the American Society for the Prevention of Cruelty to Animals (ASPCA), is generally seen as watershed in the development of *effective* anti-cruelty laws. This law and its subsequent amendment in 1867 made several important contributions to animal law (Favre, 2003). First, while previous laws were limited to the protection of owned animals with commercial value, these improvements eventually extended protection to "any living creature." The law included both intentional and negligent acts and expanded the list of illegal actions. Therefore, the early laws that were meant to protect owned animals with commercial value were expanded, and time and precedent eventually succeeded in finding a secure place for the treatment of companion animals under law. Many other states soon followed the lead of New York in the passage of similar laws to prevent cruelty to animals. States that followed rapidly on the heels of New York often copied the New York law's language quite closely. However, in the following decades, each state, including New York, modified its laws in this area, and there is now a surprising range of variation in state laws across the country. At the same time, nearly all local communities have passed their own regulations related to animals, their ownership, and their treatment. There are very limited **federal laws** related to animals. David Favre has indicated that, "There is nothing in the U.S. Constitution to suggest that the U.S. Congress is authorized to deal with animal issues" (Favre, 2003, p. 362). As a result, these state and **local laws** are the main source of regulations and legal oversight regarding animals in the United States. This chapter will examine the various laws at the federal, state, and local level that are relevant to ownership and treatment of companion animals.

FEDERAL LAWS

While Congress does not have explicit authority to legislate regarding animal issues, it does possess the power to pass laws related to interstate commerce. As a result, the limited spectrum of federal law related to animals is generally derived from this authority. The first federal law related to animals was passed to require provision of food, water, and rest for livestock being shipped interstate. In similar fashion, the application of other federal laws related to animals is typically associated with interstate commerce. This authority can be extended to cover the treatment of animals before and after they are shipped and the purpose for which they are shipped between states.

Other constitutional issues associated with companion animals are generally those associated with their status as property and rights of pet owners to due process in the protection of their rights and property. Due process accorded to an individual is protected by the Fifth and Fourteenth Amendments, protection against the seizure of public property by the Fifth Amendment and equal protection under the law by the Fourteenth Amendment (Favre, 2003). While at times frustrating for Animal Control authorities, these protections do help to ensure that laws related to the regulation of animal ownership and responsibility are enforced fairly and consistently.

Animal Welfare Act

The most significant federal law related to companion animals is the **Animal Welfare Act (AWA)**. In the early 1960s, as medical research using animals was expanding in the United States, it became a significant concern that companion dogs and cats were being stolen and sold to institutions for use as research subjects. The AWA was passed in 1966 with the express purpose to protect pet owners from the theft of their pets and to prevent the sale of stolen dogs or cats for research or experimentation. As a way to deal with the source of animals used in research, **regulations** were developed to cover the treatment of dogs and cats at breeders and **dealers** and the research facilities. The AWA has been amended several times in 1970, 1976, 1985, and 1990, expanding its scope and authority, most specifically related to care and treatment of animals in research facilities. The current primary concerns of the AWA include the following (for the full text, visit <http://www.animallaw.info/statutes/stusawa.htm>):

- The theft of pet dogs and cats being sold to research and testing facilities
- Mammals in zoos and exhibitions
- **Animal fighting** (dogs and cocks primarily)
- The breeding and wholesale distribution of some mammals
- Auctions of animals/mammals
- Animals in research laboratories (universities and private industry)
- The transportation of listed animals by other than common carriers (Favre, 2003)

The United States Department of Agriculture (USDA) through its Animal and Plant Health Inspection Service (APHIS) was given the task of enforcing the AWA. Most of the regulatory activity is accomplished through the use of **licensing** requirements, inspections, corrective actions, administrative hearings, monetary fines, and license revocation. There are three main classes of dealers and exhibitors:

- Class A: Businesses or individuals that breed and sell animals
- Class B: Businesses or individuals that buy and/or resell animals, including brokers and individuals that operate an auction sale
- Class C: Businesses or individuals who exhibit or display animals to the public—buying or selling animals may be a minor part of the business to maintain the animal collection

The AWA has specific impact on companion animals in several ways. First, the system of regulations and licensing required for animal dealers provides pet owners with some protection against the theft of their pets. Regulated research facilities may purchase animals covered by the act only from licensed breeders or dealers, who are required to document and track the sources of the animals that they sell. These licensing requirements also govern the breeders and dealers that provide puppies and kittens for sale in retail pet stores. These businesses are controversial. Animal protection groups have called them "puppy mills" and have criticized them for the treatment of the dogs used for breeding stock (Figure 5-1), failing to provide proper care and husbandry, and

FIGURE 5-1 Puppy mills are commercial breeding establishments that produce puppies for retail sale. While there is no "official" definition of a puppy mill, they are typically identified by their poor treatment of the breeding dogs, and the poor quality of the puppies. In this case, the dogs are crowded into wire bottom cages, and waste from the dogs on the top may drop onto the dogs in the bottom cage. (Dr. Randall Lockwood)

the overall quality of the puppies shipped for sale to the public (Curnutt, 2001). In 2005, Pennsylvania Senator Santorum sponsored a bill that further would have amended the AWA, providing additional regulations to protect the welfare of the animals used for breeding by licensed regulated businesses and individuals, and the puppies they produced. This bill is still pending, as various parties debate the merits of specific elements and whether and how it could address large-scale breeding operations that sell directly to public, in addition to sales to brokers and other retail outlets (<http://www.thomas.loc.gov/cgibin/query/z?c109:S.1139.IS:>).

The AWA does provide specific authority for the regulation of animals transported by commercial air carriers. For example, the law requires that animals to be shipped by a commercial carrier have a licensed veterinarian examine the animal not more than 10 days before the shipping date and provide a certificate of health. This certificate must be submitted to the commercial carrier when the animal is presented for shipping, whether or not the animal is accompanied by its owner on the flight. The carrier used for shipping must also meet standards for sturdiness to protect the animal and the airline personnel. In 2000, following a number of cases where animals either died in cargo areas during flights or escaped during the process of loading or unloading from airplanes, the Code of Federal Regulations was amended under the authority granted to the Secretary of Agriculture in the AWA, requiring that airlines maintain public records on the number of animals that die during transportation. So far, data for the first 13 months show there were 28 deaths, 12 injuries, and 6 lost animals on 15 of the 22 airlines that reported their data to the Bureau of Transportation Statistics (BTS). Full reports on each incident are available from the BTS (<http://www.airconsumer.ost.dot.gov/reports/index.htm>). In one well-publicized incident, a whippet named Vivi, returning from the Westminster Kennel Club show in New York City in February 2006, was lost at John F. Kennedy Airport when she escaped from her kennel on the tarmac. As of February 2007, she was still lost, with sporadic reports of her being sighted in the area of the airport (McGee, 2007). In much the same way that travelers may choose a particular air carrier based on its reported *on time* record, pet owners who want to travel with their pets can choose an air carrier based on the now public record for safe pet transport.

Funding for the USDA is provided through the annual appropriations process in Congress, specifically through the Farm Bill. From time to time, amendments to the Farm Bill have provided updates or changes to the AWA. One recent example in 2005 was the stipulation that the USDA evaluate the use of radio frequency identification (RFID), or microchips, for companion animal identification and establish a national standard for their use. As noted in Chapter 4, microchips have not been fully utilized by pet owners or the various agencies that could benefit by reliable identification of pets because of problems with standardization of the microchips and the readers. In this case, the USDA would be acting under the authority of Congress to regulate interstate commerce. It is interesting that the nature and language for this amendment was being considered during recovery efforts that followed Hurricanes Katrina and Rita in late summer 2005. Because of the devastation wrought by those storms, animal rescue efforts required the movement of pets out of the Gulf Coast region for foster care at animal shelters and rescue groups around the country (see sidebar The Perfect Storm in Chapter 4). The author was a party to meetings related to the Farm Bill amendment and was active during the animal rescue activities. The need for a national standard was made clear during the animal rescue process and did influence support that resulted in passage of the amendment.

The AWA does set specific husbandry requirements for dogs, cats, and other regulated animals kept in research facilities. These requirements set minimum standards for cage sizes, sanitation, and feeding, among other things. In addition, dogs must be exercised daily. While these regulations are limited to the regulated research establishments, in the absence of husbandry regulations in other circumstances, they do provide a reference point when evaluating conditions at animal shelters, boarding facilities, and breeding kennels.

Federal Regulations Related to Wildlife

Several different federal laws limit the capture, ownership, and transport of wildlife or **exotic animals**. These laws have particular significance for people interested in collecting specimens of exotic pets, including large cats, primates, reptiles, birds, and other animals.

Human Health. The Food and Drug Administration and the Department of Health and Human Services will exercise their authority to prohibit or regulate practices that have potential impact on human health. For example, in the 1970s there was concern about the spread of turtle-associated salmonellosis. *Salmonella* are naturally occurring bacteria in turtles. However, juvenile turtles are considered a greater risk for shedding the bacteria, posing a health risk to humans. As a result, in 1975 the Food and Drug Administration banned the sale of baby turtles, those with a carapace length of less than four inches (for more information visit <http://www.fda.gov/cvm/turtlereg.htm>).

Additional attention to regulation in this area developed in 2003 when monkey pox was contracted by pet owners in Wisconsin, Illinois, and Indiana, who had purchased prairie dogs. The prairie dogs had been housed next to a Gambian giant pouched rat at an animal dealer (Enserink, 2003). A number of states issued emergency orders to limit or prohibit the importation, sale, and distribution of prairie dogs and certain other rodents (Department of Health and Human Services and Food and Drug Administration, 2003). Because the animals were moving across state lines, the Centers for Disease Control and the Food and Drug Administration issued a joint order prohibiting the transportation for interstate commerce or sale of prairie dogs, Gambian giant pouched rats, and several other rodent species.

Conservation. The **Convention on International Trade in Endangered Species of Wild Fauna and Flora (CITES)** came into force in 1975, has over 150 member countries or parties, and regulates the capture, killing, confinement, or possession of wildlife designated as endangered, whether alive or dead, whole or parts. While CITES is actually an international treaty, it required federal authorization and depends on the cooperation and support of federal authorities for enforcement. CITES lists species under three appendices. Animal species listed under Appendix I are threatened with extinction and international **trade** is banned. This would include all the great ape species, a number of parrots, and marine turtles. Appendix II species are not currently in danger of extinction, but there is concern that uncontrolled trade could place them at risk. Trade in these species is permitted, but it is regulated and requires valid permits for transport. Finally, individual parties to the treaty can request that a native plant or animal be protected by having it listed under Appendix III (<http://www.cites.org>; appendices are listed at <http://www.cites.org/eng/app/appendices.shtml>).

Wild Bird Conservation Act (WBCA). During the 1980s, the United States imported approximately 700,000 birds per year for the pet trade, and about 90 percent of these were caught in the wild. In an effort to support the conservation of wild populations of exotic birds around the world, the Wild Bird Conservation Act was passed in 1992 (<http://www.defenders.org/wbcafact.html>; January 12, 2006). The act restricts importation of birds listed in the appendices of CITES as well as unlisted species at the discretion of the Secretary of the Interior (16 U. S. C. § 4907). The Secretary also has the ability to permit the import of CITES species or ban importation of all species from a particular country.

Americans with Disabilities Act of 1990 (ADA)

Under the Americans with Disabilities Act of 1990 (ADA), people who use service animals cannot be discriminated against by privately owned businesses that serve the public (ADA, n.d.). The ADA defines a service animal as *any* guide dog, signal dog, or other animal *trained* to provide an individual with a disability with assistance. Some examples would include the following:

- Assisting a person who is blind
- Alerting persons with hearing impairments to sounds
- Pulling wheelchairs or otherwise assisting persons with mobility impairments
- Assisting persons with mobility impairments with balance

While many trained service dogs wear a special harness or vest, people with service animals are not required to carry documentation of their disability or certification that their animal is trained as a service animal. However, business owners may inquire (for clarification purposes) if the animal is a service animal (for more information visit <http://www.ada.gov/svcanimb.htm>).

Pets and Public Housing. Federal law (12 U.S.C. § 1701r-1) protects the right of pet ownership for the elderly or handicapped in assisted rental housing. As a result, no owner or manager of federally assisted rental housing for the elderly or handicapped may (a) prohibit or prevent any tenant from owning common household pets or having common household pets living in the dwelling in such housing, or (b) restrict or discriminate against any person by reason of the ownership of such pets or the presence of these in the dwelling. However, owners may require removal of any pet whose conduct or condition is duly determined to constitute a nuisance or a threat to health and safety of other occupants of the housing or persons in the community.

Tax Code—Nonprofit and 501(c)(3). Nonprofit, tax-exempt charitable organizations are typically organized as corporations under 26 U.S.C.S § 501 of the Internal Revenue Law. The law specifically authorizes the formation of corporations to further "prevention of cruelty to children or animals," under 501(c)(3). These organizations do not pay tax on revenues that they collect related to their work, and contributions to the organization from members of the public are tax deductible. Groups organized as 501(c)(3) corporations are limited in their ability to be active politically. They can engage in public education related to their cause, but cannot endorse or campaign for a political candidate or make contributions to political campaigns.

STATE AND LOCAL LAWS

The majority of laws related to companion animals exist at the state and local level. These cover a wide range of animals and activities, including the types of animals that may be kept as pets, licenses for individual pets and pet-related businesses, the practice of veterinary medicine, regulations for animal shelters, and laws related to animal cruelty. Companion animals, their owners, and other people that come in contact with the animals will also be affected by laws and regulations that govern **liability** and **insurance** practice. Several Web-based resources provide summaries of state laws related to animals (visit <http://www.animallaw.info>). This is a dynamic area, with hundreds if not thousands of laws and amendments considered in state legislatures each year. Students and professionals working in the area are well advised to consult these sources, and other resources, to stay current.

Facility Licensing

Regulations related to the licenses or permits needed to operate an animal facility will vary greatly from state to state. These would include regulation of animal shelters, boarding facilities, grooming businesses, and pet stores. The regulatory authority for each situation might be found with the State Department of Health or Department of Agriculture. Placement of the oversight will give a sense of whether the intent of the regulations is directed at animal or human health issues. When regulations exist, they will generally provide standards for cleanliness and hygiene and the types of practices or procedures permitted on the premises. Inspections may or may not be required as part of the regulations. The lack of uniform practice in this area of regulations is clearly seen in Table 5-1, showing the differences in state regulations of animal shelters. While some states specifically prohibit pound seizure, other states require it for government run animal shelters. In those states that have no specific state law related to pound seizure, local laws or regulations may govern the practice.

Veterinary Medicine

All states require that veterinarians be licensed to practice veterinary medicine within that state. This will typically require possession of a valid degree in veterinary medicine from an approved veterinary medical program and passing

TABLE 5-1

Distribution of State Laws Affecting Shelter Operations in the United States *(adapted from Patronek, 2004)*

States with laws regulating animal shelters[a]	CT, FL, GA, IA, IL, KS, LA, ME, MD, MI, MO, NH, NJ, NY, NC, RI, SC, TX, VA, VT
States requiring licensing of animal shelters	GA, IA, IL, KS, MD, ME, MI, MO, NH, NY, RI, SC, TX, VT
States in which animal shelters are inspected	CT, FL, GA, IL, KS, MD, MI, MO, NH, NY, RI, SC, TX, VA
States requiring an advisory board for shelters	LA, ME, MO, TX
Training required for shelter personnel	LA, TX, VA
Training required for animal control officers	CA, FL, ME, MI, NJ, NM, VA
States with some regulation of fostering organizations	CO, VA
States in which shelter statistics must be reported	MI, VA
States with consumer protection laws applicable to pets	AR, AZ, CA, CT, DE, FL, ME, MN, NH, NJ, NV, NY, PA, SC, VA, VT
States in which pound seizure is prohibited	CT, DE, HA, MA, MD, ME, NH, NJ, NY, PA, RI, SC, VT, WV
States in which pound seizure is required for government-run shelters	IA, MN, OK, SD, UT
States in which customary farming practices are exempted from cruelty statutes	AZ, CO, CT, IA, ID, IL, IN, KS, MD, MI, MO, MT, NC, NE, NJ, NV, OR, PA, SC, SD, TN, UT, WA, WV, WY

[a]Some state statutes (for example, PA) may indirectly regulate shelters if they cover any facility that kennels dogs.

a licensing exam. In addition to the standards for the initial licensure of a veterinarian, states may set additional standards to require continuing education for an individual to maintain his or her license to practice. These regulations will be governed by the Veterinary Practice Act applicable to the individual state. The American Association of Veterinary State Boards has drafted a *Model State Veterinary Practice Act* that represents a broad consensus of current practice in the various states, as well as recommendations for change and modification (available at <http://www.aavsb.org>). It is interesting that in the model act, the purpose of the act is not related to the protection and or treatment of animals; rather it is stated, "...to promote, preserve, and protect the public health, safety, and welfare by and through the effective control and regulation of persons, residing in or out of the state that practice veterinary medicine within this state" (Section 103, Statement of Purpose of the Model Act). The practice of veterinary medicine is then defined as, "Any person practices veterinary medicine with respect to animals when such person performs any one or more of the following: (a) directly or indirectly consults, diagnoses, prognoses, corrects, supervises, or recommends treatment of an animal, for the prevention, cure or relief of a wound, fracture, bodily injury, disease, physical or mental condition; (b) prescribes, dispenses or administers a drug, medicine, biologic, appliance, application or treatment of whatever nature; (c) performs upon an animal a surgical or dental operation or a Complementary or Alternative Veterinary Medical Procedure; (d) performs upon an animal any manual procedure for the diagnoses and/or treatment of pregnancy, sterility, or infertility; (e) determines the health, fitness, or soundness of an animal; (f) represents oneself directly or indirectly, as engaging in the practice of veterinary medicine; or (g) uses words, letters, or titles under such circumstance as to induce the belief that the person using them is qualified to engage in the practice of veterinary medicine, as defined. Such use shall be prima facie evidence of the intention to represent oneself as engaged in the practice of veterinary medicine" (Section 104). Oversight of veterinary practice in a state is typically handled by the State Veterinary Board, working under the auspices of the State Education Department. The State Board is responsible for investigating allegations of malpractice or unprofessional conduct.

Veterinary technicians may or may not be licensed. In some states licensing for veterinary technicians may be voluntary. In most cases, this would require completion of an accredited program in veterinary technology and passing an exam. Licensed Veterinary Technicians (LVT; also called Certified Veterinary Technicians, Registered Veterinary Technicians, or Certified Animal Health Technicians, depending on the state) will often be permitted to perform a variety of procedures related to animal care under the supervision of a licensed veterinarian. It is likely that as veterinary medicine becomes more sophisticated, the training and requirements of licensing for veterinary technicians will become more common.

Other Professions and Activities

It is not uncommon for a state to require that dog groomers be licensed to practice. This typically reflects the desire to ensure that proper standards of hygiene are maintained and that the animal being groomed is protected from harm. It is surprising that there is limited regulation to govern dog training. Trainers do not need to be licensed or show evidence of completing any form of educational study related to being a dog trainer. Voluntary guidelines for dog training were developed by the American Humane Association (2001) and the Delta Society (2001).

States will typically limit or restrict the ability of individuals to capture, hold, or keep native wildlife as pets. In addition to the restrictions that may be stipulated on what animals may or may not be kept, the state will generally require that individuals, even including veterinarians and animal shelters that hold, care for, and rehabilitate wildlife, be licensed by the state (Casey & Casey, 2000). Holding or keeping migratory birds would require a federal license. In addition, states may issue permits to individuals to keep, hold, or exhibit other exotic species such as primates.

Dangerous Dogs

Media accounts of dog attacks that result in serious injury or fatality will often stimulate discussion of the need for "tougher" regulations regarding dangerous or vicious dogs in a state or community. There is limited accurate information on the frequency of severe dog attacks. There is no standard national, and limited local, reporting system to keep track of such attacks. In general, dangerous dog laws and regulations are based on restrictions placed on owners and dogs that have demonstrated some form of behavior considered threatening. A model law supported by the American Veterinary Medical Association follows a well-defined process to determine if a dog should be considered dangerous, appropriate steps to take to protect public safety (AVMA Task Force on Canine Aggression and Human–Canine Interactions, 2001). An alternative approach being followed by other communities are **breed-specific bans** (Bresch, 2005). Under these laws, municipalities may preemptively ban specific breeds of dogs as a prophylactic effort to protect public safety. The breeds generally covered by these bans include pit bulls (generally considered to be American Staffordshire bull terriers, Staffordshire terriers, and Staffordshire bull terriers), rottweilers, German shepherd dogs, chow chows, and Doberman pinschers, among others. Some states have specific prohibitions that prevent communities from having animal control laws that target particular breeds. Ohio has targeted pit bulls for special attention. By definition, pit bulls are automatically determined to be vicious dogs and must be penned or tied up on the owner's property (Cunningham, 2005). Florida, on the other hand, allows municipalities to regulate dangerous dogs only if done so in a way that does not target or single out specific breeds (substantial additional information on these laws is available at <http://www.dogbitelaw.com>). Breed-specific laws have been based on sparse statistical evidence that pit bulls, or any other breed, pose an increased risk of bite frequency. In fact, the AVMA Task Force on Canine Aggression (2001) concluded that the available numbers related to dog bites do not give an accurate picture of which breeds are likely to bite. Insufficient data have been collected to determine if these laws have been effective in their intent to protect public safety. Resolution in this area will probably require several rounds of court tests, redrafting of dangerous dog laws, and more accurate statistical reporting models.

Liability, Tort, and Insurance Law

Pet owners confront liability in two ways. They can be liable for damage or injury caused by their pet, or they may want to recover damages as a result of the injury or death of their pet. The growing importance of this area is reflected in the American Bar Association's formation of the Animal Law Committee through its section on Tort, Trial, and Insurance Practice in November 2004 (Gislason & Fershtman, 2006).

Dog bites are the most common way in which pet owners may be held liable for damage or injury. While there is no standard system in place for reporting dog bites, various sources suggest that dog bites are an important public health issue and do result in significant numbers of injuries and monetary loss. The Insurance Information Institute (III), an insurance industry resource, estimates that over $345 million was paid in claims in 2002 and $317 million in 2005 (III, n.d.). Traditional precedents relied on the "one-bite rule." Dogs were generally considered safe until they had bitten one time (see <http://www.dogbitelaw.com>). Owners (and the dog) would not be liable for that first bite, since it could not be predicted. However, after a first bite, the owner would be on notice that his or her dog had a history, and he or she could be held liable for damages or injury caused by subsequent bites. Practice in this area is changing and it is now more likely that owners will be responsible for damages or injury caused by a first bite inflicted by their dog. A dog owner's homeowner's insurance policy would typically pay for any damages for which the dog's owner is found liable. Some insurance companies have decided not to provide insurance to owners of some breeds that they consider a high risk (Cunningham, 2005). Cunningham has argued against this practice, citing a lack of evidence to support the practice, and the fact that it places a significant strain on families that are forced to choose between keeping their family pet and being able to get the insurance needed to keep their home. It is interesting to note that both sides of this debate point toward many of the same studies cited by the AVMA Task Force on Canine Aggression to support their position. It should be noted, however, that the authors of that report point out that their research does not provide adequate data to justify discrimination against particular breeds of dogs. Dog owners will need to be aware of the increasing attention being paid to their responsibilities in this area.

Another rapidly changing area of law is related to the damages that a pet owner can recover if their pet is injured or killed. The traditional view of the law that recognizes animals as property allowed an owner to recover damages based on the fair market value of the animal (Favre, 2003). Under this scenario, pure bred, pedigreed animals would have a greater value than non-pedigreed animals. Similarly, animals with special training or abilities would have a greater market value. However, the emotional value of the animal to the owner was not recognized and valued, and would not be part of any calculation to compensate an owner for damages. This precedent is changing based on the growing evidence that companion animals provide their owners with positive physical and mental health benefits. At the same time, greater recognition is now given to the fact that pets are considered to be part of the family where they live. As a result, both legislative and trial decisions support compensation for the emotional value of the bond that families have with their pets. In 2000, the State of Tennessee passed the *T-Bo Act* that allows a pet owner to recover up to $4,000 in noneconomic damages if his or her pet is killed or dies due to injuries caused by either the intentional or negligent acts of another (Favre, 2003). Recent court decisions have resulted in even larger awards to pet owners. In 2005, a judge in the State of Washington awarded over $45,000 to the owner of a cat that had been killed by the neighbor's dog. The award included $30,000 for the value of the cat, $15,000 in emotional distress, and additional compensation for medical expenses and the cremation of the remains (Cornwall & Welch, 2005). Additional cases are working their way through the courts across the country and it is likely that compensation awards that recognize the unique place companion animals hold in our society will become more common (visit <http://www.animallaw.info/topics/spuspetdamages.htm>).

Animal Cruelty

Many state anticruelty laws date back to the late 1800s, following the lead of New York State. While all states now have laws that in one way or another prohibit the cruel treatment of animals, there is substantial variation among the states. For example, differences in language from one state law to another will determine which species or types of animals are covered (Patronek, 2004). Some states may simply use the word "animal" while others will provide more detail by specifically including types of animals. For example, the state of South Dakota includes "mammals, birds, reptiles, amphibians and fish"; however, in the state of Georgia, the statute animal does not include "any fish nor shall such term include any pest that might be exterminated or removed from a business, residence, or other structure." In addition to these biological references to whether an animal is covered, whether the animal is owned or captive will determine if its treatment is covered under the anticruelty laws. For example, a cat may or may not be covered under a state's anticruelty law depending on whether it was "owned" by a person and kept in the house, was allowed to roam freely between indoors and outdoors, or is an unsocialized feral cat that lives on its own. The definition of ownership will also differ from state to state.

The question of how an animal is classified and whether or not it is owned has particular significance for a new development in animal cruelty laws. While animal cruelty has typically been classified as a misdemeanor crime, several states have amended their laws to classify *intentional* animal cruelty as a felony. These efforts have been based on research showing a link between cruelty to animals and other violent behavior. This topic will be discussed later in the chapter. However, it is significant that in a number of cases, felony cruelty statutes have been limited to domestic companion animals. This represents an interesting evolution from the earliest laws to prevent the mistreatment of animals that largely ignored companion animals and focused on owned livestock and animals that had *economic* value.

Defining Animal Cruelty. Animal cruelty laws typically address both *intentional* acts that cause unwarranted or unnecessary pain or suffering to an animal, and the *failure* to act in providing necessary food, water, shelter, or medical care. The former is generally classed as animal **cruelty** or **abuse** and the latter as **neglect**. The various states differ to the extent that they enumerate prohibited acts, practices that might be exempted, and as noted above, even the definition of animals pertaining to the law. Common agricultural procedures, biomedical research, hunting, fishing, and trapping are among the practices that are frequently exempted from state anticruelty laws. In addition to these two broad categories of animal mistreatment, several specific examples will be discussed: **hoarding**, animal fighting, and bestiality/zoophilia.

Animal Neglect. Animal neglect is failure to provide an animal with appropriate, nutritious food, adequate potable water, shelter from the elements, and medical and other care needed for the animal's well-being. The lack of widely accepted husbandry standards for companion animal care can complicate the investigation and prosecution of these cases (Patronek, 2004). Authorities may reference the standards defined in the Animal Welfare Act, written standards, or guidelines published by fanciers or breeders, among other sources. Judgment frequently requires documentation that the animal's condition falls outside the normally accepted standards for body condition or health. The Tufts Animal Care and Condition (TACC) scales were developed to provide an objective standard to evaluate an animal's body condition, physical health, and environmental conditions (Patronek, 1998). The scales provide visual examples and clear descriptions of the animal's physical condition related to various stages of neglect (Figure 5-2). Further evaluation of the condition of hair coat, teeth, and claws or nails, and evidence of discharge or dirt in the eyes, ears, nose, and anus may also be used to help demonstrate neglect (Figure 5-3). It is important to recognize that untreatable medical conditions must be ruled out as the cause of the condition observed. For example, a dog or cat with a metabolic disorder that compromises the absorption of nutrients from food may appear malnourished due to neglect. In cases like these, the owner may be asked to provide evidence that the pet has been seen by a veterinarian, the nature of the condition, and the type of treatment being followed. If a case is not severe, cruelty investigators may use their discretion to issue a Notice to Comply, and require the owner to provide better conditions, food, or medical treatment within a designated time frame. The investigators will then monitor the situation to ensure that the welfare of the animal improves.

Animal Abuse. Animal abuse is intentionally mistreating or injuring an animal. The range of such acts is stupefying. Cruelty investigators regularly investigate and report cases where animals are cut, stabbed, strangled, set on fire, or thrown from the roofs of buildings. Dogs are the most common reported victims of cruelty (Donley, Patronek, & Luke, 1999). However, it is uncertain whether this is due to the fact that they are abused more often, or because there is less interest or concern for cats. Anecdotal observations do suggest that stray animals, especially cats, are likely targets for abuse. Other companion animals are abused, but since they are less common, they appear less frequently in cruelty reports.

Abuse or cruelty cases are now more likely to require forensic analyses that approximate those expected in the prosecution of crimes against humans (Miller & Zawistowski, 1998; Munro, 1998; Munro & Thrusfield, 2001a; Sinclair, Merck, & Lockwood, 2006). This is especially true given the development of felony status for crimes of intentional cruelty toward animals. The defense will be more aggressive and the prosecution must be well founded, and the evidence presented needs to be carefully documented.

Animal Fighting. Dogfighting is illegal in all 50 states and cockfighting is illegal in 49 (and will be banned in Louisiana in 2008, making it illegal in all 50 states). In a number of states participating in a dogfight is a felony, and in several being an observer can also be a felony. Transportation of dogs or cocks across state lines is a federal offense. Pit bulls are the breed of choice for dogfighting. They are generally bred as carefully for their fighting skills, strength, and endurance as fanciers of other breeds will breed for herding, retrieving, or other qualities. Dogfighters can generally be categorized in one of three ways:

- Professionals—people who make a substantial investment and earn money through prizes, stud fees, gambling, and the sale of puppies, services, and products to other dogfighters. They may have a highly structured breeding and training program, including the use of performance-enhancing drugs. They will typically participate in well-organized fights that include promotional materials, refreshments, and security to guard against law enforcement.

FIGURE 5-2

The Tufts Animal Care and Condition Scale provides a standard method of reporting the mistreatment and neglect of animals in cruelty cases. *(Courtesy Gary Patronek)*

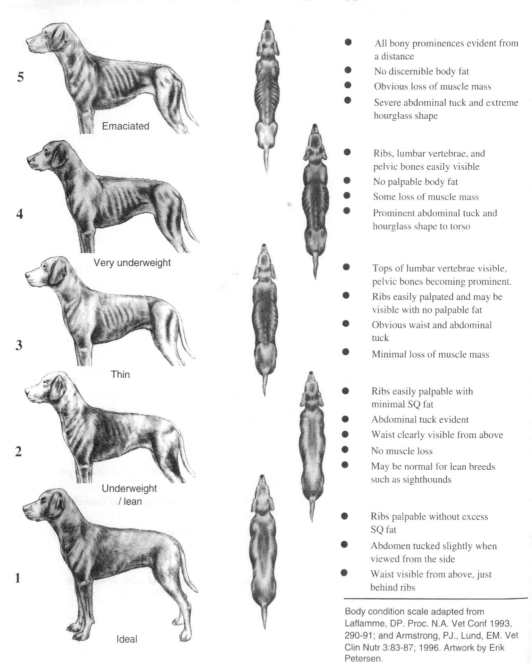

FIGURE 5-3

Cases of severe neglect such as this are not an uncommon sight for animal cruelty investigators. In cases like this, veterinarians treating the dog will document weight gain and other improvement in health and condition over time to rule out illness or metabolic disorders as the reason for the poor condition of the dog. *(Courtesy Mark McDonald)*

- Hobbyists—people who enjoy the "sport" and the dogs. They may not have the same level of knowledge or time to spend to investigate breeding lines or training. A large part of their interest may be social.

- Street fighters—people who may be involved with a range of other illegal activities. Their dogs may receive little or no formal training. Fights are often spontaneous. Their dogs may also be used for protection or to guard drugs or money and as a result may be more likely to be aggressive toward people. (Dinnage, Bollen, & Giacoppo, 2004)

Investigations of dogfighting often rely on undercover investigations, and finding evidence of paraphernalia associated with training dogs for fighting such as treadmills, needles used to inject drugs, or ropes hanging from trees. The dogs are encouraged to jump up and grab the rope and hang by their jaws to strengthen the muscles used for biting. Cruelty investigators in a number of locations are now working closely with police department antigang units to identify dogfighting activities.

Animal Hoarding. The image on the news is often striking (Figure 5-4): a dilapidated home, dozens if not hundreds of animals in poor physical condition, and a disheveled older woman worried that someone was going to take away the animals that she cared so much about. People who were once called collectors, and now termed hoarders, are not just eccentric *do-gooders* who got in over their heads (Stone, 2006). Professionals are now advocating a broad-based multiagency approach to what they describe as a significant public health and animal welfare issue. In a study of 54 case reports of animal hoarders, the typical person was an older woman, who was single, divorced or widowed, and living alone (Patronek, 1999), but women and men of many ages, some even living with family, have been identified as well. The most frequently found animals were dogs and cats, though birds, livestock, and other animals were found. The hoarder had a median number of 39 animals, though several had over 100. In many circumstances the situation may have been known to a variety of agencies, but still fell through the cracks in jurisdiction. At the same time, agencies are often loath to act since the scale of the response required overwhelms the system. Animal agencies are not sure how to deal with a person who seems clearly in need of some sort of counseling, and human social service agencies have no idea how to judge the condition of the animals or what should

FIGURE 5-4

Investigating and dealing with an animal hoarding case will often require the cooperation of several different agencies in a community. In addition to the animal shelter and cruelty investigators, public health, environmental protection, and mental health services, among others, may be called upon. The recidivism rate is high among hoarders so a monitoring system is frequently needed to ensure that the problem does not happen again. *(Courtesy ASPCA)*

happen to them. All of this is further complicated by sensational media coverage that may at times portray the hoarder in a sympathetic fashion.

The **Hoarding of Animals Research Consortium** (**HARC**, <http://www.tufts.edu/vet/cfa/hoarding.html>) recommends a task-force approach that includes the following:

- Animal control
- Public health
- Mental health
- Child and adult protective services
- Zoning boards
- Fire prevention
- Veterinary assistance (HARC, 2000)

In addition, counseling services will be required for the hoarder, and authorities will need to continue to monitor the individual since recidivism is common. While there would seem to be clear mental health issues involved, no clear psychological model has been fully resolved. At this time, some variant of obsessive compulsive disorder has been offered as an explanation.

It is important to recognize that it is not simply the number of animals that means someone is a hoarder. HARC has developed the following definition to help identify hoarders:

- Accumulated a large number of animals, which has overwhelmed that person's ability to provide even minimal standards of nutrition, sanitation, and veterinary care
- Failed to acknowledge the deteriorating condition of the animals (including disease, starvation, and even death) and the household environment (severe overcrowding, very unsanitary conditions)
- Failed to recognize the negative effect of the collection on his or her health and well-being and on that of other household members (Patronek & HARC, 2001)

The consortium does point out that each case is unique and must be evaluated and responded to in a fashion that provides best for the person, the animals, and the community.

Bestiality/Zoophilia. Of all animal topics, bestiality/zoophilia is the one most likely to make people feel uncomfortable. As a result, it is an area where we have very limited information on its prevalence and impact on people, animals, and society. **Bestiality** is sexual contact with animals. Miletski (2005) points out that bestiality has been a part of human behavior throughout history. Depictions in art and literature have been common worldwide. The behavior has enjoyed varying levels of acceptance through the ages. This has ranged from open celebration to condemnation, with the death penalty for both the person and the animal involved. The practice has generally been prohibited in the United States, combined in statutes with other sexual acts that were considered deviant. In some cases, broad state laws related to sexual behavior were liberalized in recognition of homosexuality as an acceptable lifestyle, resulting in a loophole for the practice of bestiality. In most cases, actions are being taken to again make bestiality illegal. Due to the combination of legal and social disapproval, the prevalence of the behavior is difficult to evaluate. Kinsey, Pomeroy, & Martin (1948) in a study of male sexual behavior noted that about 8 percent of men reported having had sexual contact with animals. The frequency was higher in rural males than in urban populations and this was presumed to be partly a function of available animals. Beetz (2005), however, has shown in more recent research that dogs are the most frequent targets of sexual contact. This may be related to the expanding companion animal population.

Zoophilia has been defined as an exclusive or predominant desire for sexual contact with animals, including an emotional involvement (American Psychiatric Association, 1994). Individuals describe themselves as *zoos* and have used the Internet to great benefit to share information on how to engage in sex with animals, avoid authorities, and offer emotional support to one another (Miletski, 2005). This continues to be a turbulent area of discussion, though constrained by the prevailing social taboos. In the meantime, Munro and Thrusfield (2001b) have described physical evidence of sexual assault for pets. Several pets brought to the ASPCA Bergh Memorial Animal Hospital in New York City have been examined with the aid of *rape kits* to document evidence of suspected sexual assault (personal communication with Dr. R. Riesman).

THE CYCLE OF VIOLENCE

The recognition of what has come to be called the Cycle of Violence, or **The Link** between the mistreatment of animals and humans, has a long history in Western culture. For much of that history, however, the primary concern was not about the animals, but that the mistreatment of animals could harden an individual to the cruel treatment of other people (Serpell, 1999). It is therefore interesting that empirical investigations of this observation were uncommon until the latter half of the twentieth century. MacDonald published one of the seminal papers in this field in 1963. Based on his study of sadistic psychiatric patients, he described a triad of common elements in their childhood such as fire setting, bedwetting (enuresis), and the torture of small animals (MacDonald, 1963). Anthropologist Margaret Mead concurred and, based on her observations and research on a variety of cultures, indicated that one of the worst things that can happen to a child is to abuse an animal and get away with it (Mead, 1964). Hellman and Blackman (1966) provided additional resolution when they suggested that the critical element here was that the abuse or mistreatment was directed toward animals generally associated in a positive way with humans, such as dogs, cats, and other companion animals. The American Psychiatric Association (1994) has included animal abuse as a part of the diagnosis for conduct disorder since 1987 and the FBI Behavioral Sciences Unit has incorporated cruelty to animals into its procedures to identify people who may perpetrate **violence** against other people (Lockwood & Church, 1998). Merz-Perez and Heide (2004) are among the authors who have investigated the histories of serial killers and found that these individuals frequently had an early history of violence directed toward animals. They summarize a variety of studies that have shown that notorious serial killers frequently have a history of animal abuse. For example, Jeffrey Dahmer killed, dissected, and impaled animals while a juvenile, before killing and dismembering a series of human victims. Their own research on inmates at a maximum security prison in Florida showed that those incarcerated for violent offenses were significantly more likely to have a past history of cruel acts toward animals than inmates jailed for nonviolent offenses. Arluke, Levin, Luke, and Ascione (1999) examined the criminal records of 153 individuals who were prosecuted for animal cruelty in Massachusetts and found that they were significantly more likely to be involved in other forms of criminal activity than a control group selected for similar demographic variables.

While this link between violence directed toward animals and humans appears to be well established, continued research will be needed to clarify the underlying nature of the link. Kellert and Felthous (1985) provided important information by describing nine motives for animal cruelty:

- To control an animal
- To retaliate against an animal
- To satisfy a prejudice against a species or breed
- To express aggression through an animal
- To enhance one's own aggressiveness
- To shock people for amusement

- To retaliate against another person
- Displacement of hostility from a person to an animal
- Nonspecific sadism (pp. 1122–1124)

It is clear from experience with animal cruelty cases that several of these reasons may be associated with a particular individual's behavior or a specific event.

There is specific interest in the association of animal cruelty in homes with children and the treatment of the children. DeViney, Dickert, and Lockwood (1983) found that in 53 families with a history of child abuse, abuse of animals was also found in 60% of these families. This fact has particular significance since children who witness violence are also more likely to commit violent acts. This is what helps to drive the Cycle of Violence. As Merz-Perez and Heide (2004) point out, serial killers had a history of animal cruelty and a history of being abused themselves. A later chapter will address the proposed role that humane education may have in helping to break this cycle.

DISCUSSION QUESTIONS

1. Is animal **legislation** addressed more on a federal level or on a state/local level?
2. What issues other than animal welfare and protection drive federal legislation of animal issues?
3. When and why was the Animal Welfare Act passed? Describe how it currently protects companion animals. Who is responsible for enforcing the Animal Welfare Act?
4. What are the three main classes of dealers and exhibitors as defined by the Animal Welfare Act?
5. What licensing requirements are in place for veterinarians and veterinary technicians?
6. What regulations are in place to govern dog training?
7. What regulations are in place to address severe dog attacks? What major concerns surround these regulations?
8. In what ways do pet owners become involved in liability cases?
9. What is the major distinction between animal abuse and animal neglect?
10. Using the Web site <http://www.animallaw.info>, find a law in your state that addresses at least one of the forms of animal cruelty discussed in this chapter.
11. What is the definition of an animal hoarder according to the Hoarding of Animals Research Consortium (HARC)?
12. How is zoophilia distinguished from bestiality?
13. In what states are dogfighting and cockfighting illegal? What other regulations are in place to govern animal fighting?
14. What is the Cycle of Violence?

REFERENCES

American Association of Veterinary State Boards. (2001). Statement of Purpose [Sec. 103, 104]. Veterinary Medicine Practice Act Model.

American Humane Association. (2001). *Guide to humane dog training.* Denver, CO: American Humane Association.

American Psychiatric Association. (1994). *Diagnostic and statistical manual of mental disorders* (4th ed.). Washington, DC: American Psychiatric Association.

American Veterinary Medical Association Task Force on Canine Aggression and Human–Canine Interactions. (2001). A community response to dog bite prevention. *Journal of the American Veterinary Medical Association, 218,* 1732–1749.

Americans with Disabilities Act of 1990. (n.d.). Retrieved January 12, 2006, from http://www.usdoj.gov/crt/ada/qasrvc.

Arluke, A., Levin, J., Luke, C., & Ascione, F. R. (1999). The relationship of animal abuse to violence and other forms of anti-social behavior. *Journal of Interpersonal Violence, 14,* 963–975.

Ascione, F. R., & Arkow, P. (Eds.). (1999). *Child abuse, domestic violence and animal abuse: Linking the circles of compassion for prevention and intervention.* West Lafayette, IN: Purdue University Press.

Beetz, A. M. (2005). New insights into bestiality and zoophilia. In A. M. Beetz & A. L. Podberscek (Eds.), *Bestiality and zoophilia* (pp. 98–119). West Lafayette, IN: Purdue University Press.

Bresch, D. (2005). Pit bullies: Deed not breed. *Tort Source, 7*(3), 2.

Casey, A. M., III, & Casey, S. J. (2000). A study of state regulations governing wildlife rehabilitation during 1999. In D. R. Ludwig (Ed.), *Wildlife rehabilitation* (pp. 173–192). St. Cloud, MN: National Rehabilitators Association.

Cornwall, W., & Welch, C. (2005, May 9). Judge awards $45,480 in cat's death. *Seattle Times,* Retrieved August 9, 2007, from http://seattletimes.nwsource.com/cgibin/PrintStory.pl?document_id=2002268301&slug=yofi09m&date=20050509.

Cunningham, L. (2005). The case against dog breed discrimination by homeowners' insurance companies. *Connecticut Insurance Law Journal, 11,* 1–65.

Curnutt, J. (2001). *Contemporary legal issues—animals and the law: A sourcebook.* Santa Barbara, CA: ABC–Clio.

Delta Society. (2001). *Professional standards for dog trainers.* Renton, WA: Delta Society.

Department of Health and Human Services and Food and Drug Administration. (2003). Communicable diseases: Restrictions on African rodents, prairie dogs, and certain other animals. *Federal Register, 68*(213), 62353–62369.

DeViney, E., Dickert, J., & Lockwood, R. (1983). The care of pets within abusing families. *International Journal of the Study of Animal Problems, 4,* 321–329.

Donley, L., Patronek, G. J., & Luke, C. (1999). Animal abuse in Massachusetts: A summary of case reports at the MSPCA and attitudes of Massachusetts veterinarians concerning their role in recognizing and reporting abuse. *Journal of Applied Animal Welfare Science, 2,* 59–73.

Enserink, M. (2003). Monkey pox startles Midwest. Retrieved August 9, 2007, from http://sciencenow.sciencemag.org/cgi/content/full/2003/609/2.

Favre, D. (2003). *Animals: Welfare, interests and rights.* East Lansing, MI: Michigan State University—Detroit College of Law.

Gislason, B. J., & Fershtman, J. I. (2006). Recent developments in animal tort and insurance law. *Tort Trial and Practice Journal, 41*(2), 153–180.

Hellman, D. S., & Blackman, N. (1966). Enuresis, fire setting, and cruelty to animals: A triad predictive of adult crime. *American Journal of Psychiatry, 12,* 1431–1435.

Hoarding of Animals Research Consortium. (2000). People who hoard animals. *Psychiatric Times, 17,* 25–29.

The Insurance Information Institute. (n.d.). Retrieved August 9, 2007, from http://www.iii.org/media/hottopics/insurance/dogbite/.

Kellert, S. R., & Felthous, A. R. (1985). Childhood cruelty toward animals among criminals and noncriminals. *Human Relations, 38,* 1113–1129.

Kinsey, A. C., Pomeroy, W. B., & Martin, C. E. (1948). *Sexual behavior in the human male.* Philadelphia: Saunders.

Lockwood, R., & Church, A. (1998). Deadly serious: An FBI perspective on animal cruelty. In R. Lockwood and F. A. Ascione (Eds.), *Cruelty to animals and interpersonal violence* (pp. 241–246). West Lafayette, IN: Purdue University Press.

Lockwood, R., & Hodge, G. (1986, Summer). The tangled web of animal abuse: The links between cruelty to animals and human violence. *The Humane Society of the United States News,* 10–15.

MacDonald, J. M. (1963). The threat to kill. *American Journal of Psychiatry, 8,* 125–130.

McGee, W. (2007). The beasts in the belly. Retrieved January 25, 2007, from www.concierge.com/cntraveler/articles/detail?articleId=10583.

Mead, M. (1964). Cultural factors in the cause and prevention of pathological homicide. *The Bulletin of the Menninger Clinic, 28,* 11–22.

Merz-Perez, L., & Heide, K. M. (2004). *Animal cruelty: Pathway to violence against people.* New York: Altamira Press.

Miletski, H. (2005). A history of bestiality. In A. M. Beetz & A. L. Podberscek (Eds.), *Bestiality and zoophilia* (pp. 1–22). West Lafayette, IN: Purdue University Press.

Miller, L., & Zawistowski, S. (1998). A call for veterinary forensics: The preparation and interpretation of physical evidence for cruelty investigations and prosecution. In P. Olson (Ed.), *Recognizing and reporting animal abuse'a veterinarian's guide* (pp. 63–67). Denver, CO: American Humane Association.

Munro, H. C. (1998). Forensic necropsy. *Seminars in Avian and Exotic Pet Medicine, 7*(4), 201–209.

Munro, H. C., & Thrusfield, M. V. (2001a). Battered pets: Non-accidental physical injuries found in dogs and cats. *Journal of Small Animal Practice, 42,* 269–290.

Munro, H. C., & Thrusfield, M. V. (2001b). Battered pets: Sexual assault. *Journal of Small Animal Practice, 42,* 333–337.

Patronek, G. J. (1998). Issues and guidelines for veterinarians in recognizing, reporting, and assessing animal neglect and abuse. In P. Olson (Ed.), *Recognizing and reporting animal abuse—a veterinarian's guide* (pp. 25–39). Denver, CO: American Humane Association.

Patronek, G. J. (1999). Hoarding of animals: An under-recognized public health problem in a difficult-to-study population. *Public Health Reports, 114,* 81–87.

Patronek, G. J. (2004). Animal cruelty, abuse, and neglect. In L. Miller & S. Zawistowski (Eds.), *Animal shelter medicine for veterinarians and staff* (pp. 427–452). Ames, IA: Blackwell Publishing.

Patronek, G. J., & Hoarding of Animal Research Consortium (2001). The problem of animal hoarding. *Municipal Lawyer, 19,* 6–9.

Serpell, J. A. (1999). Working out the beast: An alternative history of Western humaneness. In F. R. Ascione & P. Arkow (Eds.), *Child abuse, domestic violence and animal abuse: Linking the circles of compassion for prevention and intervention* (pp. 38–49). West Lafayette, IN: Purdue University Press.

Stone, K. (2006, March). I love you to death. *Cat Fancy,* 50–51.

Sinclair, L., Merck, M., & Lockwood, R. (2006). *Forensic investigation of animal cruelty.* Washington, DC: Humane Society Press.

CHAPTER 6

Pet Care Industry

I am as vigilant as a cat to steal cream.

–William Shakespeare (*Henry IV, Part 1*)

KEY TERMS

American Pet Product Manufacturers Association (APPMA)
Pet Industry Joint Advisory Council (PIJAC)
pet food industry
Pet Food Institute (PFI)
Association of American Feed Control Officials (AAFCO)
kitty litter
James Herriot
Mark Morris
applied animal behaviorists
veterinary behaviorists
dog training
Association of Pet Dog Trainers (APDT)
veterinary medicine
grooming
National Dog Groomers Association of America
boarding kennels
American Boarding Kennel Association
pet sitting
media
nonprofit

The growing interest in companion animals is fueling an expanding industry of businesses, professions, and organizations associated with their care and treatment and that provide services to pet owners. These include pet foods, medical care, media, and charities. It is not simply a case of more of the same things such as food, collars, and leashes. There is also a greater range of products available than ever before (Guthrie, 2005). Agrawal (2006), for example, points out that there are over 800 bakeries in the United States that specialize in producing treats for pets. Tupperware-style parties are held to introduce people to new toys and products in a convenient fashion. These parties also provide a social support network that implicitly endorses the purchase of these products (Masters, 2005). Spending on pet care supplies outsells baby care supplies $8.5 billion to $6.2 billion (Glater, 2005). Overall, the $36 billion pet care industry total is in the same range as the motion picture and sound industry, $47.3 billion, and furniture and related products at $31.4 billion (Smith & Lum, 2005). The design, manufacturing, and sale of these products are creating a wide variety of new jobs in the various pet care industries. Pharmaceutical research related to companion animal care is a growth area that is creating new job opportunities for biologists (Agres, 2003). Spending on companion animals, especially veterinary care, is steady, even during tough economic times (Richardson, 2006). This chapter will introduce the major categories of the pet care industry, their histories, and their current status and practice.

PET PRODUCTS

A visit to a modern pet supply super store such as PetSmart or PETCO is a revelation. They are the two dominant retailers in the marketplace, with over 1,500 locations between them and thousands of employees. They boast large stores with a bewildering array of products and choices (Figure 6-1). The consumer is presented with an enormous range of choice in terms of foods, treats, toys, and other needs, and they are welcome to bring their pets to the store to shop with them. These stores are a long way from the small "mom and pop" pet shops that typified the field until the late 1900s (Figure 6-2).

Through the mid-1800s, pet care products were rare. They were often homemade, or from a local craftsman. Pets typically ate leftovers from whatever their family had for dinner, and home *doctoring* was the most common form of medical care. In the latter half of the 1800s, as interest in pets continued to grow, a number of entrepreneurial companies were founded to meet the needs of pets and their owners. In some cases, these early pioneers in the business not only

FIGURE 6-1

Large, modern pet stores offer a dizzying array of products for pets and their owners. Some of these stores have decided not to sell dogs or cats, and instead, partner with local animal shelters and rescue groups to promote adoption of homeless pets.
(Courtesy PetSmart)

FIGURE 6-2

The pet industry started small, but this small mom and pop from the mid-twentieth century shows a small part of the range of pets and products that would expand in the years after World War II. Tropical fish, bird cages, collars, leashes, and foods are all evident. *(Courtesy Collection of Katherine C. Grier)*

FIGURE 6-3

Sergeant's was one of the early pet product companies that has grown and adapted to changing consumer and pet needs. Many of these companies produced booklets and pamphlets that were available for free or low cost to assist pet owners. *(Courtesy Joel Adamson)*

invented the product, they invented the *need* for the product as well. As they offered their wares, they also tracked the development of national consumer brands.

Sergeant's Pet Products is a good example of how a national brand of pet care products developed. Polk Miller was a druggist in Richmond, VA, and the founder of the Polk Miller Drug Company. He was also an avid hunter and sportsman who loved his hunting dogs. He compounded a variety of medicines and remedies for his own dogs and those of friends and relations. In 1868 he founded a new company named for his favorite hunting dog, Sergeant. In addition to medicines, Polk Miller's advice on dog care was distilled into a series of booklets on the subject (Figure 6-3). The cover of the booklet featured Polk Miller himself and his dog Sergeant. Each section of the booklet provided a description of a canine affliction, along with symptoms to watch for, and a treatment, typically a product offered by the company. Purchasing a product entitled the consumer to free advice from a company department run by a veterinarian. The booklets also featured *Senator Vest's Tribute to a Dog* (see sidebar). The sentiment expressed in the tribute, and Polk Miller's use of the story in materials distributed with his products, is an early sign of how the pet industry would exploit the human-animal bond in marketing (Glater, 2005).

Sergeant's also now owns the Geisler brand. Geisler was another of the early companies that provided a range of pet care products. Founded by Max Geisler in 1888, the company began importing and selling small caged birds such as canaries. Eventually Geisler offered a line of specialty foods for the birds. Similar to Geisler, Hartz Mountain also started with the importation of canaries in the 1920s and later added a wide range of other products. As these products represented a growing source of company income, Hartz Mountain stopped selling live animals in the 1950s (Grier, 2006).

Senator Vest's "Tribute to a Dog"

Senator Vest, of Missouri, was attending court in a country town, and while waiting for the trial of a case in which he was interested, he was urged by the attorneys in a dog case to help them. He was paid a fee of $250 by the plaintiff. Voluminous evidence was introduced to show that the defendant had shot the dog in malice, while the other evidence showed that the dog had attacked the defendant. Vest took no part in the trial and was not disposed to speak. The attorneys, however, urged him to make a speech or else their client would not think he had earned his fee. Being thus urged, he arose, scanned the face of each juryperson for a moment, and said the following:

"Gentleman of the Jury: The best friend a man has in the world may turn against him and become his enemy. His son or daughter that he has reared with loving care may prove ungrateful. Those who are nearest and dearest to us, those whom we trust with our happiness and our good name may become traitors to their faith. The money that man has he may lose. It flies away from him, perhaps when he needs it most. A man's reputation may be sacrificed in a moment of ill-considered action. The people who are prone to fall on their knees to do us honor when success is with us may be the first to throw the stone of malice when failure settles its cloud upon our heads. The one absolutely unselfish friend that man can have in this selfish world, the one that never deserts him, the one that never proves ungrateful or treacherous, is his dog. A man's dog stands by him in prosperity and poverty, in health and sickness. He will sleep on the cold ground where the wintry winds blow and the snow drives fiercely, if only he may be near his master's side. He will kiss the hand that has no food to offer; he will lick the wounds and sores that come in encounter with the roughness of the world. He guards the sleep of his pauper master as if he were a prince. When all other friends desert, he remains. When riches take wings and reputation falls to pieces, he is as constant in his love as the sun in its journey through the heavens.

"If fortune drives the master forth an outcast in the world, friendless and homeless, the faithful dog asks no higher privilege than that of accompanying him, to guard against danger, to fight against his enemies, and when the last scene of all comes, and death takes the master in its embrace, and his body is laid away in the cold ground, no matter if all other friends pursue their way, there by the graveside will the noble dog be found, his head between his paws, his eyes sad, but open in alert watchfulness, faithful and true even in death."

Senator Vest sat down. He had spoken in a low voice, without any gesture. He made no reference to the evidence or the merits of the case. When he finished, judge and jury were wiping their eyes. The jury returned a verdict in favor of the plaintiff (Polk Miller Products Corp., 1927).

PET INDUSTRY ORGANIZATION

There are two major organizations of companies in the pet industry. The **American Pet Product Manufacturers Association (APPMA)** was founded in 1958 (<www.appma.org>). Its 830 members include product manufacturers, importers, and livestock suppliers. Its primary purpose is to promote and advance pet ownership, and it is responsible for an annual survey of pet owners that is discussed elsewhere in this book. APPMA also has a government relations division that monitors pet ownership laws being considered at the state and federal level. Among other groups, APPMA helped to mobilize its members during the response to Hurricanes Katrina and Rita, shipping millions of dollars worth of food, cages, and other critical materials needed during rescue operations.

The **Pet Industry Joint Advisory Council (PIJAC)** is a second major organization in the field (<www.pijac.org>). The Pet Industry Joint Advisory Council includes pet stores and provides training and certification programs for animal care and husbandry in pet stores. Similar to APPMA, PIJAC has a very active government relations division that monitors federal, state, and local initiatives that could have an impact on the industry. Together these two organizations play a significant role in protecting the business interests of the pet industry.

PET FOODS

For most of their history as human companions, dogs and cats subsisted on a diet of table scraps, leftovers, and homemade foods. The first important development in the **pet food industry** came from James Spratt in the 1860s. Spratt was an Ohio electrician selling lightning rods in London when he observed stray dogs on a dock eating scraps of the biscuits and hard tack that sailors carried on voyages for food. These biscuits had the advantage of being easy to pack, store, and had a reasonably long shelf life. Spratt developed a recipe based on wheat meal, beet root, meat, and vegetables and was soon marketing *Spratt's Dog Cake* (Campbell, Corbin, & Campbell, 2005). Spratt had a talent for promotion and marketing, anticipating many of the strategies used today for the sale of pet foods and other products. He promoted his foods as a *brand* that represented high quality for people who cared about their dogs. He utilized logos and various promotions to ensure high visibility of his product line (Figure 6-4). He associated his food with top-quality dog shows by sponsoring events and printing kennel club registration papers for dogs, with a promotional advertisement for his dog foods on the back. His first employee, Charles Cruft, was instrumental in this strategy, and one of the world's premier dog shows, Crufts in England, is named after him. Spratt even anticipated the eventual development of specialty foods and foods for different life stages of dogs.

Spratt's success spurred the development of many other companies in the pet food arena. Many of these were small milling companies that added a pet food line to the other cereals and products they produced for human and livestock consumption. In 1894 George Rolenson and William Danforth started an animal feed business in St. Louis, MO. They added whole wheat breakfast cereals for humans in 1898. These breakfast cereals were endorsed by Dr. Ralston, a well-known health food guru at the time, promoting the benefits of high-quality, pure foods that had limited processing. The company name, Purina, was derived from the advertising slogan *Where Purity Is Paramount*. Eventually the company name was changed to Ralston Purina. The familiar red-and-white checkerboard square design was added to the company's product packaging in the early 1900s to provide a consistent, easily identifiable look. Dog food was offered as a meal through dealers of Purina Livestock feeds (Ralston Purina Company, 1994). In 1933, Admiral Byrd needed a light, convenient dog food that would not freeze to feed his dog teams during his trek to the South Pole. Purina provided dog food for Byrd's teams and would

FIGURE 6-4

Spratt's, the first commercial dog food company, anticipated many of the promotional strategies that would be at the core of the pet industry for decades to come. Dog statues like this would have stood in a prominent place to let people know that the store carried Spratt dog foods. *(Courtesy of McKissick)*

FIGURE 6-5

Celebrity endorsements have long been a part of the pet food industry. Admiral Byrd used Purina to feed his sled dogs on the way to the South Pole, and would later appear in promotional materials for the brand. Obviously, if the food was good enough to keep dogs fit and healthy in the extreme conditions of the Antarctic, it would be just what your own dog needed. *(Courtesy of Nestle Purina Petcare)*

eventually use Byrd and his dogs in advertising promotions for the company (Figure 6-5).

Purina would revolutionize the pet food industry in the 1950s, when they introduced Purina Dog Chow, the first pet food product produced by high-pressure extrusion. A Purina employee, Doug Hale, had watched a local St. Louis company bake corn curls with an extruder that compressed and heat-treated the corn. He then worked with several other Purina employees to adapt the process to the manufacture of dog food (Ralston Purina Oral History Archives, n.d.). The extrusion method of production offered a number of advantages:

- The final product had a more appealing and substantial appearance when compared with pelleted meals.
- It provided the ability to coat the surface of the food with fats and flavors to increase the palatability of the food.
- It enhanced the shelf life of the food, opening up the grocery story consumer market.

Purina Dog Chow soon became a dominant brand, and the company could not keep up with demand. Purina eventually added Cat Chow in 1962 and introduced the first semi-moist foods in 1971.

The Iams Company also arose from the livestock feed sector. Paul Iams was a student of nutrition and experimented with developing a high-quality dog food. He formed his company in 1946 and introduced his first food, Iams 999, in 1950. Paul Iams also experimented with developing supplements for mink food that would improve the quality of their fur. When he eventually developed a food that accomplished this, he discovered that the dogs that guarded the mink farms also ate the food and were healthy with glossy coats. He continued his research and in 1961 introduced Iams Plus, the first nutritionally complete diet for dogs (Iams Company, n.d.). Purina was always part of a large consumer products company that included cereals, restaurants, and even Eveready Batteries. Iams remained an independent pet food company until it was acquired by Proctor and Gamble in 1999. The acquisition resulted in an expansion of marketing for the Iams Company pet foods, including introduction to grocery store shelves.

Canned foods for pets developed out of the introduction of canned meats for human consumption in the mid-1800s. Animal parts not used for human consumption could be processed and canned for pets. At the turn of the twentieth century, there were many regional slaughterhouses located near cities. Canning plants were often opened up nearby or in conjunction with the slaughterhouse to process and preserve meats that could not be immediately shipped for sale fresh. In the years following World War I, as electric trolleys and motorized tractors accelerated the retirement of many horses that worked in both cities and on farms, they were frequently slaughtered and provided a cheap, high-quality meat for the pet food industry (Dorosz, 2006). This led to competing claims between companies that manufactured feeds from meats and grains and those that processed and sold canned meat pet foods. Each company made claims about the quality of its food and frequently employed well-known celebrities, both human and canine, in their promotional efforts.

Canned foods are higher in moisture (about 75%) than dry kibble foods (about 10%) and are pressure-sterilized to kill bacteria. As a result, canned foods have a longer shelf life than dry foods, which will deteriorate and lose nutritional value over time. Most pets find canned foods to be highly palatable because of the high levels of protein and fats. Modern processing methods have resulted in canned foods that appear very similar to human foods. Many owners find this appealing and prefer to feed canned foods even though the higher moisture level in canned foods results in a higher cost per meal. The observation that their pets seem to really enjoy canned foods and may prefer them to dry kibble will influence the owner's decision on what to feed their companion. It is not uncommon for people to feed a combination of canned and dry food, adding a couple of spoonfuls of canned food to the kibble at each meal.

Treats, Leftovers, and Raw Foods

In addition to the many commercially produced diets that are available, people also tend to feed their pets a wide range of treats and leftovers from their own meals. Treats are not unexpected given the affection of that people have for their pets. However, much as for humans, too much snacking can lead to obesity, a significant health problem among America's pets (Case, 2003). In addition to obesity, feeding excessive amounts of treats, which do not need to be nutritionally balanced or complete, and leftovers can result in animals not receiving adequate levels of required nutrients. It is generally recommended that treats and leftovers not exceed 10% of the daily caloric intake of pets. Another concern that must be considered when feeding leftovers or other human foods to companion animals is that a number of different food items eaten by people can be toxic to pets. Chocolate is one well-known example.

More recently it was discovered that raisins and grapes can be toxic to dogs. An extensive list of other foods that can sicken pets is available at the Web site of the ASPCA Animal Poison Control Center (<www.aspca.org/apcc>).

Some pet owners have also started to make their own homemade foods. While this method can provide pets with an adequate diet, it does require a substantial amount of work to ensure that the final product includes a complete nutritional balance. It is important to recognize that some preparation methods can reduce or compromise some of the nutrients that were present in the food, and without appropriate preservation, they can deteriorate rapidly over time. Individuals wishing to use a homemade diet should consult with their veterinarian for an acceptable recipe.

Raw foods have also become popular as a more *natural* way to feed companion animals. These diets may sometimes be referred to as BARF (bones and raw food) diets. Care must be taken with these diets, whether prepared at home or purchased commercially (Case, 2005). Muscle meat alone, while a good source of protein, does not provide adequate and complete nutrition for pets. Additional foods or supplements need to be given to provide needed calcium, phosphorus, iron, vitamins, and other nutrients. Raw diets also expose companion animals and owners to an increased risk of foodborne illness and parasitic infection. In addition to being dangerous to the pets, *Salmonella, Escherichia coli, Listeria,* and other pathogens are also zoonotic and place the pet owner at risk of infection as well.

Pet Food Industry Organization

The **Pet Food Institute (PFI)** was founded in 1958 and represents about 97% of the industry. It provides education and consumer research for its member companies. It also lobbies legislative bodies on issues of importance to the industry (for more information, visit <http://www.petfoodinstitute.org>).

The **Association of American Feed Control Officials (AAFCO)** is a nonregulatory, nongovernmental group that conducts feeding trials to verify the nutritional claims made by pet food companies. The Association of American Feed Control Officials defines a *complete and balanced diet* as one that provides all essential nutrients plus ingredients that provide energy at needed levels. It recommends minimum levels required for essential nutrients and maximum levels permitted for ingredients that may be toxic if present in too great an amount. On the basis of this information, AAFCO proposes model regulations for state and federal agencies to ensure consistency among nationally marketed foods (AAFCO, 2007). Nutritional claims for a food can be substantiated in two different ways. The most thorough evaluation is made when a company formulates and manufactures a food and then conducts feeding trials. Feeding trials demonstrate the digestibility and bioavailability of all nutrients at the proper level following the manufacturing process. Acceptable foods meet the needs and support the health of the intended species and life stage of the animals. A second type of evaluation would be for the company to show that the food is formulated with the proper nutrient profile through ingredient composition and laboratory analysis of the nutrient content of the finished product. A drawback to this method is that it does not address the nutrient availability, digestibility, or palatability of the final product.

Pet Food Recall 2007

In late winter and early spring of 2007, the pet food industry was rocked by a massive recall of products stemming from the contamination of dog and cat foods. Menu Foods, a Canadian company, produced packaged pet food for a great many different brands and companies. They began to receive complaints that foods it manufactured were sickening pets in February. By March, veterinarians working for Iams were receiving complaint calls about renal failure in cats that had eaten a "cuts and gravy" product produced for Iams by Menu Foods. They conveyed their concerns to Menu Foods. In the meantime, Menu Foods conducted a palatability study during which nine cats died of acute renal failure before the test ended. Menu Foods then notified the United States Food and Drug Administration on March 15, and a recall of pet foods produced by the company was announced on March 16. The recall included over 60 million packages of cuts and gravy-type foods produced for more than 100 brands of pet food.

The nature of the contamination took some time to work out. Eventually it was discovered that wheat gluten, purchased from a Chinese supplier and used to thicken the gravy and increase the protein levels of the food, had been contaminated with melamine and cyanuric acid. Melamine and cyanuric acid are both moderately toxic; however, it was found that in combination they formed crystals in the kidneys of pets that had consumed the foods. These crystals plugged kidney tubules, reducing and then blocking kidney function, resulting in kidney failure. It is still unknown how many pets died as a result of the contamination, though estimates run into the hundreds and thousands.

FIGURE 6-6

Few things changed the culture of pet keeping as dramatically as the development of kitty litter. Keeping cats indoors was a dirty, stinky endeavor without ready access to an absorbent, easy-to-clean substrate. Kitty litter made it possible to keep cats indoors most of the time, and was especially welcomed by people living in apartments, several stories above street level. *(Courtesy of McKissick)*

KITTY LITTER

Kitty litter may be one of the most significant developments in our history with companion animals. Until the development of kitty litter, cat lovers used a wide variety of materials in their kitty pans, ranging from sand to dirt, shredded paper, sawdust, and ashes. The mess and odor from these litter pans helped to make cats less than welcome in the indoor home environment. In 1948 a woman in Michigan went out to the local sand pile and found it frozen. She went to Ed Lowe, whose family ran a coal, ice, and sawdust company (Magitti, 1996). Lowe had some granulated mineral clay that he offered. The woman tried it, and it worked out so well that she told all of her friends with cats about it. This created demand for "kitty litter," and Ed Lowe was soon bagging the clay and selling it (Figure 6-6). This part of the industry has grown rapidly, and with sales of $708 million a year, it accounts for one-third of dollars spent on pet supplies. It also played a major role in helping cats become a favorite indoor pet and their eventual surpassing of dogs as the most common companion animal in the United States.

Kitty litters are now available in a wide-ranging array including the various clumping litters, with and without scents. All of these "classic" type litters are still based on the same granulated clay substrate. A variety of newer products are now made with natural organic compounds, compressed recycled newspapers, and other substrates, though these are still a small part of the overall market. In addition to the development of new litter formulations, the ingenuity of inventors can be seen in a myriad of litter box variations. There are automatic boxes that will "scoop" once a cat uses the boxes and hops out, and any number of other systems that will limit our exposure to the smell and mess of cat waste.

VETERINARY MEDICINE

When people think about companion animals and their care, veterinarians are probably one of the first thoughts that come to mind. Veterinarians are well-respected professionals in our society and are generally presented in a favorable, positive manner. In addition to the positive personal experiences that people may have with the veterinarians that care for their pets, many pet owners have been influenced by the beloved books of **James Herriot** (see sidebar). Nearly every young person with a pet has probably thought about growing up to become a veterinarian at one time or another.

Origins of Veterinary Medicine

Providing medical care for animals probably dates back to the earliest days of domestication of animals and their association with humans. These early efforts were generally directed toward animals that were important for economic reasons or were significant for the interests of the ruling classes. Horses were critical to warfare and transportation. Oxen and asses provided the power needed to forge a stable, agrarian society. Dogs seemed to intersect with all levels of society. They were used for herding, guarding, and pulling carts, and they hunted with royalty. The Babylonian Code of Hammurabi (circa 1760 BCE) included rules for veterinary practice:

- If a "veterinary surgeon" performed a major operation on either an ox or ass and cured it, the owner of the ox or ass shall give to the doctor one sixth of a shekel of silver as his fee.

- If he performed a major operation on an ox or ass and has caused its death, he shall give to the owner of the ox or ass one fourth of its value. (Dunlop & Williams, 1996, p. 55)

James Herriot

James Herriot was the pen name of James Alfred Wight (1916–1995), a practicing veterinarian in Yorkshire, England. He attended veterinary college at Edinburgh and moved to Yorkshire, where he joined a small mixed practice in the years between World War I and World War II. He documented the life and labors of a country veterinarian in a series of best-selling books. His humor and sensitivity captivated readers as he described the colorful clients and patients that he and his colleagues came across in their work. These were the years when small country farms in that part of England were just starting to transition from horses to tractors, from hand milking of cows to mechanical milking, the appearance of antibiotics and vaccines for a number of diseases, and a growing interest and emphasis on companion animals. His stories were collected and first published in the United States in *All Creatures Great and Small* in 1972. The success of this publication led to *All Things Bright and Beautiful*, *All Things Wise and Wonderful*, and *The Lord God Made Them All*. The books stimulated the production of a popular television series on the BBC that was also broadcast on American public television. Wight continued to practice veterinary medicine with his partners even after international acclaim as a best-selling author. He was obviously happy in his work, and his stories take on a feel of being romantic and heroic and no doubt inspired many readers to consider a career in veterinary medicine.

FIGURE 6-7

The veterinary profession has grown and changed in many ways from its early days of concentration on horses. Livestock and horses have given way to companion animals as a primary part of private practice, and women have overtaken men as new members of the profession. *(Courtesy ASPCA)*

To provide some context, a physician treating a freeman would be paid five shekels and treating a slave would bring a fee of two shekels. Castration was probably the first common veterinary operation performed, primarily to control the behavior of stallions, bulls, and boars.

For quite some time, animal care existed somewhere between skilled stockmen, priests, and medicine. Across various cultures the treatment of animals paralleled the philosophical and religious influences on illness and medicine for humans. Spells and prayers were often combined with surgical procedures, drugs, and medicines. For example, in China the balance of the dual forces yin and yang was an important influence on medical practice for humans and was also prominent in the treatment of animal ailments. By 650 BCE, acupuncture charts were available for treating animals, along with herbal medicines and preparations (Dunlop & Williams, 1996).

Medical care for animals further developed in Greece. It was influenced by Hippocrates (460–367 BC), a Greek physician who also formalized medical ethics. Nearly all physicians and veterinarians take some form of the *Hippocratic Oath* governing their treatment of their patients. A critical element of the oath is captured in the phrase *primum non nocere*—"Above all, do no harm." The Greeks also contributed the icon associated with medical professions, the caduceus. This is the familiar image of a snake wrapped around a staff. It is derived from the staff that was carried by Asklepios, a mythological healer and son of the god Apollo. The veterinary version of this image includes a V superimposed over the staff and snake (Figure 6-7).

The vast Roman Empire was dependent on horses to transport goods and information. The care of horses was an important consideration. The term for animal caretakers was *souvetaurinarii*, and the compounds used for holding animals were known as *veterinarium*, providing a possible derivation for the term *veterinarian* (Dunlop & Williams, 1996). The decline and destruction of the Roman Empire resulted in a tremendous loss of knowledge in the Western world. The remaining knowledge was often fragmentary, and some practices were carried on with no clear sense of understanding or purpose. Various herbs and mystical potions that might be mixed with holy water from a Catholic church were mixed and delivered to both human and animal patients. Bleeding was a frequent treatment for a wide variety of ailments, often with the aid of leeches. This was so common that *leech* became something of a synonym for someone who provided medical care.

Veterinary Education

The roots of modern medicine are found in the Renaissance era. The study of anatomy, based on dissection and vivisection, and eventually experimental physiology, led to a more complete understanding of both human and animal biology. Treatments for disease evolved, became more sophisticated, and were based on scientific principles. This was the era when the practice of medicine as a scientific discipline came to separate it from its past associations with religion and mysticism. The next stage in the evolution of veterinary medicine would be its disengagement from animal husbandry through the development of a formalized education process. King Louis XV established the world's first veterinary school at Lyon, France, in 1761. A second school was established in 1765 at Alfort, France, and additional schools would follow in Austria and other areas of the continent. The Veterinary College of London would eventually follow in 1791. The first American veterinary college would not come until 1868, when Cornell was founded in Ithaca, NY.

Much of the development of formal veterinary medicine was driven by the continued significance of horses for transport and work, livestock for food and fiber in the economy, and concern for public health. The field was largely male dominated, which would continue until after World War II. Small-animal medicine was a minor part of the training and practice of veterinarians. Rabies and distemper as diseases of dogs did attract the attention of veterinary medicine at the beginning of the nineteenth century. Rabies was a significant zoonotic threat to humans, and distemper could sweep through a dog population killing hundreds of dogs, including those kept for hunting, herding, and other purposes. The profession began to organize in America with the formation of the United States Veterinary Medical Association (USVMA) in New York City in 1863. In 1898 the organization would change its name to the American Veterinary Medical Association (AVMA). The AVMA has grown to become the largest and most important voice representing the veterinary profession. It promotes the practice of high-quality medicine and professional conduct. The AVMA also evaluates and approves the curricula of veterinary colleges.

In the early 1900s, there were a number of private veterinary colleges that often existed for short time frames. A significant development in veterinary education was the Morrill Land Grant Act of 1862 (Miller, 1981). This legislation provided the infrastructure that led to the formation of Land Grant Colleges that would support education, research, and service to support agriculture and engineering. As a number of states established universities under this plan, they included veterinary colleges as a part of the campus (Table 6-1). The educational requirements for these schools are fairly consistent from one institution to another. The vast majority of students enter with an undergraduate degree in some pre-professional major such as biology, biochemistry, or animal science. Veterinary school typically takes four years to complete, and graduates receive a doctorate in veterinary medicine (DVM; except for the University of Pennsylvania, which awards a VMD). Students are required to pass a licensing exam administered by the State Board of Veterinary Medicine in the state where they desire to practice. Following World War II a number of specialties developed within veterinary medicine. These include surgery, internal medicine, toxicology, nutrition, and many others. Individuals who wish to be board certified in a specialty are required to complete an approved two- to three-year residency under the supervision of a veterinarian who is already board certified in the specialty. In addition to other requirements that may be specific to the specialty, passing an exam is also required before being admitted as a diplomat to the College of the specialty. For example, after completion of all requirements, someone who specializes in surgery would become a Diplomate of the American College of Veterinary Surgeons and could indicate this by placing the acronym DACVS following the veterinary degree after his or her name.

TABLE 6-1

Veterinary Colleges in the United States *(Courtesy ASPCA)*

Auburn University College of Veterinary Medicine Auburn University Auburn, AL 36849 www.vetmed.auburn.edu Established in 1892	**Kansas State University College of Veterinary Medicine** 101 Trotter Hall Manhattan, KS 66506 www.vet.ksu.edu Established in 1905
UC Davis School of Veterinary Medicine One Shields Avenue Davis, CA 95616 www.vetmed.ucdavis.edu Established in 1905	**Louisiana State University School of Veterinary Medicine** Skip Bertman Dr. Baton Rouge, LA 70803 www.vetmed.lsu.edu Established in 1973
Colorado State University College of Veterinary Medicine & Biomedical Sciences 1601 Campus Delivery Fort Collins, CO 80523-1601 www.cvmbs.colostate.edu Established in 1907	**Michigan State University College of Veterinary Medicine** G100 Vet Med Center East Lansing, MI 48824 www.cvm.msu.edu Established in 1855
Cornell University College of Veterinary Medicine Ithaca, NY 14853-6401 www.vet.cornell.edu Established in 1894	**University of Minnesota College of Veterinary Medicine** 231 Pillsbury Dr. SE Minneapolis, MN 55455 www.cvm.umn.edu Established in 1891
University of Florida College of Veterinary Medicine Gainesville, FL 32610 www.vetmed.ufl.edu Established in the 1970s	**Mississippi State University College of Veterinary Medicine** Mississippi State, MS 39762 www.cvm.msstate.edu Established in 1974
University of Georgia College of Veterinary Medicine Athens, GA 30602-7371 www.vet.uga.edu Established in 1946	**University of Missouri-Columbia College of Veterinary Medicine** Columbia, MO 65211 www.cvm.missouri.edu Established in 1884
University of Illinois at Urbana-Champaign College of Veterinary Medicine 2001 South Lincoln Avenue Urbana, IL 61802 www.cvm.uiuc.edu Established in 1948	**North Carolina State University College of Veterinary Medicine** 4700 Hillsborough St. Raleigh, NC 27606 www.cvm.mcsu.edu Established in 1978
Iowa State University College of Veterinary Medicine PO Box 3020 Ames, IA 50011 www.vetmed.iastate.edu Established in 1872	**Ohio State University College of Veterinary Medicine** 1900 Coffey Rd. Columbus, OH 43210 www.vet.ohio-state.edu Established in 1885

(continues)

TABLE 6-1 *Continued*

Oklahoma State University Center for Veterinary Health Sciences Stillwater, OK 74078 www.cvm.okstate.edu Established in 1951	**Tufts University Cummings School of Veterinary Medicine** North Grafton, MA 01536 www.vet.tufts.edu Established in 1978
Oregon State University College of Veterinary Medicine Corvallis, OR 97331 http://oregonstate.edu/vetmed Established in 1975	**Tuskegee University College of Veterinary Medicine** Tuskegee, AL 36088 www.tuskegee.edu/cvmnah Established in 1945
University of Pennsylvania School of Veterinary Medicine 3451 Walnut Street Philadelphia, PA 19104 www.vet.upenn.edu Established in 1884	**Virginia-Maryland Regional College of Veterinary Medicine** Blacksburg, VA 24061 www.vetmed.vt.edu Established in 1978
Purdue University School of Veterinary Medicine 625 Harrison Street West Lafayette, IN 47907 www.vet.purdue.edu Established 1957	**Washington State University College of Veterinary Medicine** Pullman, WA 99167 www.vetmed.wsu.edu Established in 1899
University of Tennessee College of Veterinary Medicine 2407 River Dr. Knoxville, TN 37996 www.vet.utk.edu Established in 1967	**Western University College of Veterinary Medicine** Pomona, CA 91766 www.westernu.edu/xp/edu/veterinary/home.xml Established in 1998
Texas A & M University College of Veterinary Medicine College Station, TX 77483 www.cvm.tamu.edu Established in 1916	**University of Wisconsin-Madison School of Veterinary Medicine** 2015 Linden Dr. Madison, WI 53706 www.vetmed.wisc.edu Established in 1983

Post–World War II Developments

Two significant developments in the years following World War II have helped to change the face of veterinary medicine. One was the increasing number of women entering the field. Aleen Cust was the first woman to graduate from a veterinary college in 1900, when she completed her studies at the Veterinary College of Edinburgh. However, she was not permitted to join the Royal College of Veterinary Surgeons at that time. She would wait until 1923 before she was allowed to take the licensing exam. The first American woman to complete veterinary school was Elinor McGrath in 1910. Women were rare in the field until the later part of the twentieth century. It is likely that women were often discouraged from entering the profession due to the perceived physical requirements of working with horses and livestock in the early days of the profession. A remarkable transition has occurred, and women now make up the majority of students in veterinary schools, and of new veterinarians entering practice. Some part of this change in gender representation may be associated with another significant change in veterinary practice. In the years following World War II, horses were eventually phased out of work on farms and elsewhere. At the same time, as noted in earlier chapters, the number of animals kept as companions began to grow rapidly. The *horse doctor* of the past was giving way to the small animal practitioner. A glimpse of this development could be seen in the formation of the American Animal Hospital Association (AAHA) in 1933 to promote high-quality small-animal medicine. It would not be until the post–World War II years, however, that the growth of this part of veterinary practice would really take hold.

Mark Morris

Dr. **Mark Morris** had a profound impact on the science of pet nutrition and on the veterinary profession. Born in Colorado in 1900, his daily commute to school included walking, riding a bicycle, and riding a horse and a horse-drawn buggy. A poor student in mathematics at first, he eventually buckled down to master the subject when he realized that it was essential to understand fully the chemistry that he enjoyed. He began his veterinary studies at Colorado State University, transferring to Cornell to complete his veterinary degree in 1926. Soon after graduation he bought a mixed large- and small-animal practice in New Brunswick, NJ. As his practice expanded he took on a partner to handle the large animal cases so he could concentrate on companion animals. This was a bold move in those days, as few veterinarians concentrated on treating pets, relying on livestock care as the primary focus of their work, and income. He built a new hospital designed for the treatment of dogs, cats, and other pets. Morris had not lost his fascination with chemistry, and his new hospital included a laboratory. It was here that he began a series of studies that would help to change the field. Working with blood and urine samples from his patients, he conducted detailed analyses to link the chemical profiles of samples from the animals with specific diseases. He combined this work with studies of nutrition and was soon mixing special diets for patients in his care. When he was confronted with a seeing-eye dog with kidney problems, Morris produced the first *prescription diet* for a specific condition. This formulation, which he canned himself, eventually became Prescription Diet k/d®. As he used the diet with other animals, he was convinced that this approach could be used as part of the treatment for a number of different ailments. Word of his success with this approach soon spread in the profession, and colleagues began to request his prescription diets for their patients. Demand overtook the ability of his small staff to mix and can the products in his small hospital. Morris would eventually negotiate an agreement with Hill Packing Company in Topeka, KS, to produce, can, and distribute his prescription diets to veterinarians. In mid-October, 1948, Morris supervised the first production run at Hill Packing Company, and prescription diets were soon available for national distribution. This humble start would lead to an immensely successful pet food company.

If all Mark Morris had done was to develop the concept and perfect the formulation of his prescription diets, he would occupy an important niche in the history of companion animal care. However, in addition to this work, he also played a key role in the organization of the American Animal Hospital Association, serving as its first president. He would eventually serve as president of the American Veterinary Medical Association in 1961. When he died in 1993, he left behind many contributions to the care of companion animals. His interest in scientific research to benefit animal care is carried on by the foundation that he founded in 1948 (Haselbush, 1984). The Morris Animal Foundation, endowed with royalty funds from the sale of prescription diets, has provided over $42 million in funding for over 1,200 studies in animal health research (Morris Animal Foundation, 2006).

Corporate Practices

While the traditional model of companion animal veterinarian is a private practice and small business owner, this is now changing. There is some consolidation in the field. A good example of this is Banfield, The Pet Hospital. Banfield started as a single small practice in Portland, OR, in 1955. In 1993 the group opened a practice in a PetSmart store. By 1999 Banfield purchased 114 veterinary practices within PetSmarts to become the sole provider of veterinary services at the stores. For a while these practices were known as VetSmart, but in 2000 all these practices were renamed Banfield, The Pet Hospital. As of 2006, there are now 600 practices within the Banfield system. They are linked by sophisticated software into a comprehensive case record/management system. This allows them to stay on top of emerging disease trends, cost management, and pricing strategy for services. There are other large corporate practices of this type, including Veterinary Clinics of America (VCA). They offer a variety of advantages and challenges for the practice of veterinary medicine. These practices can make it easier for veterinarians to balance their personal lives and schedules, receive a salary and benefits, and not bear the primary responsibility for management of what is really a small business. On the other hand, there is some concern among veterinarians that working in such a practice will limit their individual freedom to practice medicine if there is too much regimentation in service delivery. The issues will be important to monitor over the next decade, as they will influence veterinary careers and the delivery of health care to companion animals.

Veterinary Health Insurance

Veterinary health insurance has been slow to take hold in the United States and lags behind its acceptance overseas (Veterinary Practice News, 2006). The industry is expected to grow from $185 million in 2005 to $550 million in 2010. Currently 83% of policies are purchased for dogs and just 15% for cats. Infections are the most common claim for dogs (9%), and urinary tract infections for cats (7%).

ANIMAL BEHAVIOR SERVICES

An understanding of animal behavior was likely an important part of human evolution. Knowing what an animal was doing and what it meant by its actions may have been the difference between *getting dinner* or *being dinner* for our ancestors. Naturalists, explorers, and philosophers frequently described and commented on animal behaviors that they had observed or were reported to them. The systematic study of animal behavior began with Charles Darwin when he included behavior in his work on evolution and natural selection. His classic work, *The Expression of the Emotions in Man and Animals* (1872/1965), described how behavior could help a species adapt, and traced the evolutionary links and similarities between humans and animals. George Romanes followed in Darwin's footsteps and published *Animal Intelligence* in 1882 (Romanes, 1882/1970). He emphasized careful descriptions of animal behavior and, like Darwin, placed species and behaviors into an evolutionary context. A criticism raised against both Darwin and Romanes by those who continued the research into animal behavior was their tendency to anthropomorphize, or interpret animal behaviors in human terms, and place too much emphasis on anecdotes or single examples of a behavior. Darwin, for example, described how his dog played with its food and how this behavior helped it to relish its food more (Darwin, 1872/1965). A well-known example of how observation of a single animal's behavior, without experimentation or rigorous evaluation of the context and conditions of the behavior, is that of Clever Hans (Griffin 2001; Linden, 1999). Hans was a horse who could not only count but could do arithmetic. Hans and his owner traveled throughout Europe putting on demonstrations of his skills. Audience members would pose simple math problems, and Hans would tap his hoof on the ground until he had arrived at the correct answer. A special commission of scientists was established and was given the task of understanding how Hans was able to do these amazing things. Led by Oskar Pfungst, the commission discovered that Hans was not able to count or do math. If Hans was unable to see his owner when he was "answering" a question, he would not be able to provide a correct answer. Apparently, as Hans tapped his foot and approached the correct answer to the problem, his owner must have made subtle changes in facial expression or how he held his body. At that point Hans would stop tapping. While it was disappointing to many that Hans was not able to count and do math, the story makes two important points. The first is that we need to always ask questions about how and why an animal is behaving the way it is, and second, whether he could count or not, Hans had a remarkable ability to read subtle signs in his owner's behavior.

Significant contributions to the understanding of animal behavior, especially how they learn, would be made by Ivan Pavlov and Edward Thorndike. Pavlov had already made important discoveries in the study of digestive physiology, winning a Nobel Prize for his work, when he turned his attention to the study of conditioned reflexes in dogs (Pavlov, 1927). Pavlov demonstrated that reflexive behaviors, paired with a reliable signal, can over time be elicited by that signal or stimulus (Figure 6-8). Thorndike studied a variety of species, including cats, and established the *Law of Effect* (Thorndike, 1913). He observed that behaviors that are followed by pleasant or positive results were likely to occur more often, while those followed by unpleasant results would

FIGURE 6-8

Ivan Pavlov apparently used dogs that he or his assistants found in the street for his experiments on conditioned reflexes. Many of the dogs were photographed, and afterwards adopted into new homes. *(Courtesy of Tim Tully)*

decrease in frequency. Pavlov's work would form the basis for understanding classical conditioning, and Thorndike's work established the fundamental nature of instrumental conditioning. J. B. Watson (1959) and then B. F. Skinner would later expand our understanding of instrumental conditioning, providing a strong theoretical and practical framework for our understanding of learning (Skinner, 1938). Both forms of conditioning or learning are used in training animal behavior and treating behavior problems presented by companion animals (Reid, 1996).

While the study of animal learning was primarily focused on research conducted in laboratories, another group of scientists was studying animal behavior under natural conditions. Konrad Lorenz, Niko Tinbergen, and Karl von Frisch studied what they considered the natural behaviors of animals. Their studies of these behaviors fit well with an evolutionary understanding of animal behavior. This approach to the study of animal behavior came to be known as ethology. Lorenz, Tinbergen, and von Frisch shared the Nobel Prize in Medicine in 1973 for their contributions. While knowledge of companion animal behavior has been part of their care as long they have lived with humans, the integration of the scientific study of animal behavior with companion animal care and treatment is a more recent development. While there are currently three main groups of practitioners providing behavior services for pets, namely, **applied animal behaviorists**, **veterinary behaviorists**, and dog trainers, there is overlap in how they approach the issues and frequently combine learning and ethological perspectives (Zawistowski, 2004).

Dog Training

Modern **dog training** began in Germany in the early 1900s. It was based on the practices developed for training dogs for German military and police applications (Lindsay, 2001). Stories that highlighted the performance of heroic dogs during World War I fueled greater public interest in dog training. A number of German trainers came to the United States and helped further popularize dog training. Among these trainers was Carl Spitz. In addition to training family pets, he also trained dog actors, including Buck in *The Call of the Wild*. The performance of Buck, Rin Tin Tin, and other well-known canine film stars helped to establish public appreciation for the role that training could have on a dog's abilities and role as a companion. Obedience training as a competitive event within the dog fancy was a natural extension. Helen Whitehouse-Walker and Blanche Saunders and Josef Weber (Burch & Bailey, 1999) were instrumental in getting the sport started and accepted by the American Kennel Club (AKC) as a sanctioned form of competition in 1936.

At the start of World War II, the U.S. Army had a limited canine service in place. The AKC played a key role in the recruitment of dogs for training and military service (Lindsay, 2001). Many of the civilian and military trainers and handlers from the war-dog program developed dog-training practices in the years following the war. William Koehler was an influential trainer coming from this tradition. The methods that he advocated became quite controversial because of his forceful and coercive approach and potentially harmful effects on dogs. These included hanging a dog, or holding it up by its leash and collar, cutting off its air or beating the dog. These techniques are less acceptable than they were, and positive training methods are more often encouraged and used (American Humane Association, 2001; Delta Society, 2001).

In addition to the military training influence on the field, marine mammal trainers and their techniques have had a strong influence on dog training. Among these, Karen Pryor (1975, 1985) has been particularly significant. She is generally seen as the most visible and active proponent of clicker training for companion animals. This methodology is derived from the system used by marine mammal trainers. The clicker is used as a sound bridge between the behavior performed by an individual and a treat or reinforcement (marine mammal trainers typically use a whistle for the sound bridge). Adherents of clicker training are effusive in the praise of using this technique to train dogs, cats, and other animals and to correct behavior problems.

Dog Training Organizations. While dog training is a very individualistic profession, there have been several attempts to form organizations to set standards of professional conduct and promote the profession. The National Association of Dog Obedience Instructors was formed in 1965 (Lindsay, 2001). Its primary activity was to provide training for group obedience instructors. These instructors would then work with people who were preparing for competition in AKC obedience competition.

The Society of North American Dog Trainers (SNADT) formed in 1987 and continued until 1995. It involved Job Michael Evans, Carol Benjamin, Captain Arthur Haggerty, and Brian Kilcommons and other trainers well known for their books and television appearances. SNADT promoted dog training as an honorable profession that played an important role in helping dogs fit into human society. The organization also developed a two-level certification program for trainers to demonstrate their knowledge and skill, emphasized continuing education, and held monthly meetings with speakers who would address topics ranging from animal behavior research and learning theory to business practices. A variety of competing interests would eventually lead to the dissolution of SNADT. While SNADT was short-lived, it set a tone for changes in dog training. Its emphasis on professional development, ethical standards, and business practices would prove a lasting influence in the field.

Ian Dunbar, a veterinarian and behaviorist, played a key role in the formation of the **Association of Pet Dog Trainers (APDT)**. The APDT holds a well-attended annual conference featuring a wide variety of speakers. It has developed a certification program that reflects the broad range of knowledge that a professional dog trainer should have. Certification is based on an exam administered by an independent third party, and successful candidates earn the title *Certified Pet Dog Trainer* (CPDT). The certification program is now completely independent of the APDT and is administered by the Certification Council of Pet Dog Trainers (CCPDT).

Applied Animal Behaviorists

While academically trained scientists have studied and written about companion animal behavior, they seldom attempted direct application of their theories and observations to companion animals with behavior problems. Tuber, Hothersall, and Voith (1974) published an article titled "Animal Clinical

Psychology: A Modest Proposal," which set the stage for the development of a new field. While a few individuals such as Peter Borchelt took up the challenge, it would be nearly two decades before a formal program was developed to set standards for applied animal behaviorists. In 1991, the Animal Behavior Society established a Board of Professional Certification and determined the education and experience requirements for certification. There are two levels of certification: Certified Applied Animal Behaviorists have a doctorate in animal behavior and five years of experience in the field, and Associate Applied Animal Behaviorists have a master's degree and a minimum of two years of experience. The education requirements balance a background in both ethology and learning theory. Certification is valid for five years, and individuals are required to submit a recertification application to demonstrate that they have remained current and up to date in the field. In addition to working with companion animals, applied animal behaviorists also work with laboratory animals or in zoos.

Veterinary Behaviorists

The American College of Veterinary Behaviorists (ACVB) was formed to recognize the treatment of behavior problems in animals as a specialty within **veterinary medicine**. Similar to other veterinary specialties, diplomate status is achieved following completion of the doctorate in veterinary medicine, an approved residency under the supervision of a diplomate in the field, and passing a qualification examination. In addition to behavior modification methods that might be employed by an applied animal behaviorist or dog trainer, veterinarians may also prescribe drugs. At this time there are only two drugs approved by the Food and Drug Administration (FDA) for the treatment of behavior problems in companion animals—both for dogs. Anipryl is approved for the treatment of canine age-related cognitive disorder, and Clomipramine is approved for the treatment of separation anxiety. Veterinarians have the professional discretion to prescribe other drugs based on their professional judgment and experience. This extra-label usage allows them to employ a variety of the psychoactive drugs that have been tested and approved for the treatment of human disorders.

DOG GROOMING

All dogs require some sort of **grooming**. Even the "hairless" breeds need appropriate skin cleaning and care. Grooming is especially important for many of the breeds with long or thick coats of hair. Failure to groom can have a negative impact on the dog's health. While dog owners are typically able to manage the basic needs of bathing and brushing their pets, skilled help may be needed for more difficult grooming needs or for the custom cuts that are desired for some breeds, or for showing in competitions. Competence in grooming is typically acquired through training at a dog grooming school or apprenticeship with an experienced groomer (Figure 6-9).

The **National Dog Groomers Association of America** (<www.national doggroomers.com>) was formed in 1969 to promote excellence in professional

FIGURE 6-9

Professional groomers need to study the various breeds and the appropriate way to groom each. In addition to pet grooming salons, some groomers will travel to a pet owner's home in mobile grooming vehicles. "Do-it-yourself" grooming parlors provide pet owners with tubs, towels, and shampoos so that they can bathe their own dogs. *(Courtesy ASPCA)*

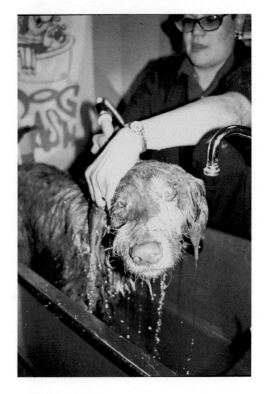

standards. The NDGAA offers a variety of accredited training workshops to enhance an individual's skills. Completion of various workshops and successful performance on an exam that includes both written and practical portions will result in certification as a "National Certified Master Groomer." The exam includes breed identification, breed standards and anatomy, use of pesticides among other areas, as well as a demonstration of grooming skills on breeds of dogs representing the various breed groups. The exam requires knowledge of cat grooming as well.

Most dog-grooming professionals work as independent businesses. In addition to grooming parlors, a number of groomers have begun to operate mobile grooming salons, coming to a pet owner's home to groom their dogs. Some of the large pet supply stores may also include a grooming service at their locations. Another area of growth in the field has been the appearance of self-service grooming locations. Dog owners can bring in their pet and use tubs and bathing facilities at the location to wash their dog themselves. This is especially convenient for people who have large dogs but live in small apartments.

BOARDING, DAY CARE, AND PET SITTING

Occasionally, people need to find temporary care for their pets while they travel, during work hours, or for other reasons. In many cases, people are able to arrange for friends or family to care for their pets. At other times, however, other arrangements need to be made. **Boarding kennels** have been available for decades as a temporary place for dogs, cats, or other animals to stay. The **American Boarding Kennel Association** (<www.abka.com>) represents the

FIGURE 6-10

Doggie daycare provides an option for people who want their dogs to have some fun running around playing with other dogs while they are working. *(Courtesy Heather Mohan)*

profession and promotes training and standards of practice. All ABKA members are required to adhere to a standard of ethical practice that respects the rights of pet owners and the welfare of the pets in their care. The organization supports this philosophy by offering training programs for pet care technicians and kennel operators. Successful completion of these training programs and passing required exams allow people to earn certification in the field. There is also a voluntary accreditation program for facilities. This certification program includes business practices, animal care procedures, sanitation, fire safety, and a number of other topics.

While boarding facilities are generally considered for some form of extended overnight care for pets, daycare for pets is becoming more common. Often called *doggie daycare*, it provides people an option for pet care for part of a day (Figure 6-10). This can be especially helpful for people with long workdays who are unable to walk and exercise their pets during the day. It can also provide dogs with an opportunity to socialize with other dogs during the day, since the dogs may be kept in a large indoor or outdoor play area with other dogs spending the day at the facility. In other cases, individuals may arrange for a regular dog walker, someone who will come in on a regular basis and take their dog for a walk. This might be with several other dogs, or alone, depending on how much the dog owner is willing to pay for the service. Once again, this is a convenient option for people who are away from home for long periods during the day and want to ensure that their pet gets some exercise and a chance to go to the bathroom outside.

Another service growing in popularity is **pet sitting**. In this case, pet owners will arrange to have someone come into their home and care for their pets when they are away. This will generally include feeding, cleaning cat litter, walking dogs, and other needs. This is especially useful for people with cats, since cats are generally more comfortable in their home than in a boarding facility. Reptiles, birds, aquariums, and other pets that would be more difficult (or impossible) to move also benefit from this sort of care option. It is important that pet owners check carefully to ensure that the pet sitter they employ is experienced and skilled in the care of their type of pets. The National Association of Professional Pet Sitters (NAPPS, <http://www.petsitters.org>) was formed in 1989 to represent people working in this field. Its mission statement addresses the needs of the professional, the public, and the pets:

- To provide tools and support to foster the success of members' businesses
- To promote the value of pet sitter to the public
- To advocate the welfare of animals

The National Association of Professional Pet Sitters supports this mission by offering training programs that include options for certification upon successful completion of an exam. They are also able to offer liability insurance to members through an NAPSS insurer.

MEDIA

Companion animals have appeared in the primary **media** of almost every era. Dogs and cats can be seen in the hieroglyphics that adorned the pyramids of the pharaohs, the pots of Rome and Greece, and the hand-copied books and scrolls of the Middle Ages. It is no surprise that companion animals are a common theme in our modern media formats. Classic books featured companion animals, and they were an integral part of early children's literature (see Chapter 3.9). Rin Tin Tin and Lassie starred in films before they went on to join Americans in their homes in early television series. These traditions continue in films such as *The Truth About Cats & Dogs* and best-selling books such as *Marley and Me* by John Grogan (2005). The expansion of television variety through cable and satellite access has led to not only television shows that depict pets, but now an entire network dedicated to animals, Animal Planet. Animal Planet is part of the Discovery television family of networks and draws heavily on the interest and affection that people have for their companion animals for viewers, sponsors, and content. Popular shows on the network have ranged from veterinary care to pet psychics and from breed profiles to cruelty prevention (Figure 6-11). Radio shows also feature companion animal topics. Jon Patch hosts a regular syndicated radio show, *Pet Talk*, which is carried in 95 markets and features a wide range of guests who provide personal stories about their life with pets and expert pet care information for listeners (Figure 6-12).

The Internet has had an enormous influence on how people entertain and inform themselves about companion animals. Any Internet search engine will generate thousands of Web sites associated with pets. The quality and nature of these sites will vary from excellent training and health care information to sharing humor (see Figure 6-13). Care needs to be taken, however, since it is as easy to broadcast false or misleading information as truthful and helpful information. Urban legends about products that might be dangerous to animals (but often are not), home remedies for sick pets (which may not work or may put the pet at greater risk), and other questionable materials can spread like

FIGURE 6-11

Animals have their own network, Animal Planet. It features a wide range of shows about wildlife, pet care, and other topics. One of the most popular shows on the network is *Animal Precinct*, which follows the real-life exploits of ASPCA special agents as investigating animal cruelty in New York City. *(Courtesy Christian Oth/Animal Planet)*

wildfire through the Internet. For the most part, however, the Internet has proven to be a great benefit for people and companion animals. Research for this book was greatly facilitated by use of the Internet. For those who are connected, it has a great impact on nearly every part of a person's life, including the companion animals that share that life.

FIGURE 6-12

Jon Patch is the host of *Pet Talk* radio, and each week provides entertaining guests and pet care information. Here he interviews Jacque Schultz, a certified pet dog trainer. *(Courtesy Jon Patch)*

FIGURE 6-13

The Internet has made it possible to share pet stories with people across the nation and around the world quickly and easily. Pet humor is one popular topic. Care must be taken however, since *urban myths* travel as quickly as factual information on the information highway. *(Courtesy ASPCA)*

How many dogs does it take to change a light bulb?

GOLDEN RETRIEVER: The sun is shining, the day is young, we've got our whole lives ahead of us, and you're inside worrying about a stupid burned out bulb?

BORDER COLLIE: Just one. And then I'll replace any wiring that's not up to code.

DACHSHUND: You know I can't reach that stupid lamp!

ROTTWEILER: Make me.

BOXER: Who cares? I can still play with my squeaky toys in the dark.

LAB: Oh, me, me!!!!! Pleeeeeeeeeze let me change the light bulb! Can I? Can I? Huh? Huh? Huh? Can I? Pleeeeeeeeeze, please, please, please!

GERMAN SHEPHERD: I'll change it as soon as I've led these people from the dark, check to make sure I haven't missed any, and make just one more perimeter patrol to see that no one has tried to take advantage of the situation.

JACK RUSSELL TERRIER: I'll just pop it in while I'm bouncing off the walls and furniture.

OLD ENGLISH SHEEP DOG: Light bulb? I'm sorry, but I don't see a light bulb!

COCKER SPANIEL: Why change it? I can still pee on the carpet in the dark.

CHIHUAHUA: "We don't need no stinking light bulb."

GREYHOUND: It isn't moving. Who cares?

AUSTRALIAN SHEPHERD: First, I'll put all the light bulbs in a little circle. . .

POODLE: I'll just blow in the Border Collie's ear and he'll do it. By the time he finishes rewiring the house, my nails will be dry.

How many cats does it take to change a light bulb?

Cats do not change light bulbs. People change light bulbs. So, the real question is:
"How long will it be before I can expect some light, some dinner, and a massage?"

ALL OF WHICH PROVES, ONCE AGAIN, THAT WHILE DOGS HAVE MASTERS, CATS HAVE STAFF!

CHARITIES AND NONPROFIT ORGANIZATIONS

In addition to the multibillion dollar pet product industry supported by pet-owning consumers, there is also a large **nonprofit** sector that is supported by charitable contributions. These range in size from the Humane Society of the United States with a budget of over $100 million and a staff of hundreds to small local organizations that are organized and run by volunteers with little or no budget. In general, charitable contributions to animal protection and environmental organizations accounts for about $8.9 billion, a small fraction of the total of the $260 billion dollars that Americans give to charity each year (Bain, 2006). It is estimated that companion animal organizations receive between $2.5 and $3 billion each year (Figure 6-14). Charitable organizations are strictly regulated under federal law, with substantial attention paid to accounting for finances. Each year they are required to file an accounting form called the 990

FIGURE 6-14

U.S. charitable giving in 2005. Total giving was $260 billion, with animal welfare and environment accounting for 3.5% or $8.86 billion. Environmental giving accounted for most of this figure, with about $3.4 billion directed toward animal welfare. This includes companion animal welfare groups, as well as groups who oppose vivisection and hunting, and hold other similar positions. *(Courtesy Dr. Tom Latrielle—Humane Society of Broward County, 2007 Walk for the Animals)*

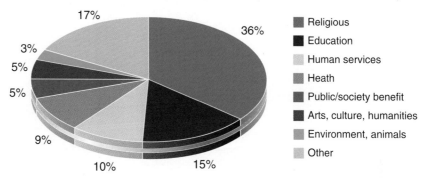

that lists revenue and expenses related to administration, fund-raising costs, and program delivery. These are public documents and charities are required to make them available to interested parties. In addition there are independent oversight groups that scrutinize the finances and activities of charitable organizations and rate them for their effectiveness and efficiency. Guidestar, for example, provides as Web site (<www.guidestar.org>) that breaks these finances down into an easy to understand format and further rates charities for their use of funds collected and used in program delivery versus expenses related to administration and fundraising.

Fundraising is done through a variety of methods. Direct mail marketing is a well-known example where solicitations are mailed directly to the public. Charities may sell or exchange their lists with other charities to expand the number of people that they are able to solicit. Bequests are another common source of gifts to nonprofits where individuals may remember a favorite charity in their will. Special events such as "dog walks" can be used to not only raise funds, but also to generate some fun and a sense of community around a shared appreciation for companion animals (Figure 6-15).

In addition to contributions from the public, nonprofit animal groups may receive support from foundations or corporate partners. Foundations can be organized in several different ways; they share in common the mandate to distribute financial assistance to individuals and groups that meet the foundation's mission-related requirements. Foundations may be set up by individuals wishing to use their financial resources to assist a cause that they believe in. Maddie's Fund is one such example. Established by Dave Duffield, the founder of the PeopleSoft Company, it has an endowment of $250 million dollars. Named for Duffield's miniature schnauzer Maddie, the foundation supports efforts to eliminate the euthanasia of companion animals in animal shelters as a response to pet overpopulation. Maddie's Fund targets large-scale, high-visibility projects. For example, in 2005 it provided $15.5 million dollars to groups in New York City over a period of five years to make New York a no-kill city. Corporations may also establish foundations that support

FIGURE 6-15

Animal groups raise funds for their groups in a wide variety of ways. Dog walks are great community events that can be used to raise funds, have fun, and educate the public. The Humane Society of Broward County attracted more than 4,500 people and their pets in March 2007 to its 17th Annual Walk for the Animals. *(Courtesy ASPCA)*

philanthropic efforts. These may be closely associated with the business and products of the corporation, or they may have limited or no direct connection to their business interests. PETCO and PetSmart both have foundations that distribute money to companion animal groups to support pet adoption programs. Funds for these two foundations come from both in-store donations by customers and contributions by employees.

In addition to charitable, philanthropic donations to nonprofit groups, corporations may also enter into *cause-marketing* agreements with these groups. Unlike the donations that a corporate foundation may make to a group, funds contributed through a cause-marketing arrangement are more closely tied to a specific marketing or product promotion effort. For example, the Meet-Your-Match pet adoption program described in an earlier chapter is part of a cause-marketing relationship between the Iams Company and the ASPCA. Program support materials include the logos of both the ASPCA and Iams, and help to support building the public awareness of both organizations. Pet adopters benefit because the program is designed to assist them in the selection of an appropriate pet, and participating shelters benefit because they gain access to expertise in adoption management, adoption support materials, and promotional materials that support their pet adoption programs. Another type of cause marketing is where a company promotes the fact that for every unit of a particular product sold, a portion of the purchase price will be contributed to a particular charitable cause. Another way in which nonprofit groups may benefit from a relationship with a corporation is through an endorsement or licensing agreement. In this case, a nonprofit animal group may endorse or lend their name to a particular product. For the consumer, this provides some confidence that the product is of good quality and will benefit them and their pet. This may be in the form of a *seal of approval*, and the organization will typically receive a royalty payment based on the sale of product units.

DISCUSSION QUESTIONS

1. What are the two major organizations of companies in the pet industry? What are one difference and one similarity between the two organizations?
2. What role did James Spratt play in the establishment of the pet food industry?
3. What is extrusion? When was the extrusion method first applied to the manufacturing of pet food, and by whom? What advantages does it offer in the manufacturing of pet food?
4. How and when did canned pet food develop? What are the advantages and disadvantages of feeding canned food as compared to feeding dry kibble?
5. What are the challenges/concerns with feeding homemade and raw diets?
6. What are two ways that a pet food can meet a nutritional claim? Which way is preferred, and why?
7. What were some substrates used as kitty litter prior to the use of clay? How did the introduction of clay litter affect the popularity of cats as companions?
8. When and where was the first veterinary school established?
9. When and where was the first American veterinary school established?
10. What species and diseases prompted the establishment of small animal medicine as a subset of veterinary medicine? Approximately when did this occur?
11. How did the gender make-up of the veterinary profession compare before and after World War II? What factors contributed to this shift?
12. What are the two levels of certification for Applied Animal Behaviorists?
13. What are three ways that the American Boarding Kennel Association promotes standards of practice in its profession?
14. Name two services that exist for pet owners who are unable to provide the desired level of daily attention for their pets.
15. Approximately what proportion of the money donated to charity each year in the United States is received by companion animal organizations?
16. Describe two methods by which corporations financially support nonprofit groups.
17. What major contributions has Mark Morris made to the companion animal industry?

REFERENCES

Agrawal, M. (2006, February 24). Dog crazy. *Life: America's Weekend Magazine.*
Agres, T. (2003). Dog drug targets push research prospects. *The Scientist, 17*(21), 52.
American Humane Association. (2001). *Guide to humane dog training.* Englewood, CO: American Humane Association.
Association of American Feed Control Officials. (2007). Pet food regulations. In *AAFCO Official Publication.* Atlanta, GA: Author.
Bain, M. (2006, July 10). Donating our dollars and hours. *Newsweek,* 65.
Burch, M. R., & Bailey J. S. (1999). *How dogs learn.* New York: Howell.

Campbell, K. L., Corbin, J. E., & Campbell, J. R. (2005). *Companion animals: Their biology, care, health, and management.* Upper Saddle River, NJ: Pearson-Prentice Hall.

Case, L. P. (2003). *The cat: Its behavior, nutrition, and health.* Ames, IA: Blackwell Publishing.

Case, L. P. (2005). *The dog: Its behavior, nutrition, and health.* Ames, IA: Blackwell Publishing.

Darwin, C. (1965). *The expression of the emotions in man and animals.* Chicago: Chicago University Press (original publication in 1872).

Delta Society. (2001). *Professional standards for dog trainers: Effective, humane principles.* Renton, WA: Delta Society.

Dorosz, Edmund, R. (2006). Commercial pet food: Yesterday and today. Retrieved January 10, 2005, from www.netpets.org/cats//reference/food/competfood.html.

Dunlop, R. H., & Williams, D. J. (1996). *Veterinary medicine: An illustrated history.* St. Louis, MO: Mosby-Year Book.

Glater, J. D. (2005, November 16). Wild demand and competition for new products. Retrieved August 9, 2007, from http://www.nytimes.com/2005/11/16/business/businessspecial/16glater.html?ex=1289797200&en=252381da1b076953&ei=5090&partner=rssuserland&emc=rss.

Grier, K. C. (2006). *Pets in America: A History.* Chapel Hill, NC: University of North Carolina Press.

Griffin, D. (2001). *Animal minds.* Chicago: University of Chicago Press.

Grogan, J. (2005). *Marley and me.* New York: Harper Collins.

Guthrie, D. (2005, February 14). Pets become bigger part of family budget. *Detroit News.*

Haselbush, W. C. (1984). *Mark Morris: Veterinarian.* New York: R. R. Donnelley and Sons.

Iams Company. (n.d.). *The history of the Iams Company.* Dayton, OH: Iams Company.

Linden, E. (1999). *The parrot's lament.* New York: Dutton.

Lindsay, S. R. (2001). *Applied dog behavior and training, vol. 2: Etiology and assessment of behavior problems.* Ames, IA: Iowa State University Press.

Maggitti, P. (1996, July). Cat litter: The inside scoop. *Pet Business, 48.*

Masters, C. (2005, May). Let's party! *Time Inside Business, A20.*

Miller, E. B. (1981). Private veterinary colleges in the United States 1852–1927. *Journal of the American Veterinary Medical Association, 178*(6), 583–593.

Morris Animal Foundation. (2006). *Morris Animal Foundation annual report, 2006.* Englewood, CO: Morris Animal Foundation.

Pavlov, I. P. (1927). *Conditioned reflexes* (G. V. Anrep, Trans.). London: Oxford University Press.

Polk Miller Products, Corp. (1927). *Polk Miller's dog book*, 12th edition. Richmond, VA: Polk Miller Products.

Pryor, K. (1975). *Lads before the wind.* New York: Harper and Row.

Pryor, K. (1985). *Don't shoot the dog: The new art of teaching and training.* New York: Bantam.

Ralston Purina Oral History Archives. (n.d.). Comments of Dick Kealy and Ron Flier on the development of the extrusion process for the manufacture of pet foods.

Ralston Purina Company. (1994). *The Ralston chronicle: 1894–1994.* St. Louis, MO: Ralston Purina.

Reid, P. J. (1996). *Excel-erated learning.* Oakville, Ontario: James and Kenneth Publishers.

Richardson, K. (2006, June 26). Bets on pets can hold up in down times. *Wall Street Journal*, p. C1.

Romanes, G. J. (1970). *Animal intelligence.* London: Gregg International Publishers (original publication 1882).

Skinner, B. F. (1938). *The behavior of organisms: An experimental analysis.* New York: Appleton Century Crofts.

Smith, G. M., & Lum, S. K. S. (2005). *Annual industry accounts: Revised estimate for 2002–2004* (p. 18–69). Washington, DC: Bureau of Economic Analysis, Department of Commerce.

Thorndike, E. (1913). *The psychology of learning.* New York: Teachers College.

Tuber, D. S., Hothersall, D., & Voith, V. L. (1974). Animal clinical psychology: A modest proposal. *American Psychologist, 29,* 762–766.

Veterinary Practice News. (2006, January). Pet insurance. *Veterinary Practice News, 18*(1), 7.

Watson, J. B. (1959). *Behaviorism.* Chicago: Chicago University Press.

Zawistowski, S. (2004). Applied animal behavior. In M. Bekoff (Ed.), *Encyclopedia of animal behavior* (pp. 138–142). Westport, CT: Greenwood Press.

CHAPTER 7

Competitions

If you're wealthy you ride to the hounds; if you're poor, you go to the dogs.

—Unknown

KEY TERMS

competition	conformation	Canine Freestyle
show	Canine Good Citizen (CGC)	Cat Fanciers' Association (CFA)
earthdog	lure coursing	American Fancy Rat and Mouse Association (AFRMA)
flyball	breed standard	
weight pulling	agility	
American Kennel Club (AKC)	trial	
obedience	Iditarod	

Over the centuries, humans have worked with creativity and scientific know-how to develop specific traits in our companion animals. We have bred animals to have certain shapes, colors, strengths, and characteristics to help us in our work—the Australian cattle dog has been bred to crave the opportunity to weave and dash to keep a herd moving together, the homing pigeon can find its way home from hundreds of miles away to carry messages, and the cats that live and work on the family farm seem pleased to catch unwanted rodents. Many of these species also live in our homes, away from their intended work. They no longer need to work for their supper; yet they keep the same behaviors their relatives have as working animals. Are there ways for the cattle dog living in the suburbs to use its intelligence and stamina? What about the Ragdoll cat? It has most likely been hundreds of generations since one of its relatives had to catch a mouse. Sometimes in the absence of their original occupations we have found ways to provide an outlet for our companions to demonstrate their innate behaviors and physical abilities through the sports and work companions can share with us.

Humans are very social beings. We enjoy the company of others, and if the individual is not another human, our companion animal becomes critical to our welfare. We like to join other people to share interests, skills, and information. As youngsters, we may have been members of the Scouts; in college, a sports team or chess club; and later in life, a professional society or social club. We also enjoy sharing experiences with our friends, and who is better than our favorite companion animal? We know that our pet is incredibly special, much like parents feel that their child is unique and no other baby can be quite so beautiful or intelligent. It is likely that the first time two people showed up on market day with their dogs, a discussion developed over who had the best looking, fastest, strongest, best hunting, or best herding dog. This spirit of **competition** has since given rise to a bewildering array of competitions and **show** for all sorts of companion animals. Over centuries, humans have developed many different ways to demonstrate to others that their pet is the best. By using their animal's natural behaviors or physical characteristics we have designed competitions for our pets. The competitions can be for prestige, money, or just for the sheer fun of it. They may be for young competitors, experienced professionals, or a family out for some wonderful recreation. But no matter what the goal, or who the competitors may be, through shows and sport we heighten the relationship between pet and guardian. We travel and train together, thereby spending hours working as a team, getting to know each other's ups and downs—always strengthening the bond.

We have all heard of frog jumping contests, 4H competitions for young people to show the care and knowledge they have put into raising a pig, goat, or calf, horse races, and cat shows but the greatest variety of competitive sport has been designed for canines. The following is a list of the many activities and sports designed for dog sports (<wikipedia.org/wiki/list_of_dog_sports>).

- Agility
- Bikejoring
- Canicross
- Carting
- Dog hiking
- **Earthdog** trials, or ground trials
- Field trials
- **Flyball**
- Disc dog
- Musical freestyle
- Mushing
- Obedience training

- Protection sports
- Pulka
- Sootering
- Sled dog racing
- Skijoring
- Sheepdog trials
- Sighthound racing
- Tracking trials
- **Weight pulling**

And there are certainly more waiting in the imaginations of people in the United States and around the world. We will not address all of them in this book, but rather focus on a few that explore the variety of competitive sports for dogs and their human partners, and we will consider a few activities for other pets.

SHOWS (CONFORMATION)

Organized events where people can show their farm or domestic animals for fun or as a way to enhance their value as breeders of their species are immensely popular in the United States. There are organized events in almost every city and small town across the country—anywhere where people wish to compete by exhibiting the animals they love and care for.

Judging is based on how well the individual conforms to the accepted standards of the breed, or can be more playful and judged as being the most beautiful, unusual, or most similar in appearance to the owner. The Cat Fanciers' Association (<www.cfainc.org/shows.html>) recognizes 39 breeds of cat for their national shows. As with all official shows, there are classes the animal passes through, beginning with the kitten class for cats under eight months old, and moving through several levels to reach the champion status. Probably a weekend does not pass in the United States when there is not a rabbit show going on somewhere. It is a perfect sport for children to participate in as they care for and show their pet rabbits. But it is also a very serious undertaking for people who are in the business of breeding animals, are professional handlers, or just love the show as a hobby (<http://www.rabbitweb.net>).

The showing of dogs is a most popular competitive undertaking by dog fanciers around the world. Perhaps the best known competitive show in the United States is hosted by the Westminster Kennel Club every year in New York City (Figure 7-1). The **American Kennel Club (AKC)** was established in 1884 to promote the study, breeding, exhibiting, and advancement of purebred dogs. Through its registry and stringent rules, the AKC makes certain that breeds stay pure and their features are true to the characteristics of the breed. This is important for owners of the dogs for they can be assured of the dog's quality when they buy, sell, or show their dog. The registry assures that the highest quality canines are maintained. The AKC is the largest nonprofit purebred dog registry in the country.

FIGURE 7-1

The Westminster Kennel Club show is held in February each year at Madison Square Garden in New York City. It attracts thousands of entries, all of them already champions, tens of thousands of spectators, and an international television audience in the millions. *(Lisa Croft-Elliott/ Westminster Kennel Club)*

For a dog to be AKC registered as a certain breed, it is essential that there is written record of its family tree going back three or more generations. If a dog is registered with the AKC, it must mate with an AKC registered bitch (female canine) in order for their offspring to also be registered with the AKC.

The AKC approves and maintains official records of over 15,000 sanctioned and licensed events involving registered dogs per year. The AKC has about 500 member clubs and over 4,000 affiliated clubs across the country. These clubs arrange and host dog shows, field trials, and **obedience** events. They also provide educational activities to benefit the community; giving presentations at schools, fairs, libraries, shelters, hospitals, scout groups, and health clinics.

Showing dogs combines the thrill of competition with the joy of seeing beautiful dogs—those that conform closest to the standards of their breed. Dog shows (**conformation** events) are one of many types of AKC dog events in which AKC registered dogs can compete. These events draw nearly two million entries every year. The dog's conformation, or overall appearance and structure, are judged as an indication of the dog's ability to produce quality puppies. In addition to conformation events, AKC registered dogs participate in competitions of strength, tests of instinct, and the breed's trainability. Trainability is tested specifically in obedience trials, **Canine Good Citizen (CGC)** tests, field trials, agility trials, **lure coursing**, rally, hunting tests, herding trials, tracking tests, and coonhound and Earthdog events. The size of dog shows can range from all-breed shows with 3,000 dogs entered to small local specialty club shows.

All-breed shows are open to over 150 breeds and varieties of dogs recognized by the AKC. *Specialty shows* are restricted to dogs of a specific breed or to a variety of one breed. The Bulldog Club of America is only for bulldogs, while the Poodle Club of America allows three varieties of the breed: standard, miniature, and toy poodles.

Group Shows are limited to dogs belonging to one of the seven groups recognized by the AKC: hounds, sporting, working, terrier, toy, nonsporting, herding (the newest group that broke away from the hunting group in 1983), and miscellaneous. Miscellaneous means a transitional class for breeds attempting to advance to full AKC registration. These dogs have clubs specifically for their breed, there are breeders across the country, and consistent standards are being met for or the breed. To be eligible to compete in an AKC show, a dog must meet the following criteria:

- Be registered with the AKC
- Be six months or older
- Be a breed for which classes are offered at a show
- Meet any eligibility requirements in the written standard for the breed

Spayed and neutered dogs are not eligible to compete in conformation classes, because the purpose of the competition is to evaluate dogs of quality to continue a line of excellence for the breed.

Each breed has "an official standard." These are characteristics that allow the breed to perform the function it was bred for, including structure, temperament, and movement. The judges are experts on the breeds they evaluate. During the competition, the judge goes to the standing dog to examine and evaluate each individual dog against the **breed standard**, awarding points for how close it comes to the idealized standard for its breed. The judge runs his or her hands over the dog's body to see if the muscles, bone structure, coat texture, teeth, ears, and eyes conform to the breed's expectations. They watch the dog move to see its gait and observe if it is physically pleasing and up to standard. Each dog is also viewed in profile—*stacking,* or *positioning*, the body so that head, torso, and legs are lined up properly for the breed.

The dogs participate in the competition to win points toward their AKC championship. It takes 15 points, including two "*major*" wins of three, four, or five points awarded by at least three different judges to become an AKC "*Champion of Record.*" This is an honor, and also assures that puppies will be in great demand and bring greater prices. The number of points awarded at a show is based on the number of dogs competing, with five being the maximum number points possible.

There are six classes—each one divided so males (dog) and females (bitch) compete separately.

- Puppy—for dogs between 6 and 12 months, not yet champions
- 12–18 months—for dogs of that age who are not yet champions
- Novice—for dogs aged six months and over that have not won three first prizes in the novice class, a first prize in *Bred-by-Exhibitor, American Bred,* or *Open classes*, nor one or more points toward their championship
- *Bred-by-Exhibitor*—for dogs that are exhibited by their owner and breeder, not yet champions
- *American Bred*—for dogs born in the United States from a mating that took place in the United States, not yet champions
- *Open*—for any dog of the breed, at least six months old

During the show, each dog has its turn to come before the judge. If it is a small dog, it will be placed on a table so the judge gets a good look and can easily touch the dog. Larger breeds learn to stand still, in the most perfect position. You will often see the handler reposition a leg or hold up a tail to make certain the dog is seen in the most flattering and breed appropriate way. This is *stacking*. The judge evaluates each of the following points for every breed.

Finally, the dog is invited to run around the ring in order to demonstrate their gait or way of moving, and how well they meet the standard for the breed. The dogs and handlers always look proud and agile, almost floating in front of the judge and audience.

With all of this information in mind, the judge may ask for all of the dogs to come out again or go back to reexamine one aspect of the dog's body before making the final evaluation. As the judge points at the dog and handler, they step forward to take the blue, red, yellow, or white to indicate first, second, third, and fourth place, respectively.

After the various classes are judged, all the dogs that win first place compete again to see which is the best of all the classes. Only the best male and the best female receive championship points and a coveted purple ribbon. Then the winning dog (male) and winning bitch (female) that were awarded the blue and white ribbon compete with the champions for the Best of Breed award. At the end of the Best of Breed competition, three awards are usually given: Best of Breed, Best of Winners, and Best of Opposite Sex.

Dog shows are a process of elimination. Only the *Best of Breed* winners (a purple and gold ribbon) advance to compete in the *Group* competitions. Four ribbons are awarded in each group, but only the first-place winner advances to the *Best in Show* competition. This dog will be judged to be the best of any of the other dogs competing in that show—the best of group, either sex, and any breed. In being judged, it is important to keep in mind that these dogs are not being compared to one another. They are being judged against the standards for their breed. The essential question is whether the beagle is closer to being the "perfect beagle" or is the golden retriever closer to being the "perfect golden retriever." When the judging is over, however, win or lose, they can go back to their crate for a well-deserved rest, the Red, White, and Blue ribbon hanging on the kennel door.

Certainly, it is the dog, its genetics, and care that make it a fine example of its breed, but the handler can bring out the best in his or her dog to create a champion. The handler, or person who shows the dog to the judge, may be the owner, a friend or family member, or an individual who is paid to present the dog in the best possible way, much like a jockey in horse racing. The handler will travel to the show—sometimes down the block for a "match" show that is informal with no championship points awarded, or across the country to one of the most prestigious competitions, like the Westminster Kennel Club dog show. The dog will have gone through much preparation for the event. There is training, grooming, excellent diet, and exercise. And, just before the competition begins, the dog will be groomed with brushes, hair dryers, and even gel to assure every hair is in its place, providing the best possible appearance. The dog must be in good "sound," meaning in excellent mental and physical shape. In order to be good competitors, the dogs must enjoy the competition and be well behaved, as well as an excellent example of the breed.

Of course, there are other less formal competitions that are not AKC sanctioned. They are sponsored by clubs or designed to show off the funniest, biggest, or smallest dog in a community. There are professional handlers and children who find this a wonderful way to bond with their dog and share a sense of fun and challenge. Dog shows judge the standard and appearance of the dog, but there are other forms of competition that demonstrate the innate traits of the dog and are far more athletic.

AGILITY

Agility is a competitive sport emphasizing teamwork between the handler and dog as they negotiate a timed series of obstacles. Agility trials are open to all dogs, of any size, which are physically and temperamentally suited to the sport. Cats, not to be outdone, have recently begun to have agility competitions, but here we will focus on our canine companions.

The dog that competes in the agility trials must be quick, intelligent, and willing to respond correctly to every command given by the handler. Using their natural instinct to follow the pack leader, the dog cues in to the human's shoulder and arm positions to instruct him to slalom in and out of poles, jump through rings, run into a tunnel and out, and after going full speed—stop, lie down for a count of five—then jump up to race through the remainder of the course. Each set-up is very different, depending on the age, size, and ability of the dogs. They look like fabulous play spaces for children, colorful, multi-dimensional, and fun.

The agility course looks a great deal like the obstacles in a horse show, and this is not an accident, but based on the beginnings of dog agility trials. In order to entertain the audience at the Crufts dog show in 1978 in England, between the obedience and conformation competitions, a dog trainer, Peter Meanwell, constructed a largely jumping style course resembling something from the equestrian world to demonstrate a dog's speed and agility. The audience loved it, and the sport of agility was born. (There are reports that other places at other times had developed the sport. But it is widely believed that the event in 1978 was the first major presentation.) People wanted to see more, and have their own dogs participate. The sport went on to grow locally, nationally, and internationally. By 1979, several British dog training clubs were offering training in the new sport of dog agility, and in December of that year the first Agility Stakes competition was held at the International Horse Show at Olympia in London. By 1980, the British Kennel Club became the first organization to recognize agility as an official sport with a sanctioned set of rules. The competition then traveled to the United States (<www.wikipedia.dog_agility>, August 14, 2007).

One of the many wonderful things about agility trials is that they are open to all kinds of dogs—purebred or not—and all kinds of humans. The American Kennel Club admits only purebred dogs to their competitions, but many other associations welcome all dogs, including mixed breeds, as long as they are ready for the sport of the agility course. It is quite something to see Australian cattle dogs alongside golden retrievers, rat terriers, brown spotted "who knows what" dogs, and even basset hounds as they mill about the grounds before going to the competitive group for their size and ability. The most common breeds to enjoy the challenge are those who are quick on their paws, like border collies, Shetland sheepdogs, retrievers, and Australian shepherds. Dogs as large and cumbersome as Newfoundland's have been known to impress the crowds. Very serious competitors are breeding hybrid specialists for competition. One popular cross is the *border-border*, a border collie crossed with a border terrier. Agility trials are also open to all kinds of people: experienced handlers, children, persons with disabilities, and casual competitors who simply enjoy the challenge and time with their canine friend.

The object of the agility course is to *run clean*, meaning to run through the course and take all obstacles correctly in the proper order. There can be no dropped bars on jumps, all contact points have to be hit, and the dog and handler together must finish on or under the time designated for the run. It is fast and complex. There are no standard courses. The judge sets every course before the competition. The handler can walk the course before the **trial** begins, memorizing each obstacle, distances between them, difficult obstacles for their dog, and then remember them during the competition while running full tilt and instructing the dog at the same time. The dog, on its part, must be able to "read" the handler's instructions while also running at top speed and executing jumps, turns, and stops. Being a team, understanding each other perfectly is paramount to their success. The dog must be in sync with the trainer, understanding a new language of hand signals, posture, and words, because there is no other means of control. The dog must want to follow directions, and the handler must be clear in giving them. The dog knows if it is being asked to weave in and out of poles or rush through the tunnel, repeat the hoop jump, or stop at the yellow line on the way down the A frame. The successful team is made of a dog and human that love the challenge of the agility course and are up to the challenge, both intellectually and physically (Figure 7-2).

FIGURE 7-2

Pamela Reid, Ph.D. and Certified Applied Animal Behaviorist, having fun running with one of her agility dogs. Pam has trained and competed in agility with several dogs, including Eejit, winner of multiple championships and titles. *(Courtesy Dr. Pam Reid)*

Just like any athletes, both human and dog must be in excellent physical shape. From puppyhood on, the dog is conditioned. Beginning at four months of age, the dog starts to take longer and longer walks, and works up to some little obstacles, a ramp, balance beam, or jump. As with young human athletes, care must be taken to assure that growing bones and fragile joints are not damaged by too much activity too soon. For the humans, the workout is also intense, with lots of running and quick turns, often resulting in knee injuries.

To get a true understanding of the pace of these competitions, imagine: On a "jumpers" course there are only jumps and tunnels (the AKC includes weave poles). There are many twists and turns and traps. A "trap" is an opportunity for the dog to go off course without proper attention from the handler. Obstacles are placed between 15 and 20 feet apart. A fast dog can cover a 140-yard jumpers course with multiple twists and turns, 20 jumps, and two tunnels in about 22 seconds. This is like you running 100 meters in 18 seconds where each hurdle is set at 6'6".

The North American Dog Agility Council (the other organizations have similar, if not the same system) has a number of classes: *Regular, Jumpers,* and *Gamblers*. Within these there are skill categories that include novice, open, elite, and champion. The team moves up through the various categories by acquiring points. Points are awarded for the time to complete a course, how well each obstacle is completed, with points taken away for mistakes (knocking off a bar, not stopping on time, refusing the tunnel). Judges call out the points at each hurdle, while others keep track of the points.

Agility trials are a growing sport. The fun and excitement of the competition is spreading to clubs, backyards, and formal competitive circles around the world. This activity brings out the best in the dogs and humans as they strengthen the bond between them, working as a team.

EARTHDOG EVENTS

The sport of agility trials, though not exclusively for herding dogs, is well suited to their intelligence and physical ability. But, what of the dogs who are low to the ground, bred to hunt game above and below the earth? The small terriers and dachshund are excellent at tracking animals that live underground, tunneling after them and/or barking like mad to alert the human hunter that something has been cornered. Now that most of these dogs no longer have jobs as hunters, a sport was devised that allows these breeds to demonstrate their skill and speed. There are four levels of Earthdog competition:

Level One: *Quarry*. The dog is introduced to a 10-foot tunnel with one right angle turn. At the end of the tunnel there is a cage with rats (safe from any danger from the dogs). The dog follows the scent of the rats to the end of the tunnel, navigating the turn and barking at the rats. He is then removed from the tunnel.

Level Two: *Junior Earthdog*. The dog can earn a title at this level. Now it gets trickier. The dog has to navigate a 30-foot tunnel with at least three right angle turns, in 30 seconds, bark at the rats for 60 seconds, and then allow the handler to take him out of the tunnel. No one should get hurt in the process. If the dog can do this twice under two different judges, it is awarded the title of J.E. (Junior Earthdog).

Level Three: *Senior Earthdog*. Now it gets even trickier. The tunnel is 30 feet long, with three right angle turns, but there is a false unscented exit and unscented bedding with used rat bedding at the end of it. The dog has 90 seconds to travel the tunnel and get to the rats. The rats are barked at for 90 seconds until their cage is taken out of the tunnel with them in it. The dog has to get out of the tunnel and back to the handler within 90 seconds. Successful completion of the course under two different judges awards the dog the title of S.E. (Senior Earthdog).

Level Four: *Master Earthdog*. To become an M.E. (*Master Earthdog*), the dog, with a companion, has to find his way above ground to the den for 100 to 300 yards—a true "hunt." He has to investigate a "fake" empty den, and then both dogs have to find the real entrance, mark it, and make it through the tunnel. This tunnel is a bit more complicated than the senior den. There are barriers that resemble tree roots and rocks. Again, the dog "works" the rats for 90 seconds, and has to allow the handler to remove him within 15 seconds. During these three minutes, the other dog has to wait his turn for the hunt, with minimal barking or agitation. Once his teammate comes back out, it is his turn to find the rats, make sure the handler knows they are there, and get back out to the entrance of the burrow within the given time frame and least amount of agitation.

Perhaps the Earthdog trials do not appear to be as much of a human/dog team as the agility trials, but the dogs are able to demonstrate their innate ability, while still working with the handler for a successful finish.

⊙ FLYBALL

Flyball is a relatively recent sport. Conceived in the late 1960s, the sport took off after the first tournament in 1984 and a demonstration on the Johnny Carson show. It is a team sport, matching two teams of four dogs each. Tournaments are divided into divisions of dogs of equal ability. Currently there are 16,000 dogs registered in the North American Flyball Association (NAFA).

Each team runs a relay race over a series of hurdles spread along a 51-foot course. The hurdles accommodate the smallest dog on the team. The hurdles are set at the height of the shoulder (rounded up or down) minus 4″. So, a dog that stands 15 inches at the withers will set the team hurdles at 11 inches (Figure 7-3A–D).

After a quick start and taking one hurdle after the next, the first dog makes it to the end of the course and triggers a flyball box. This sends a ball flying. The dog retrieves the ball, returns up the course leaping over the hurdles, and back to the start. Once over the start/finish line the next dog races over the hurdles, releases the ball, retrieves it, returns over the hurdles to the start so the third dog can begin. After the fourth dog finishes, time is called. The team to finish the fastest, with the least errors, wins. Amazingly, the fastest team on record has run all four dogs through the course in less than 16 seconds!

FIGURE 7-3

In flyball, tag teams of dogs need to race over a series of jumps; a dog collects a ball at the far end of the course and runs back to its team so that the next dog can race off. *(Courtesy Tom Schaefges Photography)*

(A)

(B)

(C)

(D)

LURE COURSING

Sight hounds, the regal breeds of greyhounds, Ibizan hounds, Irish wolfhounds, salukis, whippets, and Afghan hounds, to name a few, have been admired for their dignity, speed, loyalty, and agility for centuries. Sleeping at the feet of kings, they were resting from the hunt. They joined in the chase of game, using keen eyesight and bursts of speed to bring down prey. In today's world, they have little opportunity to exercise in a way that allows their bodies to move at full speed. Nor can they use their intelligence and hunting ability. Lure coursing was developed for the pleasure of the owners and the passion of the dogs. The American Sighthound Field Association (ASFA) was started in 1972. Every weekend, one or more of the 120 ASFA member clubs around the country organize a field event to provide an opportunity for the hounds to test their ability to hunt by giving chase to the lure that represents a racing rabbit—although it is often a white plastic bag.

Generally three dogs race at once, wearing different color jackets. They give chase to an artificial lure that moves like game trying to escape the hunt. The dogs are judged for their ability to follow the lure, their enthusiasm, their agility, speed, and endurance. The dogs are given points for each of those categories. The hounds run the course twice. They are awarded placements and points based on where they finished, first, second, or third place, and the number of hounds they competed against. Points from both runs are totaled to get their score.

Always excited to run, the winners may go on to compete for *Best of Breed* and then the *Best in Breed* may run the course to be named the *Best in Field*. Lure coursing leads to titles, but most importantly it keeps their natural abilities honed and bodies and minds at peak performance.

The sport of greyhound racing came from lure coursing. The dogs chase competitively around a track in pursuit of an artificial lure. Many dogs are bred for the sport, but if they are not suitable or do not prove to be winners they are kept out of the races. Some are adopted as pets; others may be euthanized.

HERDING

The American Herding Breed Association (AHBA) is serious and proud of herding breeds and their ability to keep other animals together and out of harm's way. The organization serves as a resource to people who are interested in learning more about canine behaviors related to herding, herding breeds, training, and practical herding. In addition, the *AHBA Herding Trial Program* provides dogs of different levels of ability to show their skill in ranch or farm work. The dogs are not competing but rather showing how well they work and how well their owners have cultivated the dog's instincts to keep hoof stock, sheep, or fowl in control (Figure 7-4).

FIGURE 7-4

Border collies still work in many places to keep herds of sheep moving, and herding trials are held to test the breeding and training of the very best. *(Courtesy Dr. Patricia B. McConnell)*

In addition to the AHBA there are also organizations for specific herding breeds such as the Australian Shepherd Club of America (ASCA). The ASCA holds competitions that lead to a certificate and title of *Stockdog*. The AKC states, "The purpose of competitive herding trial program is to preserve and develop the herding skills inherent in the herding breeds and to demonstrate that they can perform the useful functions for which they were originally bred."

OBEDIENCE TRIALS

Dogs appear to be happy when they please their human partner. With kindness, patience, and clarity, most dogs will quickly learn what is expected of them. As pack animals, they look for direction from their leader. An eagerness to learn, coupled with a particular breed's ability to use their physical ability to assist humans, makes it possible to be extremely obedient and well trained. In their work, in competition, and as members of our society it is important for our dogs to be controlled, friendly, and safe.

The CGC program is an excellent way to teach human and dog to behave in such a way that they are well mannered in public, allowing safe and pleasant interactions with the community, on the street, on meeting a child, or on a visit to a hospital. In the program, the basic commands are learned as well as skills such as walking comfortably in a crowd, allowing strangers to touch him, and stay relaxed in any situation without becoming distracted. These are the basic skills for a companion dog and for pursuing other sports, or becoming active in the community. To make the training more fun and challenging, the CGC has developed a program so dog and trainer can earn certificates.

If the partners want more of a challenge after accomplishing the goals of the CGC program, they can enter the *Obedience Trial* competition. Obedience trials test a dog's ability to perform a prescribed set of exercises. In each exercise the dog must score more than 50 percent of the possible points and get a total score of at least 170 out of a possible 200 points. With each qualifying score, the dog has earned a leg toward an obedience title. After earning three legs, the dog has earned an obedience title. The team can then advance through three more levels to win more advanced titles.

Rally Obedience is a bit less stringent than obedience trials. The dog and handler complete a course designed by the Rally judge. The team moves through a course of 10–20 stations at their own pace. Each station has a sign providing instructions regarding the next skill to be performed. The dog and handler move as a team. Unlike the obedience trials, the competition is much less strict. The team can communicate through vocal commands and signals, but there can be no touching or physical corrections.

The Rally provides a step for the dog/human team and handlers from training for the CGC program to obedience and agility competitions.

IDITAROD: *THE LAST GREAT RACE ON EARTH*

Mushers and dogs race over some of the Earth's roughest terrain... frozen rivers, endless tundra, deep forests, and treacherous mountain ranges. The dogs learn to work together, and the team works with the human to survive the below freezing temperatures, the wind whipping snow so powerfully that they are barely able to see, and a grueling, dangerous course of 1,200 miles (the distance is often quoted as 1,049 as a symbolic figure referring to a thousand mile race in the 49th state).

Each Musher has his or her own strategy for running the race: how to care for their dogs, which dogs run best together, when to rest, and how to train. The dogs, usually northern breeds such as Siberian huskies, Alaskan malamutes, or Samoyeds, are eager to run. Nonetheless, it takes years of training and running for hundreds of miles to become a good sled dog and part of a great sled team. The Musher must not only care for dozens of dogs but also organize to carry enough supplies for all the dogs and himself for up to three weeks on the trail. They must have enough to sustain them, but not so much weight that it becomes more difficult for the sled team to negotiate the terrain at a good speed or become too unstable to follow a narrow trail.

Musher comes from the French word *marche*, which means to walk. During the gold rush days, the word was used and then made into a stronger command, "Mush" to drive the dogs.

The rules of the **Iditarod** are stringent. The dogs must have booties to protect their pads from being cut on the ice, veterinarians look them at each of the 20 checkpoints, and rest times are enforced. Each Musher has 2,500 pounds of food for the dogs that is distributed by volunteers at the various check points. Despite stops, exhaustion and accidents happen so that dogs must sometimes be airlifted to a clinic. Dogs have died during the race and many animal rights groups protest the treatment of the dogs before and during the race.

In the city of Anchorage, where the annual race starts on the first Saturday of March, the human population is excited for the event to begin. For most of the people along the route, from tiny village to metropolitan area, the Iditarod is a celebration to be proud of. Local schoolchildren, foreign visitors, film crews, people from all socioeconomic levels, and corporate sponsors all do what they can to be involved. The event is so important that the Mushers and dogs are well-known celebrities.

The dogs are also excited to begin the race. They strain on their leads, barking, eager to begin a race that may last up to 17 days! The dogs pull as a team; their coordination is critical as they speed over a route that will take them from the mountains to the seacoast. As they rush to the finish line, they are always vigilant. Any turn taken too quickly or misjudged by the Musher may lead to disaster. An accident or the decision to take the wrong fork on the trail will cause the team to lose valuable time, but more importantly, will risk injury in a wilderness cloaked in snow, a region covered in almost constant darkness, and temperatures so low that the breath of the dog and the Musher turns to ice. Winds along the river blow mercilessly and temperatures can dip to –60°F.

This Iditarod trail was once the only route for mail and supplies from coastal towns to mining camps in the interior and out to the harsh Western territories of Alaska. Without roads or sea planes it was only possible to bring a bit of civilization in to parts of Alaska by dog sled. The sleds were not empty going back to the coast; rather they were filled with the gold that made many men rich and even more just cold and hungry. The Iditarod is celebrated to remember this part of Alaska's history and the bravery of the men and dogs who helped settle the land. Of greatest significance was the "Great Race of Mercy to Nome." In 1925, sled dogs saved the citizens of Nome from an outbreak of

FIGURE 7-5

A well-matched sled team moves with grace and speed. Many people enjoy the thrill and satisfaction of working with the dogs even if they do not compete in races. *(Courtesy Dr. R. P. Coppinger)*

diphtheria. A package of antitoxin had to get to the dying population and the only way to get there was by sled dog. The Mushers and their teams made the 674 miles in less than six days, saving lives and making history.

Those who complete the race are still connected to the hardy people and loyal dogs who settled this wild land. Although they may come from countries all around the world to compete for the monetary prize, they are also there to test themselves. If they come in first or last, they have competed against other teams and against some of the harshest circumstances nature can produce. Human and dog, as fragile as they seem, race over one thousand miles through the grandeur and danger of Alaska's wilderness, and survive. The Iditarod may be the most well known of the sled dog races, but there are many other races and simple excursions that are less intense that people are able to enjoy with their dogs (Figure 7-5).

WEIGHT PULLING

It is likely that sled dogs that have participated in the Iditarod would make good competitors in a very "athletic" sport, *weight pulling*. The object of this activity is for a dog to pull the greatest weight. The dog pulls a cart or sled that is weighted down for 16 feet within a 60-second time period. The competition can be held on dirt or on snow. Sometimes the path is fenced in so the dog is sure to pull in a straight line.

To begin the race, the dog stands at the starting line and waits while the handler goes to the finish line. If the dog crosses the start line before he is called, it is a false start. Two false starts and the dog is disqualified. Any dog can compete if they are fit and enjoy the challenge. There are 12 different weight classes, so the competition is between dogs of similar weights, if not breeds. Dogs as diverse as mastiffs and poodles eagerly compete.

The sport was first organized in 1984. Since then it has spread to ten regions across Canada and the United States. Dogs compete against each other and themselves. There are three certificates available: *Working Dog* (WD), in which the dog pulls 12 times their weight (five times if they pull on snow) at four different events; *Working Dog Excellent* (WDX), in which the dog pulls 18 times their weight (10 times on snow) at four different events, and *Working Dog Superior* (WDS), in which the dog pulls 23 times their weight (15 times if on snow) at three different events.

Dogs that are successful at this sport are true athletes. They must be physically fit and have great stamina and the determination to pull as much as 2,000 pounds across the finish line. No food rewards or handling are allowed. The winning dogs have a strong desire to please their trainer, straining on the harness, struggling to pull the sled or cart just to hear the happiness in the human's voice. It is the natural ability, genetic makeup, and determination that allow these athletes to cross the finish line.

CANINE FREESTYLE

Perhaps on the other extreme of *weight pulling* is the canine competitive sport of Musical **Canine Freestyle**. The World Canine Freestyle Association defines canine freestyle as "A choreographed musical program performed by handlers and their dogs. The object of musical freestyle is to display the dog and handler in a creative, innovative, and original dance, using music and intricate movements to showcase teamwork, artistry, costuming, athleticism, and style in interpreting the theme of the music." The sport expands the scope of training by adding an artistic element. It is creative and extremely entertaining for an audience.

As a team, human and dog create a masterpiece of movement and entertainment. They must be able to understand one another and be proficient in obedience training. The dog probably does not help with the costumes or the selection of the music, but the human must have a sense of what he and his canine partner can accomplish together.

Very much akin to freestyle is *Heelwork-to-Music*. This art form incorporates the art of dressage with traditional dog obedience, dance, and costuming with an emphasis on nonstandard obedience movements. Both Heelwork-to-Music and musical freestyle should create a visually exciting display that is enjoyable for the person and dog as well as the audience.

It could well be that people have danced with their dogs throughout history. Dogs following their handlers as they practiced dance steps or ice skating forms, children joined by their pet dogs as they danced about in their own revelry. Who knows? But formally it was in Canada in 1989 that Val Culp in British Columbia developed the sport. Soon after, the sport caught on in England and then soon followed in the United States. The idea was to have fun and promote responsible pet ownership through obedience training. By 1995, there were two different styles—in Canada, the routines were very theatrical with ostentatious costumes, and in the United States, the performance and demonstrations were more controlled with most of the attention to the dog and little to costuming. There are different levels of competition: starters, novice, intermediate, and advanced. Performances last from two and one-half minutes to four minutes and should cover the entire 40-foot-by-50-foot rink.

The sport has continued to grow, with competitions around the world. By the year 2000, two new divisions were added for testing: Sassy Seniors for dogs over 9 and handlers over 65 and *Handi-Dandi Dancer*s for the creatively challenged dogs and/or handlers. The sport is especially nice because both senior humans and dogs can continue to compete as they age. Only the steps become slower and there are not so many athletic moves, but the joy and bond remains for both partners.

The Canine Freestyle Federation, Inc. (CFF) is an international organization dedicated to defining and developing the sport and providing the structure for demonstrations and competitions. Another organization, the *Musical Dog Sport Association*, also seeks to advance canine freestyle and to "Share the joy of the canine/human bond achieved through positive training, enhanced by the artistry of music and choreography."

There are probably more activities waiting to be formalized into competitive sports, just as throwing a Frisbee for a pet dog has developed into the competitive sport of *Disc Dog*. And who was the water-loving retriever that raced off a dock into the water after a stick, causing some humans to create Dock Jumping. The next time you are playing with your dog and he invents a new game with a ball, behavior, or instinct, perhaps you will create a new canine competition.

CAT SHOWS

The **Cat Fanciers' Association (CFA)** currently recognizes 37 breeds in the championship class in its shows, along with three provisional breeds and one miscellaneous. Similar to dogs, each breed is judged against a standard that is maintained by each breed council (<www.cfa.org/shows>). Cats entered in the *Championship Class* must be pedigreed, unaltered cats over eight months of age. In addition to the championship class, cat shows may often have several other classes:

- *Kitten*—unaltered or altered, pedigreed kittens between the ages of four and eight months
- *Premiership*—altered, pedigreed cats over the age of eight months
- *Provisional*—breeds that have not yet achieved championship status
- *Miscellaneous*—breeds not yet accepted for Provisional status, but accepted for registration and showing in the Miscellaneous class
- *Veteran*—any male or female, altered or unaltered, not younger than seven years on the opening day of the show that if chosen to could otherwise be shown in the championship or premiership classes
- *Household Pet*—all random bred or non-predigreed cats

The household pet class is of particular interest since it gives anyone with a companion cat an opportunity to participate in a cat show. The class is open to any cat over four months of age. Any cat over the age of eight months must be either spayed or neutered, and entries cannot be declawed. Since there can be no standard for this class, the cats are judged for their uniqueness, pleasing appearance, unusual markings, and sweet dispositions. Each household cat that exhibits good health and vitality receives a Merit Award (Figure 7-6).

FIGURE 7-6

The Cat Fanciers' Association welcomes non-pedigreed house cats at its shows, where they are judged on their health, condition, and sweetness. *(Courtesy Emily Manos)*

RATS AND MICE

The **American Fancy Rat and Mouse Association (AFRMA)** was founded in 1983 to promote and encourage the breeding and exhibition of fancy rats and mice for shows and pets (<www.afrma.org>; March 14, 2007). For shows, rats and mice are divided into various classes based on coat types, color, and color patterns. Rats should have a long racy body in good weight, with large bold eyes, a long, clean head showing breadth and length, large ears, a long, tapering tail, and an average body length of 8–10 inches. Mice should have a long, slim, racy body, large bold eyes, a long, clean head showing breadth and length, large expressive ears, a long tapering tail, and an average length from nose to tail tip of 8–9 inches. Both rats and mice must be tractable and easy to handle, with no evidence of physical defects or unsteady temperament. It is interesting that the AFRMA makes a point of *not* having a position on culling, reptile keeping, or feeding of rats and mice to reptiles, nor does it have a position on animal rights.

DISCUSSION QUESTIONS

1. What are the seven breed groups recognized by the American Kennel Club (AKC)?
2. What is a breed standard? How is it assessed in a conformation competition?
3. How do agility and conformation competitions differ in terms of athleticism and the types of dogs allowed to participate?
4. Which breed types are most likely to be involved in the following competitions? How does the associated activity relate to the dogs' original working roles? Lure coursing, Iditarod, Earthdog events.
5. What is the purpose of the *Canine Good Citizen* (CGC) program?
6. Is it more likely to find neutered animals competing in a Cat Fanciers' Association (CFA) cat show or an AKC conformation competition?

REFERENCES

Show
www.akc.org
www.akc.org/kids
www.akc.org/breeds
www.cfainc.org.shows.html
www.grca.org Golden Retriever Club of America
The American Kennel Club. *The complete dog book* (20th ed.). (2006). New York: Ballantine Books.
Evans, M. (1992). *ASPCA pet care guides for kids: Rabbits*. New York: DK Publishing.
Searle, N. (1992). *Your rabbit: A kid's guide to raising and showing*. Pownal, VT: Storey Communications.

Agility
www.akc.org American Kennel Club
www.canismajre.com/agility
www.cleanrun.com

www.nadac.com North American Dog Agility Council
www.usdaa.com United States Dog Agility Association
Daniels, J. (1991). *Enjoying dog agility: From backyard to competition.* Wilsonville, OR: Doral Publishing.
Simmons, J. (1992). *Agility training: The fun sport for all dogs.* Somerset, NJ: Howell Book House.
Hobday, R. *Agility fun the Hobday way*, Vol. 2. South Hadley, MA: Clean Run.

Earthdog
www.akc.org.events/earthdog
www.dirt-dog.com. American Working Terrier Association

Flyball
www.flyball.org North American Flyball Association

Lure Coursing
www.akc.org/events/lure_coursing American Kennel Club
http://www.asfa.org/coursing American Sight Hound Field Association

Herding
www.ahba-herding.org The American Herding Breed Association
www.akc.org/events/herding, www.asca.org/programs+/stockdog, www.asca.org Australian Shepherd Club of America
http://www.workingdogs.com Working Dogs

Obedience
www.akc.org/events/obedience

Rally Obedience
www.akc.org/events
www.apdt.com/rally Association for Pet Dog Trainers

Iditarod
www.alaskanet.com Iditarod Dog Sled Race – Facts & Figures
www.dogsled.com
www.iditarod.com
www.njsdc.com New Jersey Sled Dog Club

Weight Pulling
Dog Fancy Magazine. (2006). Training secrets for bully breeds. *9.*

Freestyle
Jaskiewicz, A. (2005). *Getting started with freestyle.* Retrieved August 14, 2007, from http://www.canine-freestyle.org/articles_gettingstarted.asp.
Sullivan, M. (2006). *New steps for old bones.* Retrieved August 14, 2007, from http://www.canine-freestyle.org/articles_fandf_newsteps.asp.
http://worldcaninefreestyle.org. World Canine Freestyle Organization
http://musicalsportassociation.com Musical Dog Sport Association

Assistance Dogs

CHAPTER 8

A dog teaches a boy fidelity, perseverance, and to turn around three times before lying down.

—Robert Benchley

KEY TERMS

assistance dogs
service dog
Americans with Disabilities Act (ADA)
service animals
social/therapy animals
companion animal
Delta Society
animal-assisted activities (AAA)
animal-assisted therapy (AAT)

Assistance Dogs International, Inc. (ADI)
hearing dogs
Seeing Eye
Leader Dogs
therapy dogs
Boris M. Levinson
Green Chimneys Children's Services and School of Little Folk
search and rescue
tracking
trailing

patrol dogs
water rescue
narcotics dogs
evidence dogs
Schutzhund
sentry dogs
messenger dogs
mine dogs
conservation dogs
Working Dogs for Conservation Foundation (WDCF)

For thousands of years, dogs and humans have worked side by side, each depending on the other for their livelihood and protection. The Yorkshire terrier of the twentieth century, with a ribbon in her hair and sequined collar, is descended from fearless, rugged wolves that survived by their instincts. Over the centuries, a vast array of breeds have developed to serve a multitude of responsibilities designed by man—from retrieving game to sitting quietly on a person's lap.

Sled dogs

Sled dogs were bred and used by native peoples of the coldest regions of the world to help them survive. The Alaskan malamutes were bred by a group of Eskimo people known as the Mahlemiut. The dogs were large and capable of pulling immense weight—hauling food back to the villages along the Anvik River in Alaska. Siberian huskies originated with the Chuckchi people of northeastern Siberia. They used the dogs to herd reindeer and pull loads. Huskies are smaller and faster than the malamutes. Both breeds were sought after during the gold rush and have continued to work with humans to pull sleds for work and for pleasure.

In the twenty-first century, most of the innate qualities of dog breeds are no longer useful to the majority of people, so competitions have been designed as an outlet for dogs to practice their natural physical and mental skills. But it is not all play. There are canine professions where dogs use their skills to work. Humans have also found unexpected ways for dogs to provide service to humans.

Of the many commendable abilities our canine helpers possess, their work as **assistance dogs** is most diverse. By partnering with a person with disabilities, the assistance dog provides a specific service as well as provides their handler with greater independence and companionship. The lives of thousands of people are enriched, thanks to these dog's abilities to serve as the eyes of the blind and visually impaired, the ears of the deaf and hearing impaired, and helpers to people with other disabilities. The phrases assistance dog and **service dog** are used in slightly different ways by different organizations, although the dogs do the same work, using their abilities to help people with disabilities. In this section we will learn the different definitions.

Service dogs were once considered to be only those dogs trained to lead the blind. This is no longer the case. Dogs reliably perform a huge variety of tasks for an almost limitless range of needs the disabled have, thereby allowing these people to function independently and in safety.

Among the jobs service dogs perform are as follows:

- Leading a person with a visual impairment around obstacles, to destinations, and alerting them to danger
- Sound discrimination to alert a person with hearing impairments to the presence of specific sounds
- General assistance
  - Mobility—helping persons balance, pull a wheelchair, rise from a wheelchair or from a fall, pulling a wheelchair, opening/closing doors, finding another person.
  - Retrieval—getting items that are dropped or out of reach, carrying items in mouth or backpack.
  - Sense and alert—some dogs are able to sense when a person is going to have a seizure—though to date it is unknown how—and thereby make certain that the person is in a safe position and prepared. (Some dogs are also able to detect cancer in a person before there is any visible evidence.)

According to the **Americans with Disabilities Act (ADA)** (ADA, 1990), a *service dog* is one that has been *individually trained to do work or perform tasks for the benefit of a person with a disability*. To be considered a service dog, it must be trained to perform the tasks directly related to the person's disability. Also as defined by the ADA, a disability is *a mental or physical condition which substantially limits a major life activity* such as

- Caring for oneself
- Performing manual tasks
- Walking
- Seeing
- Hearing
- Speaking
- Breathing
- Learning
- Working

Dogs may also help with disabilities that are not visible, such as deafness, epilepsy, and psychiatric conditions.

Other animals are trained to provide assistance to humans (e.g., monkeys, horses, parrots), but the most common animal employed to help people is the dog. There are numerous labels for the work that dogs perform for people with disabilities and the definitions are not standardized. For example, a dog trained to help a person walk may be referred to as a *mobility dog*, a *walker dog*, or a *support dog*. Additionally, many service dogs are cross-trained to perform more than one category of work (guide and mobility for a person who is blind and is wheelchair bound). The dogs are not only for adults with disabilities, but there are organizations, like Loving Paws Assistance Dogs, that provide dogs for physically disabled children. Most of the children have spinal cord injuries, but may also have conditions like muscular dystrophy, cerebral palsy, and other impairments that require assistance and a loving companion.

Service animals are legally defined (ADA, 1990) and are trained to meet the disability-related needs of their handlers. Federal laws protect the rights of individuals with disabilities to be accompanied by their service animals in public places. Service animals are not considered to be *pets*.

Social/therapy animals do not have a legal definition. They are often animals that did not complete service animal or service dog training due to health, disposition, trainability, or other factors, and are made available to people with disabilities to have as pets. The animals might or might not meet the definition of service animals.

Therapy animals may also be defined as those animals that visit with people in hospitals, nursing homes, or prisons to bring some of the feelings of well-being to people who might otherwise not be able to have a pet. Federal laws have no provisions for people to be accompanied by therapy animals in places of public accommodation that have *no pets* policies. **Companion animal** is not legally defined, but is accepted as another term for pet.

The **Delta Society** is the national organization best known for its work pertaining to the human-animal bond, supporting research and formalizing training and ethics surrounding service animals. Established in Portland, Oregon, in 1977, the founders of the organization, Michael McCulloch, M.D., and Leo K. Bustad, D.V.M., Ph.D., wanted to understand the quality of relationships between pet owners and pets. At that time, pets were not considered of any importance to the health and well-being of humans. Delta Society spent the early years after it was established providing funds for research on why animals are important to the general population and specifically how they affect health and well-being. Indeed, the studies found that there was scientific evidence that nonhuman animals greatly enhanced human lives. With this credible research, the organization proceeded to see how animals could enrich the lives of people who are ill and disabled.

By the early 1980s, Delta Society was able to disperse educational materials to apply the scientific information to everyday life. Pet owners and a broader community joined the veterinarians and scientists who had formed the organization. In the 1990s, Delta Society built on its scientific and educational base to provide direct services, including comprehensive training in **animal-assisted activities (AAA)** and therapy to volunteers and health care professionals. With the development of the standards of practice in animal-assisted activities and **animal-assisted therapy (AAT)**, Delta Society was able to provide guidance in the administrative structure of AAA/T programs, including animal selection, personnel training, treatment plan development, and more. Today, the organization continues to develop standards-based training materials. The Professional Standards for Dog Trainers: Effective, Humane Principles provides guidelines for all dog training developed by Delta Society. This is being followed with a comprehensive curriculum for service dog trainers.

Delta Society's current programs implement the mission of incorporating pets into the lives of the following:

- The general population to improve health (health benefits of animal activities)—The health benefits of animal activities focus on the role of companion animals in the everyday lives of people. Delta Society educates the public on the ways a strong human-animal relationship helps to maintain good health, grow and develop in positive ways, and cope with life's challenges.

- People who are ill to improve healing (AAT Services and Pet Partners Program)—These programs help people heal through the comfort and motivation from a specially trained pet.

- People who are disabled to improve independence (National Service Dog Center)—The center provides referrals and resources for people with disabilities with service dogs to overcome problems in their environments and achieve greater independence. The effect of the information extends beyond the family, workplace, and community so that there is greater understanding and acceptance of the role that dogs have in the lives of the disabled. The National Service Dog Center also provides widespread education about service dog issues and referrals to resources.

Service dogs may not be for everyone with a physical or emotional disability. Before an individual decides to team up with a service animal, many things must be seriously considered. These issues include learning about the abilities and limitations of service dogs, evaluating the person's needs, the impact a dog will have on the person's lifestyle, and all of the issues surrounding responsible care and treatment of the animal.

If the decision is made to apply for, train, and partner with a service dog, it is also critical to assess the person's ability to function on a daily basis. Will the person have better stamina if a dog can perform some tasks for him or her? Would having a service dog help him or her get more exercise? Allow him or her to be more mobile? Would a dog help him or her become more social? Would the dog's presence alleviate some of the concerns of the human caregivers of the person with disabilities, knowing his or her safety is being looked after? How difficult are the activities of daily living and could a dog assist with them?

This is a partial list of the types of disabilities a person may have and benefit from partnering with a service dog. Certainly, there are many other situations where a trained companion will enhance quality of life for an individual and his family.

- Spinal cord/head trauma (injury, stroke)
- Visual or hearing deficits
- Arthritis
- Ataxia/poor balance
- Multiple sclerosis
- Cerebral palsy
- Muscular dystrophy
- Spina bifida
- Seizure disorders
- Cardio/pulmonary disease
- Arteriovascular disease
- Psychiatric disabilities

Karen Allen and Jim Blascovich (1996) completed a two-year study and found that people with service dogs scored higher for psychological well-being, self-esteem, community integration, and the amount of control they could exert over their environment. In addition, the number of personal human assistant hours required for care decreased by an average of 78 percent, representing a huge savings in health care costs. In addition to performing the tasks the dogs are trained for, they provide the person with disabilities all of the benefits of having a companion animal. Plus, there is substantial research that indicates having a companion animal in one's life

- lowers blood pressure
- moderates stress

- improves motivation
- decreases serum cholesterol
- relieves the effects of loneliness

The Delta Society and the United States government define all dogs that *assist* humans as service dogs. This is in contrast to the organization **Assistance Dogs International, Inc. (ADI)**, which has set standards for the assistance dog industry since 1987. It is a coalition of nonprofit organizations that trains and places all types of assistance dogs. The ADI has three categories: guide dogs, **hearing dogs**, and service dogs (those that help in all different ways). Together, they call all of these trained dogs **assistance dogs**. The ADI is working to establish consistent terminology internationally.

Through the ADI, the industry can share ideas, work on issues pertaining to the legal rights of people with disabilities, and establish guidelines and ethics for the training of the dogs. ADI's mission is to

- establish and promote standards of excellence in all areas of assistance dog acquisition, training, and partnership;
- facilitate communication and learning among member organizations; and
- educate the public to the benefits of assistance dogs and ADI membership.

The ADI wishes to create standard terminology worldwide as well as to create minimum standards for all dogs, service dog centers, and assistance dog partners. The following are the standards for training service dogs that respond/alert to a seizure:

1. A minimum of 120 hours of training over a period of no less than six months is required and must take place under the supervision of a program's trainer. During that time at least 30 hours of regularly scheduled training must be devoted to field trips and public exposure.

2. The dog must be able to perform both on- and off-leash obedience skills with voice and/or hand signals. These commands include: sit, stay, come, down, and heel. Dogs must demonstrate absolute control on and off leash.

3. The dog must show social behavior skills of no aggression, no inappropriate barking, no biting, no snapping, no growling, no inappropriate jumping on strangers, no begging, and no sniffing of people. Dogs must be unobtrusive and have good household skills.

4. The dog must be trained no less than one *appropriate response skill* (i.e., vocal alert, physical contact alert, activation of an emergency medical alert system, etc.).

5. The training time with the student prior to placement must be a minimum of 60 hours. Instruction must include obedience skills, proper behavior, and implementation at home.

6. The training facility must require the recipient to complete a follow-up progress report once a month for the first six months following the placement. Personal contact will be done by qualified staff or program volunteer within 18 months of graduation and annually thereafter.

7. Identification of the seizure response/alert dog will be accomplished with a laminated ID card with a photo of the dog and partner and names of both recipient and dog. In public the dog must wear a harness, backpack, or slicker with a logo that is clear and easy to read and identifiable as a service dog.

8. At the onset of training, every dog will be spayed or neutered and will have a thorough medical evaluation to determine that the dog does not have any physical problems that would cause difficulty for a working dog.

9. It is the program volunteer's duty to educate the client of the responsibility to inform in advance their support system of the proper response to the seizure alert/response dog.

The following descriptions are for dogs that help provide a better quality of life for humans with disabilities.

GUIDE DOGS

Perhaps the most commonly known work of assistance dogs is as guide dogs. Guide dogs are trained to partner with blind or visually impaired people so the handler has greater independence. With the dog guiding him or her, the handler can be guided safely through daily activities at home, at work, and anywhere he or she wishes to go. With a well-trained, willing dog with him or her, there is freedom and security for the person needing assistance. The dogs are commonly, but incorrectly, called **Seeing Eye** dogs. *The Seeing Eye* is the name of only one of many guide dog training schools.

The first guide dog school was founded in Germany during the First World War to provide a better quality of life to returning soldiers who were blinded in combat, mostly by coming in contact with lethal gas. It was at the school that an American woman named Dorothy Harrison Eustis, who was living in Switzerland, saw the full potential of the German shepherd dog. She was training German shepherds for the Swiss Army and European police units, but after seeing how the German dogs gave new life to the soldiers who had been blinded, she returned to America to found the Seeing Eye, Inc., in 1929, the first training program for guide dogs in the United States.

In February 1929, the first class saw two students graduate. By the end of that year, 17 blind men and women had the opportunity to regain their independence and dignity with their own Seeing Eye dogs. The school moved from Tennessee to New Jersey in 1931 and since then has matched over 13,000 specially bred and trained Seeing Eye dogs with 6,000 blind men and women from around the world and from all socioeconomic levels and professions. They have all learned how to care for and work with their dogs at home, on the job, on public transportation, and in all public places (<http://www.seeingeye.org>).

The Seeing Eye trains their dogs to guide blind people, breeds and raises the dogs, teaches instructors the science and technique of training dogs as guides for blind people, instructs them in the proper use and care of the dogs, and educates the public about the role of guide dogs and the capability of blind people for independent living.

The Seeing Eye has its own breeding program, keeping information on every sire and dam in a database, making it easier to select parents who will produce puppies with the qualities that make an excellent guide dog (intelligence, health, moderate size, sound hips, and gentle disposition). Mostly the breeds are German shepherds, Labrador retrievers, golden retrievers, and occasionally boxers or mixed breeds. Since the founding of The Seeing Eye, Inc., in the United States, many other organizations have been founded to meet the demand for these very special dogs. **Leader Dogs** for the Blind was the second American school for guide dogs, founded in 1939 by Lions Club members. Leader Dogs is a nonprofit organization and provides all training, expenses, and the dogs free of charge to the visually impaired. Their dogs are mostly purebred retrievers and German shepherds and graduates of the program are called *Leader Dogs*, much as The Seeing Eye's dogs are named for the organization that trains them.

Guide Dog Training

Potential guide dogs come from various sources. Some organizations breed and raise their own puppies, while some rely on *foster families* to raise the puppies until they are ready for formal training. When dogs become old enough to start training, most guide dog *schools* will conduct a test to analyze the dog's potential to be a guide dog. If the dog passes this test, it continues on to harness training, where it learns to help a person move around safely—navigating curbs and avoiding overhead obstacles. The dogs may be taught additional skills, such as retrieving items for their handler. Schools may vary in their training techniques. The following is the curriculum for Seeing Eye dogs (<http://www.seeingeye.org>).

When a dog is about 18 months old, it enters a four-month course with a sighted Seeing Eye instructor. Seeing Eye instructors have completed a rigorous on-site three-year apprenticeship. They educate the dogs to respond to a system of rewards and corrections. The reward often is no more than a loving pat, and the correction, a verbal reprimand. Instructors work in teams of four. Training schedules are staggered so as one team works with an incoming class of students, another has dogs in the final stages of training, ready to go out in the world.

Each training cycle begins at The Seeing Eye campus, where the dogs are taught basic obedience commands and to pull with a harness on. They then progress to a residential street where they are taught to lead and stop at curbs. As the dogs advance, the route gradually changes from quiet side streets to the busier downtown areas where there is heavy car and pedestrian traffic. Finally, the dogs learn *intelligent disobedience* to disregard a command if it would lead their person to danger. The dogs use what they have learned but also must think and use their intelligence in order to make decisions for their partners, sometimes a contradictory one to the command. After four months, the dogs are ready for their final exam. Instructors wear blindfolds and walk with their dogs while a training supervisor evaluates each dog. That is what it takes to train the dogs—then come the people. Human students are trained individually and then with their own guide dog for two to three months with an instructor. When the newly created team has graduated, they are certified and go out on their own to live and work together. Depending on the organization, follow-up training to ensure the dog is still doing his or her job may or may not be required.

The average working life of a Seeing Eye dog is about 8 years. Many live and work to the ages of 12, 13, or older. The bond between handler and dog is very strong, for the dog is a constant companion and family member. Therefore, an individual may choose to keep the aging dog as a pet, have him placed with a new owner as a pet, or retired at The Seeing Eye, which will then find the dog a good home. A graduate who decides to retire his or her dog may return to The Seeing Eye to be trained with a new partner.

Hearing Dogs

Hearing dogs, also called *signal dogs, sound alert dogs, hearing ear dogs,* or *hearing assistance dogs*, are a category of assistance dogs that are specially selected and trained to help people who are deaf or hearing impaired. It can be very lonely and alienating not to be able to hear, and a dog provides increased independence, more confidence, and a feeling of security for someone who may not be able to receive auditory cues if there is danger. Imagine going to bed at night, alone in the house, knowing that if a fire alarm sounded, or someone knocked at the door, or even screamed at you, you would have no idea. It would be hard to be at ease. There is excellent equipment for the hearing impaired, but for many people it is awkward and certainly does not provide the same comfort as a dog, curled up at your side, especially one with perfect hearing.

The primary responsibility of the hearing dogs is to help their handlers become aware of important sounds such as a ringing telephone, alarm clock, a baby's cry, or the door bell. When the dog hears one of those sounds it will go to its handler, alert him or her by putting a paw on him or her, and then lead him or her to the telephone, door, or other sound signal (Figure 8-1).

All of these sounds are important, but there are others that can save a life. A hearing dog will alert its handler to emergency sounds by touching with its paw and then lying down in an alert position that means *danger*. The dog's response to smoke alarms, fire alarms, carbon monoxide alarms, or burglar alarms provides the hearing impaired person with time to make sure he or she and his or her dog are able to escape.

Of course, there are many other sounds outside the home that are critical to the safety and social well-being of people: sirens, machinery, cars backing up, people calling out to you—the world is full of auditory warnings that if not heard can be dangerous or at the very least unpleasant.

Hearing dogs are generally small in size, energetic, quick learners, and very comfortable in a variety of situations. Before entering a training program they are tested for proper temperament, sound reactivity, and willingness to work. There is no need for hearing dogs to be purebreds, or of a certain breed. In fact, The Hearing Dogs for Deaf People, an organization in England, obtains 75 percent of the dogs they train and provide hearing impaired population from shelters.

Once the dogs are selected and admitted to a training program, they are started on socialization training—usually in the home of a volunteer.

FIGURE 8-1

Hearing dogs are trained to alert hearing-impaired people to a wide variety of sounds, including a telephone ringing. Telephones may also be hooked up to a nearby light so that it is on when the telephone rings. At the San Francisco SPCA homeless dogs are selected and trained as hearing dogs. Once training is complete, they have a new job and a home. *(Courtesy Francis T. Metcalf)*

They learn to be comfortable with a variety of people and learn basic obedience. During these months of training, the dogs are also exposed to things they will face in public such as elevators, traffic, other dogs, all kinds of situations, even the occasional trip to the veterinarian. If they complete this period of socialization with no problems, then they are trained specifically in sound alerting to identify certain sounds and how to respond to them. The third and final stage of training involves the hearing dog and the hearing-impaired person working in partnership. This training usually takes place in the home of the person who will handle the dog. Dog and human are supervised and coached in the home, on walks, and on other outings away from home, until trainer, dog, and human feel at ease.

In the United States, Title III of the Americans with Disabilities Act of 1990 allows hearing dogs, along with guide and other service dogs, access to anywhere the general public is permitted to go; state laws also provide for access. There are fines and also criminal penalties for interfering with a hearing dog team or denying access to a hearing dog. The same penalties apply to a person trying to disguise a dog as a hearing dog illegally. In order to identify these assistance dogs and their hearing impaired partner, the dogs often wear an orange collar and leash and even a cape or jacket.

Therapy Dogs

Therapy dogs are not assistance dogs with the same rights, privileges, or responsibilities of dogs that help one individual with a disability become more independent. Therapy dogs interact with people, and usually more than one

FIGURE 8-2

Therapy dogs may visit people in hospitals, senior homes, or schools. Several programs encourage children to read, by having them practice reading to a therapy dog. The dog is non-judgmental and seldom corrects the child's pronunciation. *(Child at Gillen Brewer School)*

person, only when they are invited. They come for a visit, not for a lifetime. By law (the ADA) service dogs are allowed access to a hospital or school along with their handler. If a therapy dog wishes to visit patients in a hospital, or a school for the handicapped, he and his handler would have to receive permission, present papers assuring their health status, and sign waivers of responsibility (Figure 8-2).

Therapy dogs bring comfort to people in a variety of settings including hospitals, long-term care facilities, nursing homes, mental health centers, children's residential facilities, prisons, and domestic abuse shelters. It is difficult to secure real scientific evidence regarding the effect they have on people, but there is certainly a great deal of anecdotal evidence that they can brighten anyone's day. Some general effects from a therapy dog visit are that they:

- Promote a feeling of well-being (elderly, children, hospitalized patients).
- Provide unconditional affection (prisons, domestic abuse shelters).
- Improve focus (Alzheimer patients and people suffering from clinical depression).
- Interact with people who have difficulty communicating (some psychiatric patients).
- Stimulate memory functions.
- Encourage and aid speech functions (stroke patients).
- Motivate simple physical activities for the mobility impaired (brushing, stroking the dog).

- Provide practice for specific physical therapy functions (throwing a ball).
- Provide a model for perseverance (because many therapy dogs came from shelters, they have been through some very tough times. It comforts some people to know that not only did the dogs survive, but now they are helping others and are very useful).

The most important characteristic of a therapy dog is that he is very comfortable around people, with a steady, friendly temperament. The excellent therapy dog must be friendly and outgoing but very well behaved so it does not jump up on a frail, elderly person, rush about the hospital, or lick someone who is uncomfortable with that kind of attention. The dog may come across some very unusual and sometimes threatening situations—someone shouting, throwing things, or acting out—but it must remain calm and at ease. Not every person in the facility will wish to be visited by a dog, so the team must be respectful and aware.

Training includes not only good citizen training, but also a familiarity with many different situations dog and handler may encounter. There are national organizations and local clubs to train with and obtain certification. However, there are no national standards.

Boris M. Levinson and Pet-Oriented Psychotherapy

In the days following the terrorist attacks on September 11, 2001, thousands of emergency responders clambered over the rubble piles looking first for survivors, and then for the remains of victims. Close at their side were dozens of trained SAR dogs. These images were captured by the media and linger in our memories. At the same time, other trained dogs were also hard at work in a place that few people were able to see. On a pier jutting into the Hudson River an emergency center was established for families waiting for word on lost loved ones. It was here that over 300 dogs certified for animal assisted therapy greeted families and emergency workers, soothing jagged nerves. A fireman, just in from ground zero, still dusty and worn, might sit with a paper cup of coffee in one hand, the other hand dropped to his side, caressing the head of a golden retriever. Staring into the distance, he might mutter or mumble about what he had seen or felt our there. Perhaps he was talking to the dog, or maybe the dog's human partner kneeling nearby. Another swig of coffee and the fireman would be off to the next job. The dog and partner would then wander down to the family area and play with several children. Distracting them while their father provided a description of his wife, and a small knot of hair found in the comb she had used the morning before going off to work at the World Trade Center. The pet therapy teams worked in relays, and were there every day until the family assistance center closed. Their presence and benefits were readily accepted by everyone. The therapeutic value was in little doubt and was a testament to the vision of a man who foresaw this opportunity some 40 years earlier.

Boris M. Levinson, born in 1907, was a Lithuanian immigrant who had arrived in the United States in 1923. His family lived at the very edge of New York City in East New York, Brooklyn. There was a dairy farm just down the road from their home, and Levinson enjoyed seeing the many animals he could see living around him. He did well in school and in 1947 earned a degree in clinical psychology from New York University. He went to work as a psychoanalytically trained child psychologist. Students recall that as a teacher he was austere and tough, but marveled at his gentleness when dealing with the children. In 1953, while working with a severely withdrawn child, he first began to use his dog Jingles as part of his therapy sessions. His success with other children encouraged him to expand his efforts, and Jingles became an important part of his therapeutic strategies. In 1961 he presented a paper at the annual meeting of the American Psychological Association on his work, and what eventually called Pet Therapy. Levinson would continue to refine and promote the use of animals as part of psychotherapy until his death in 1984. The Delta Society, dedicated to efforts in this area, presents the Jingles Award each year to animals that have shown exceptional service in the therapeutic settings (from the preface of Levinson, 1997).

Part of training for the human partner is that he or she understands enough about the health conditions of the patients and people he or she and his or her partner will visit to be helpful and cause no harm. It is also important for the human to be honest about the kinds of conditions that may be uncomfortable for him or her to be around. Perhaps being with old people in a nursing home is not easy because he or she thinks of his or her own aging parents, but he or she and his or her dog are great together when visiting prisoners. Everyone has to find their niche so they bring great comfort to the many people who might otherwise be shut off from the healing qualities of a fine dog.

A facility that uses therapy animals on a daily basis, not just to visit the youngsters, is **Green Chimneys Children's Services and School of Little Folk**. Located in Brewster, NY, it provides care for children who cannot live at home because of their mental or physical disabilities. Believing that all living things must be cared for, the founders, Myra and Sam Ross, bring domestic and farm animals into the everyday lives of the children. The facility has a working farm with dogs, cats, and smaller animals that play a role in the therapeutic setting (Figure 8-3). The animals are part of the curriculum and

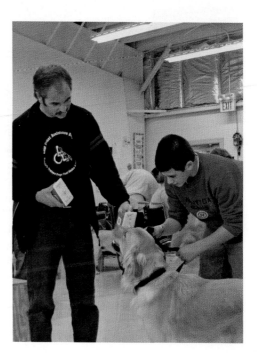

FIGURE 8-3

The children at Green Chimneys participate in programs to train service dogs. The patience and consistency that the children need to master to be successful with the dogs is an important part of their therapy.
(Courtesy Green Chimneys)

are often in the classroom. The children draw them, write about them, and find ways to incorporate them into all aspects of their day. For many children, it is easier and more healing to have a dog to talk to than a human and bridge the way to having a healthy interaction with other people.

Merely having animals present in an environment may make a huge difference in the lives of people with emotional problems, but they can also play a part in actual therapy. The animal does not replace the human therapist, but for a nonverbal or autistic child the presence of a gentle dog or cat may help open up an avenue of care not available simply with a trained adult. Children will often share thoughts or experiences with pets, even if they do not feel comfortable speaking with a human.

Green Chimneys

Green Chimneys provides care and treatment for at risk and emotionally injured children. It was founded in 1947 by Samuel B. Ross, Ph.D., and members of his family in Brewster, NY, on a 200-acre farm campus. Satellite programs are located in Danbury, CT, New York City, and Westchester County, NY, but it derives its name from the iconic green chimneys on the buildings of its Brewster campus (Figure 8-4). Green Chimneys is an international leader in AAT for children. Their main campus provides residential facilities for about 100 students and approximately 75 day students. Green Chimneys offers a full array of psychological, therapeutic, and counseling services for children. These traditional therapies are interwoven with multiple opportunities for children to work on a farm with animals and plants, and assist with a licensed wildlife rehabilitation program.

The children who are referred to Green Chimneys may be deeply disturbed. They can be angry and withdrawn. In many cases traditional therapies have failed to resolve their problems. A large number of the children come from urban environments, and being taken to Green Chimneys literally transports them to another world. Once on the farm they confront sights, sounds, and smells that they have never encountered. Anger and acting out are not very helpful when you need to move a horse or cow from its stall. Working with the animals requires them to stop and listen, pay attention to the subtle behaviors (and some not so subtle) of the animals, and learn to communicate in a calm and effective fashion. Their *guides* on the farm may be a psychologist in boots and bib overalls, or another child who has been there longer and already learned how to do one job or another. The children will often develop a special interest and relationship with animals on the farm that were rescued from abusive situations by agents of the ASPCA. They seem to recognize and appreciate that these animals may need an especially gentle touch. In these cases, children and animals very much help to cure one another.

FIGURE 8-4

The Ross family bought an old farm to build a new kind of place for children. The green chimneys provided a name and a landmark. *(Courtesy Green Chimneys)*

SERVICE DOGS

We have all seen films or read books where bloodhounds are in hot pursuit of a criminal newly escaped from prison. The dogs bay, their noses on his trail, they drag their handlers through the woods, up and down mountains until they get their criminal. Without the amazing sense of smell the dogs possess (they have 20–40 times more scent receptors than humans depending on the breed) the escape might have been successful. Lost in an avalanche, there would be little hope of survival if it were not for dogs that are able to smell a human several feet beneath the snow. Brave dogs worked tirelessly, searching for the living or the dead following the destruction of the World Trade Center buildings on September 11, 2001, and little is more handsome or authoritative than a German shepherd guarding a government building. Dogs work side-by-side with humans in the service of **search and rescue**, doing other forms of police work and in the hazardous conditions of war.

Search and Rescue

Dogs trained in SAR techniques can find lost people whether they are living or dead. They can search under water, through snow, collapsed buildings, or the deep woods. They may also use their abilities to track crime suspects or escaped criminals (Figure 8-5).

Tracking and Trailing

There are two major ways for a dog to follow the trail of a person, although they are really on two ends of a continuum. **Tracking** is when the dog follows the exact path the person has taken. **Trailing** is when the dog can pick up the

FIGURE 8-5

Search and rescue dogs are now an important part of disaster response teams. Trained dogs and their handlers respond to both natural and man-made disasters. *(Courtesy ASPCA)*

person's scent, even when it is dispersed by wind, rain, and temperature and the actual path is unclear. Dogs can even trail people in cars, from the scent that blows out of the window or through the vents of the car. It is not true that dogs will lose a scent if it passes through water. If the water is still, the scent will remain on the surface. But even if the scent has dissipated on the water's surface, the dog will find the scent on the other side of the stream or river.

The best way for a dog to get a solid, uncontaminated scent is to smell an article of clothing, preferably underwear, a shirt, or socks that have been close to the person or at least have been handled by the person they are seeking. The dog will then pursue the same scent. Often the material has been handled by others, contaminating the scent. However, even then the dog can decipher which smell belongs to the person it is meant to track, because the scent is no longer around it and the smells of other people (e.g., police officer, parent of the missing child) can still be smelled. It will then follow the one smell that is no longer in the space with him.

There are some terms to be used for tracking. A *track solid* dog follows a track, usually the newest one. A *track sure* dog will follow the track associated with the scent he started with, and will not follow a track laid by a different person as long as the second track was laid at a different time. A *track clean* dog will follow the correct trail even if it crosses other trails laid at the same time. For example, for disaster work (finding victims in rubble), dogs lead their handlers toward any human scent. This is tracking solid. A bloodhound given a scent article will track clean, finding that same individual regardless of whatever crosses the track it is following.

Since the sixteenth century, bloodhounds have been selected as the best dogs for performing difficult and long searches. They are capable of covering a great distance and the loose skin on their face allows them to cup and catch even the faintest scent. Their stubborn and patient nature allows them to stick to trails that are miles long or take days to cover. Shepherds and Labradors do not do as well on old or long trails, but they work off leash better than bloodhounds, so they can work more quickly if there is a need during a search or when the dog must scramble over debris as in a building collapse.

When no scent article is available, as when someone is lost in a wilderness area and there is nothing available that belonged to the person, some dogs can *air scent*. This means they can detect human scent in the air and pursue that odor, hopefully to find the missing hikers. **Patrol dogs** can also use the technique to find anyone hiding in a confined area such as a building.

The United States Department of Agriculture (USDA) uses beagles to search for contraband fruit, produce, or other agricultural products at airports and other ports of entry. Small, agile, with excellent noses, the beagles scramble over suitcases with their jaunty tails waving. Travelers are seldom put off by these cheerful hounds as they go about their work.

Some search and rescue dogs are trained to search through rubble for people looking for any human scent. Avalanches and airplane and train accidents are all examples of sites where these kinds of trained dogs are employed. Search and rescue dogs are also adept at finding cadavers, new or old. They can even help at archeological digs to locate old graves or collect all the bones scattered in an area (<http://www.K9web.com>).

Air scenting search dogs are also able to locate victims of drowning (<http://www.ardinc.org>). They are most effective in those cases where a large area of water that requires searching must be narrowed down. It is not clear how scent reacts in the water, but it is assumed that the body gases are carried to the surface in oxygen bubbles. Underwater currents, temperature, and the wind most likely have effects on the scent that is available from the submerged person. But even with all the uncertainties, dogs usually are able to narrow the search area to within 100 feet or less of the drowning victim's location. In cases where dogs have *alerted* and subsequently missing people have been found, the depth of water ranges from 4 to 60 feet deep and the time since the drowning has been from a few hours to up to three weeks. Once the dogs have identified a general location, divers or sonar devices are helpful in pinpointing the exact location of the body so it can be retrieved. Interestingly, it is easier for a dog to locate a body in water 35 feet deep or if it has been submerged for 48 hours. This may be as a result of built-up body gases.

Some dogs are trained as **water rescue** dogs. They retrieve from below the surface of the water. The dogs will jump into a lake, ocean, or river, locate the person in distress, allow them to hold on to their side or the feathers on the tail, and tow them back to land. Some times the dog can bring a life ring and rope to the person in trouble and bring them back to shore.

Another subset of SAR is the detection by **narcotics dogs** and **evidence dogs**. These dogs are trained to use scent to alert to the presence of controlled substances at customs or at border crossings. The dogs can search through luggage, cars, buildings, and even have been known to outsmart smugglers by finding illegal drugs in rotting food, gasoline, or other strong smelling material. The olfaction of the dogs can be so acute that they are able to smell small quantities of material as a car passes by (as they might through a border crossing). With such an alert, police can stop the vehicle later so the dog and handler can keep up with their work.

Drugs are certainly important contraband to locate and keep off the streets, and dogs do a great service by helping customs officers and police. They are also very effective, especially in today's environment, to alert if there is the presence of a bomb or other explosive device.

Training these dogs is a challenge because the illegal materials, drugs, explosive devices, and other dangerous materials are not readily available. Only certified law enforcement personnel can train the dogs by obtaining small quantities of the contraband. This is also true of cadaver search dogs; the parts of cadavers must be obtained legally. If the dog does not have a résumé of training by law enforcement, and there is no record that the training material was legally obtained, then evidence cannot be accepted in a court of law.

BOMB-SNIFFING DOGS

Dogs do not actually sniff a bomb; they are smelling molecules of any number of explosive materials they have been trained to identify. The dogs, German shepherds, Labrador retrievers, vizlas, or Belgian Malinois, can *sniff* within a five-foot radius. Most of the techniques used to train these dogs are classified by the United States government and the majority of the dogs are trained at Lackland Air Force Base in San Antonio, TX. Before a dog and handler can be put to work at an airport, sports stadium, department store, subway, or any other location there might be the threat of an explosive device, there is a long period of training. First the human is sent for a three-month course. There, his classroom training is supplemented by training with experienced dogs. After the introductory period, the officer is assigned a dog and the new partners are schooled together. After almost a year, they are ready to go out to protect the public. Even after years on the job, they are tested on an annual basis to make sure they are sharp.

For Bara, a bomb-sniffing dog who is a member of the Massachusetts State Police Canine Explosive Detection Program, the work day is ten hours long. On an average day she and her trainer may be asked to check five or six situations at the airport. And then, at the end of the work day, there is even more training so both handler and dog stay fresh and skilled. Like many bomb-sniffing dogs, Bara is a *passive responder*. This means that if she should detect an explosive she will go sit in front of it and wait quietly for her reward (<http://www.Petplace.com>).

Dogs can also use their amazing sense of smell to fight our battles with human disease. There is a growing body of evidence that dogs are able to detect some cancers even before modern medical devices can (Smith, 2004).

Patrol Dogs

We often see police and soldiers standing guard or patrolling city streets, airports, government buildings, and train stations with a patrol dog partner at their side. The dog is generally a powerful dog, like a German shepherd, Doberman Pinscher and Rottweiler. Their size alone, and the knowledge that impressive teeth and jaws await any human aggressor, makes them a strong psychological deterrent, as well as a physical one.

Any dog working for the police is a *police* or patrol dog, including all of the categories of SAR just covered. But usually the term *police dog* refers to the dogs trained to chase suspects in a crime and hold them for arrest. Of course, dogs can be trained, and are, to multitask. They can smell illegal drugs, track a person, and then hold the person. Most dogs will go to *bark and hold*. This means that the dog will bark at the suspect and not attack unless the person tries to escape or to hurt the dog. There is controversy over this; many officers believe that it is best if the dog attack immediately. No matter how the dog reacts, the handler must have full control of his dog partner and call him off with one command (<http://www.k9web>).

The organized use of patrol dogs is fairly recent; the first police department to use them was the Orange, NJ, department in 1907. Others followed, including the Royal Canadian Mounted Police K-9 Section, established in 1937 (Chapman, 1990).

There is no standard for protection work, but the National Association of Protection Dogs is trying to establish national standards for protection work, and educate the general public about these dogs.

SCHUTZHUND

The **Schutzhund** (or *protection dog* in German) is used to describe a German shepherd who is trained to utilize and maintain the qualities in the breed that make him a useful, safe, and happy companion. In addition, Schutzhund also refers to a very stringent series of tasks designed to test tracking, obedience, protection abilities, and courage. In Germany, one is not allowed to breed a German shepherd and register the puppies without a *degree* in Schutzhund. The Schutzhund trials are a competition to demonstrate the dog's obedience and courage. The training is extremely difficult, requiring both a very skillful trainer and motivated, intelligent dog. And although the most common dog is a German shepherd, other breeds may pass the trials and therefore can be called a Schutzhund. Since the 1920s, other dog breeds also participate in the Schutzhund trails, including Doberman pinschers, Bouvier des Flandres, rottweilers, boxers, giant schnauzers, and other larger working breeds that have the personality and intelligence to master the demanding curriculum that will earn them the title of Schutzhund.

The Schutzhund training and breeding program was developed in the 1920s in Germany by the Shaeferhund Verein (German Shepherd Dog Club) to maintain the excellent traits of ability and temperament required of the breed. There are three major degrees awarded—SchH I, SchH II, and SchH III, in order of increasing difficulty. The traits that make a good Schutzhund candidate are mostly innate characteristics that must be bred for. Even among dogs from Schutzhund bitches and dogs, few have the necessary drive, intelligence, and strength to achieve a SchH III title. In addition to family lines, early development is key to success. All training and play has to be positive, with no strong corrections and no domination by another dog. So, just as it is important to train a dog of the proper temperament it is also critical to make sure a trainer uses the proper techniques and does not train with violence and fear.

Many people believe that a Schutzhund is trained only for protection. This is often true, but since the training and the completion of the program require many skills to be mastered, the dog will make an excellent protection dog for the work place, the home, or as a companion for the humans who understand this quality of the canine and how to care properly for them.

Protection dogs need not be aggressive by nature. To the contrary, excellent protection dogs must be of the best temperament: stable and trustworthy around people. A dog must have confidence and a steady disposition, not a ferocious angry personality. In the same vein, a shy dog is also a poor candidate for protection work or Schutzhund training because the stress of the training will prove to be too much for the dog. The best and safest protection dog will have a true understanding of what is being asked of it, will be trained with consistency and kindness, and will have the innate personality and breeding to have a balanced temperament, just as a person who is absolutely sure of himself or herself need not *show off* their weapons or fighting skills. They just know that if put to the test, they will prevail. The good protection dog carries himself with confidence, knows what to do, and has no need to be aggressive. Their bite is worse than their bark, so they can just stand quietly.

MILITARY

Among the least appreciated working dogs are those that serve in the military. Since Roman times, when razor sharp collars were put on the dogs so they could attack the enemy to bite and cut, dogs have been soldiers fighting with their army. During the Revolutionary War, dogs served as pack animals, and in

World War I their job was primarily to kill rats in the trenches. It was not until the Second World War that K-9 units were formally organized so their strength, loyalty, intelligence, and superior sense of smell could be used to full advantage (<http://www.defenselink.mil/news/Sep2004>).

In 1942, the first training program was established at Front Royal, Virginia, to prepare dogs to help in World War II. Established by the American Kennel Club and a new organization, *Dogs for Defense*, dogs were obtained and trained to serve in the conflict oversees. Without a breeding program or established program, the American public was asked to donate dogs they felt would be suited to the purpose of war. A patriotic country sent over 19,000 dogs of 30 breeds to be evaluated and trained. Only about half of those dogs were considered suitable. At first it was thought only about 200 dogs would be necessary, but soon it became evident that they were saving the lives of soldiers and the program was very successful. Soon the war Dog Program was training their own dogs and also for the Navy and Coast Guard. Training lasted between two and three months before they were sent into battle. They were trained in basic obedience as well as to ride in military vehicles, to work during gunfire, and to wear muzzles and gas masks. After this period of *basic* training they were sent through specialized training. Then and today they serve the following functions in the military.

Sentry Dogs

These are worked on a short leash and taught to give warning by growling, alerting, or barking. They are especially valuable for working in the dark. They accompany a military or civilian guard on patrol and give warning of the approach or presence of strangers.

Scout or Patrol Dogs

In addition to the skills listed for **sentry dogs**, scout patrol dogs are trained to work in silence in order to aid in the detection of snipers, ambushes, and other enemy forces.

Messenger Dogs

The most desired quality in **messenger dogs** is loyalty, since the dog must be motivated by the desire to work with two handlers. They learn to travel silently and take advantage of natural cover when moving between the handlers.

Mine Dogs

Mine dogs called the M-dogs or mine detection dogs are trained to find trip wires, booby traps, and metallic and nonmetallic mines (Figure 8-6).

Of the 10,425 dogs that were trained during World War II, 9,300 were for sentry duty. Most were issued to the Coast Guard for beach patrols guarding against enemy submarine activities. Dogs on combat patrols often could sense the enemy 1,000 yards ahead. The dog would raise his hackles, prick his ears, and stiffen the body. His ability saved many lives since the chance of ambush was greatly reduced. This was especially important and effective in the islands

FIGURE 8-6

Land mines are one of the lingering dangers of modern warfare. Trained mine detection dogs easily outperform any mechanical or electronic device yet created. The ideal dog is careful and controlled in its search for the buried explosives. *(Courtesy Dr. Randall Lockwood)*

of the Pacific. Units with dogs suffered far fewer casualties as a result of ambush in the jungle or surprise attacks. After the war, the dogs were returned to their original owners. If they were not wanted, then they were adopted to other families or stayed in the military to train other dog handlers.

Today, the 341st Training Squadron (341TRS), Lackland Air Force Base, San Antonio, TX, is the premier Military Working Dog School. The mission of the Department of Defense Dog Center is to provide trained military working dogs and handlers for the Department of Defense (DoD), other government agencies, and allies through training, logistical, veterinary support, and research and development for security efforts worldwide. An estimated 2,300 dogs are part of the DoD Military Working Dog Program. The 341TRS provides basic working dog handler instruction to over 525 students annually. Additionally, they train over 300 working dogs for all of the DoD and Federal Aviation Administration (FAA). Its four major sections provide logistic, dog training, handler training, and medical support to the entire DoD. Most K-9 teams are deployed for six months rather than a year after it was discovered that after that amount of time the dogs do not perform as well, many of them suffering from the stress of combat (<http://www.defendamerica.mil>).

Throughout their careers, the working dogs receive excellent veterinary care, with checkups every six months. But, as with any athlete, the day comes when they must retire. Only recently, in the year 2000, a law was passed allowing retired military dogs to be adopted by former handlers or other qualified people. Ronny was the first dog to be retired to his former handler, moving to North Carolina from the Lackland Air Force Base. Prior to this, military dogs were used to train new human handlers or adopted to law enforcement facilities.

The number of dogs in the military is currently increasing steadily. They serve an important role in the war on terror, detecting explosives, working in the field, and serving as a great psychological deterrent to the enemy. But, even for the dogs, the stress of combat is intense and they are deployed for only six months. After that period of time in active duty they are no longer sharp and need a rest. There are several hundred military dogs in the Middle East as well as those handled by private contractors. The dogs do their jobs well, always eager to please their handlers, and the military is eager to assure the health of this excellent resource. Research is underway to design an effective gas mask for the dogs as well as body armor and a pill to help military working dogs survive a nerve agent attack. It seems that the invention of a device to replicate the olfactory ability of a dog will be a very long time in coming, and the ability of the dog to be a great soldier will never be surpassed.

CONSERVATION DOGS

New areas are constantly being found where dogs can work with human partners. A more recent example is the training of **conservation dogs**. Through **Working Dogs for Conservation Foundation (WDCF)** dogs are helping to obtain data on hard to track endangered species. Using their keen sense of smell to track wildlife, dogs have joined environmental research teams. Even when the animal the research team is trying to study cannot be found because it is scarce, or the terrain is difficult, the biologists can still learn a lot about the species' range and behavior from their tracks, hair samples, and particularly their scat (fecal matter). With technological advances the analysis of scat can provide information about the abundance of the species, the sex of the animals, parasite load, and often familial relationships.

WDCF has trained dogs to help locate a variety of species (seals, fox, turtles, snakes, black-footed ferrets, bears, and termites). Even in ecosystems that are hard to navigate for humans, the dogs are able to find burrows, follow trails, and discover other scent markers left by the target population. The conservation dogs are currently being used in projects in North America, Asia, Africa, and Micronesia. In the past several years they have provided assistance around the world in a variety of conservation efforts. In Kenya, Africa, teams provided assistance in scat detection of wild dogs and cheetahs for the Samburu-Laikipia Wild Dog Project and the Tanzania Carnivore Project. In Russia they do research on discrimination scent testing in collaboration with the Amur Tiger Scent Dog Monitoring Project and Save the Tiger Fund, and here in the United States the dogs have joined with the United States Fish and Wildlife Service Law Enforcement, and the Montana State Fish, Wildlife and Parks to search for illegally killed wolves. In California they search for kit fox dens and bobcat and pine snakes in New Jersey (<http://www.workingdogsforconservation.org>).

DISCUSSION QUESTIONS

1. What is the Delta Society? How do its programs contribute to the lives of people who are ill or disabled? How do its programs contribute to the lives of the general population?
2. Where and why was the first guide dog school formed? What was the first training program for guide dogs in the United States, and when was it established?

3. What breeds are commonly used as guide dogs? Is this different from the breeds commonly used as hearing dogs?
4. What characteristics must a therapy dog possess? What training does a therapy dog undergo?
5. What is the difference between tracking and trailing?
6. During which world war were dogs formally organized and trained to serve? How was the American public involved in this effort? Is the number of dogs currently used in the military increasing or decreasing?

REFERENCES

Allen, K., & Blascovich, J. (1996). The value of service dogs for people with severe ambulatory disabilities. *Journal of the American Medical Association, 275,* 1001–1006.
Americans with Disabilities Act of 1990. (1990). Retrieved August 14, 2007, from www.ada.gov/svcanimb.htm.
Chapman, S. G. (1990). *Police dogs in North America.* Springfield, IL: Charles C. Thomas.
Levinson, B. (1997). *Pet-oriented child psychotherapy* (2nd ed.). Revised and updated by G. P. Mallon. Springfield, IL: Charles C. Thomas.
Smith, P. (2004). Dogs may be able to detect some cancer tumors. *Pittsburgh Post-Gazette,* Retrieved October 26, 2004, from http://www.postgazett.com./pg/04300/401479.stm.

Service Dogs
<http://www.cofc.edu> Service and Therapy Dogs
<http://www.workingdogs.com/book001.htm> Working Dogs Book Store
<http://www.dogplay.com/resources/books> Dog-Play: Books and Resources

Therapy Dogs
<http://www.k9web.com/dog-faqs/working.html>
<http://www.ardainc.org/watertss.htm> Water Sniffing Dogs
<http://ask.yahoo.com> Bomb Sniffing Dogs
<http://www.petplace.com>
<http://seeingeye.org>
<http://guidingeyes.org>
<http://www.hearingdogs.org.uk>
<http://www.ecad1.org>
<http://www.adionline.org> Assistance Dogs International

Military Dogs
<http://www.defenselink.mil>
<http://www.militaryworkingdog.com>

Therapy Animals
<http://www.deltasociety.org>
<http://www.greenchimneys.org> Green Chimneys Children's Services, Inc.
<http://www.workingdogsforconservation.org>

CHAPTER 9

Companion Animals in the Home

You enter a certain amount of madness when you marry a person with pets.

–Nora Ephron

KEY TERMS

acquisition	labradoodles	Harmony Neighborhood Charter School
planners	cockapoos	counseling
impartial	humane education movement	biophilia
smitten	George Angell	
designer dogs	Bands of Mercy	
puggles		

More often than not, people welcome companion animals in their homes to be members of their families (Friedman et al., 1984; Hirschman, 1994). Americans acquire dogs, cats, and other companion animals from a wide range of sources. It may be surprising, but pet stores and animal shelters are not the most common source of dogs and cats in American homes (Table 9-1). The most common source of dogs (32%) and cats (43%) as pets are friends and relatives, usually for little or no cost. Almost a third of dogs are purchased from breeders (31%), while very few cats come directly from breeders (3%). Animal shelters account for 16% of dogs and 15% of cats, and many cats (34%) end up in homes after being taken in as strays (APPMA, 2005). These data indicate that a large number of dogs and cats are acquired with limited opportunity to provide education on responsible pet care. Birds, fish, small animals, and reptiles are more frequently purchased at a variety of retail outlets, including pet stores, pet superstores, and specialty stores. Where friends and relatives acquired the pets they passed on to the individuals surveyed is not known from these data. However,

TABLE 9-1

Common sources of companion animals (APPMA 2005, p. XXIII)

SOURCE	DOG %	CAT %	TOTAL FISH %	BIRD %	SMALL ANIMAL %	REPTILE %
Adopted from a pet superstore	1	3	—	—	—	—
Animal shelter/humane society	16	15	—	2	—	7
Breeder/hatchery/animal farm	31	3	1	16	7	5
Discount store	—	—	15	—	7	—
From a friend/relative	32	43	10	34	28	17
Gift	6	3	6	8	—	—
Internet/online	<.5	—	1	—	1	1
Newspaper/private party	13	8	—	5	4	3
Pet store	6	4	27	19	34	38
Pet superstore	—	—	20	8	16	9
Previous owner/from own pet/bred at home	5	15	7	5	27	9
Rescue group	5	3	—	—	—	—
Specialty store (fish or bird)	—	—	33	22	—	—
Stray/caught or found outside/flew into yard	9	34	2	5	4	17
Veterinarian	>.5	2	—	1	—	—

New et al. (2004) note that both dog- and cat-owning households produce large numbers of unplanned litters, yielding puppies and kittens that can be passed on at little or no cost to new owners. The question of **acquisition** cost is significant because dogs and cats acquired for little or no cost are at increased risk for relinquishment to an animal shelter (Patronek et al., 1996a, 1996b; Salman et al., 1998). If you look at both the frequency with which new cats are acquired by taking in a stray and the frequency with which cats leave households by wandering away, it is hard to escape the conclusion that some part of the cat population is cycling between various homes.

Further analysis of relinquishment data has focused on behavioral issues of pets (Salman et al., 2000), human health and personal issues (Scarlett et al., 1999), and moving (New et al., 1999). Taken as a group, these analyses reveal that relinquishment of pets to an animal shelter results from a complex mix of reasons. Behavior problems are the most common reason for the relinquishment of dogs and the second most common reason for cats. Among the common behavioral reasons for cat and dog relinquishment were biting and other forms of aggression, household destruction, house soiling, and problems with other pets. These data are consistent with the data collected from a free behavior help service run by the American Society for the Prevention of Cruelty to Animals (ASPCA). Another key element noted in the relinquishment studies was the presence of other animals in the home. Multiple-pet homes were more likely to relinquish a pet to an animal shelter.

Pet behavior may have an effect on the level of attachment that pet owners have for their companion animals (Serpell, 1996). Weakly attached owners are more likely to be dissatisfied with their pet's behavior. Understanding the relationship between pets and people and its influence on retention in the home is further complicated by studies that suggest that owner personality and behavior are associated with dog behavior problems (O'Farrell, 1997; Podberscek & Serpell, 1997). Among other things, owners of aggressive dogs were found to be more tense, emotionally less stable, shy, and undisciplined (my personal experience working with owners to correct behavior problems is consistent with these reports). Additional research will be required to elucidate these relationships more fully. Especially important will be determining the extent to which owner personality and behavior influence that of a pet, and how or whether human variables influence pet choice.

Analysis of health and personal issues indicates that households in flux are at risk for relinquishing a pet. Allergies, illnesses, new jobs, and births and deaths of people in the household were all identified as reasons for relinquishment of dogs and cats. Overall, if these data are combined with an evaluation of moving as a reason for relinquishment, one can develop a picture of households that are unstable for a variety of reasons and conclude that relinquished pets are the victims of instability. These households were more likely to add pets to their homes, and often kept one or more pets, even when they were relinquishing one or more to an animal shelter. Young adults are over-represented among those relinquishing pets to animal shelters. This is not surprising since this is the time in life when people are most likely to move (U.S. Department of Commerce, 1995), as well as change jobs, get married, and have children.

When relinquishment data are combined with the acquisition data, it is clear that many companion animals change hands through friends and relatives or as strays in an informal network outside of the framework of retail sales, direct purchase from breeders, or adoption from an animal shelter. This has implications for both animal welfare and pet product commerce. Limited opportunities will be available for education on responsible pet care or product sales at the point of acquisition.

HOW PETS ARE CHOSEN

Leslie Irvine is a sociologist who conducted research at an animal shelter. She was struck by how people connect with animals and developed three general descriptions of the process (Irvine, 2004). The first group that she describes was the **planners**. These are people who know what they are looking for in terms of breed, size, and color. They may be trying to replace a companion animal from the past, or continuing a family history with a particular type of pet. The second group was the **impartial**. These people have generally had animals in the past but have few explicit preferences. They are open to whatever might be available, just looking for something that will be a good pet and member of the family. The third group was the **smitten**. People in this group describe an irresistible pull to a particular animal. The dog or cat may be nothing like what they had been looking for when they came into the shelter. These animals may have some distinctive markings or distinguishing characteristic. While the research was conducted at an animal shelter, her observations would probably apply to other venues for acquisition. It would not be difficult to see how these ways of choosing a pet could be played out under other circumstances. It would certainly seem that people who go directly to breeders would be planners; they know what they want and what they are looking for. Strays may come into someone's life when they are ready to accept a new companion, with no particular type in mind; they are impartial. The smitten could, of course, find a pet in any of the various sources. It is indeed a special moment when something just feels right and you know that this dog or cat is the one for you. It will be interesting to someday follow up on people who acquire pets in these different ways, and know whether there are differences in how likely the pet is to stay in the home for the rest of its life.

Trends

Hal Herzog and colleagues have examined trends in the popularity of dog breeds. Herzog et al. (2004) examined the AKC registration numbers between 1946 and 2001. While these registrations do not represent all dogs, and not even all purebred dogs since not everyone registers a purebred dog that they purchase, they do provide a reasonable picture of dog breed popularity. They found that breed popularity has followed a random copying model similar to that used to study other forms of cultural transmission (Cavalli-Sforza & Feldman, 1981), based on the random drift model of population genetics. The popularity of dog breeds is largely the result of random chance. People are likely to choose dogs that copy the choices of friends and others. This is similar to how changing tastes and fads in clothing and music are transmitted through a community or population. Someone wears their baseball cap with the brim turned backwards, someone else sees it, thinks it looks cool/hot/phat (the choice of descriptors here has also followed a similar drift model over the past 25 years!), and copies the style. As a particular style or choice becomes more common, it is more likely to be copied and maintained. In the same way, when someone shows up at the dog run with a breed that has not been seen, other people seeing that dog will ask about it, and may copy the choice when they acquire a new dog. The authors also note that there have been cases where cultural events have had a profound influence on the popularity of a dog breed. One example is the spike in registrations of Dalmatian registrations in the years following release of the Disney film *101 Dalmatians* in 1985. In 1985 there were 6,880 new Dalmatians registered. By 1993 there was a greater than sixfold increase in registrations, peaking at 42,816. By 1999, however, the registrations for Dalmatians had declined to their historic levels, with 4,652 new registrations.

Herzog and Elias (2004) further evaluated the effect of a high-profile and popular dog event on breed popularity, the Westminster Kennel Club Dog Show. Westminster is the nation's oldest dog show, and enjoys a national television audience of over five million viewers, as well as substantial coverage of the winner in the popular press, news, and television. They examined registration numbers for the four years before and five years (counting the year of winning) after a breed won Best in Show. They found that winning the Best in Show has no impact on breed popularity. The clear conclusion is that the winners of Best in Show do not reflect American's preferences for pet dogs.

More recently, a variety of **designer dogs** have become popular (Kelly, 2006). These are intentional hybrids of two different purebred dogs. It is thought that these types of mixes will avoid the inherited diseases and problems sometimes seen in different dog breeds. These dogs also provide a unique and different look. Common hybrids are **puggles**, which are a mix of pug and beagle, **labradoodles**, a mix of Labrador retriever and poodles, and **cockapoos**, a mix of cocker spaniel and poodle. The fact that a number of celebrities have acquired one or another type of designer dog has no doubt helped to encourage their popularity.

CHILDREN AND ANIMALS

Children and animals seem to be a natural combination. There is a long history of literature and folklore that links children with animals, and how this can help to promote kindness in children (Zawistowski, 1998). This is reflected in the earliest roots of children's literature. In the late 1700s and early 1800s, when literature specifically created for children first began to appear, animals were frequently included as part of the story. These early stories would also include a moral theme. One such example, published in 1798, "The Life, Adventures, and Vicissitudes of a Tabby Cat" included a description of a cat having its ears cut off with a pair of scissors by a terrible young man. Other stories included stealing eggs from birds' nests, and other forms of cruelty to animals (Figure 9-1). Most important, however,

FIGURE 9-1

The Story of the Robins is representative of early children's literature that combined stories about animals with moral messages for children. *(Courtesy Dr. Stephen Zawistowski)*

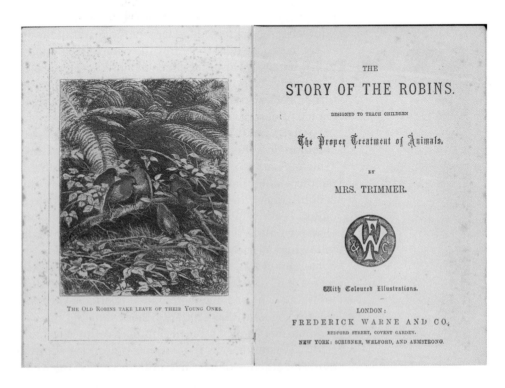

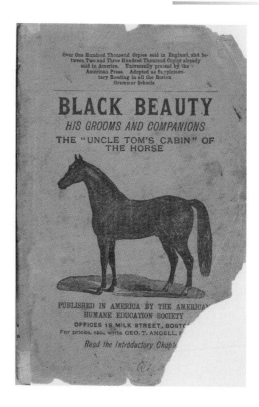

FIGURE 9-2

Black Beauty was considered the *Uncle Tom's Cabin* of the horse because it brought the rampant mistreatment of horses to wide public attention. George Angell, founder of the Massachusetts SPCA, sponsored the publication and wide distribution of the book in the United States. *(Courtesy Dr. Stephen Zawistowski)*

was that as part of the story something terrible would usually happen to the children or people who were cruel to animals, and the children who were kind were usually rewarded. The most famous such story was *Black Beauty* written by Anna Sewell in 1877 (Figure 9-2). Eventually this approach would coalesce into the **humane education movement**. **George Angell**, founder of the Massachusetts Society for the Prevention of Cruelty to Animals, was the leader in this effort (Angell, 1884). Angell's father had been a minister but died when George was a young boy. His mother supported her son and herself as a schoolteacher. When George was a young man, he would also teach school while working his way through college and law school. This had a profound impact on his approach to preventing cruelty. He strongly believed that the most productive approach would be to teach kindness to children. His approach found fertile ground in the educational establishment of the time (Good, 1956; Spring, 1985). *McGuffey's Newly Revised Eclectic Reader* was published in 1843 and included many stories about animals and nature. True to the prevailing theme of children's literature, these stories also reflected moral themes. In this same era, the *common school* philosophy of Horace Mann dominated thinking in the education field. This approach emphasized the importance that public education could have on well-informed citizens who shared a common knowledge of morals and culture. The public school would be the place where children of different backgrounds would come together to learn and appreciate the American experience. There was a strong belief that this type of education would be able to help solve many of the social problems that faced the nation.

In 1882, Angell began to organize **Bands of Mercy** in the schools in Boston. These after-school clubs encouraged children to learn about animals and how to care for them. Meetings would include recitation of a prayer, testimony about animals the children had helped in the past week, and singing the Bands of

Mercy hymn. The idea spread rapidly across the country. By 1883, there were 600 Bands of Mercy found in schools across the country, with a membership of 70,000. In 1889, Angell formed the American Humane Education Society to promote work in this area. In 1923, there were over 140,000 Bands of Mercy with over four million children enrolled. Over time, the Bands of Mercy dwindled and disappeared. Changes in the education system and new and different demands resulted in schools discarding these quaint relics of another era.

These early humane education efforts did leave a legacy, enshrined in state laws across the country that require the teaching of humane education (Antoncic, 2003). Thirteen states currently have humane education laws (California, Florida, Illinois, Louisiana, Maine, New Jersey, New York, North Dakota, Oregon, Pennsylvania, Tennessee, Washington, and Wisconsin). In Louisiana, Maine, New Jersey, and Oregon these laws reflect a legislative intent regarding the importance of humane education, but they do not make the teaching of humane education mandatory. In the other nine states, teaching humane education is mandatory, though no additional funding is provided for this purpose. As an example, the New York State law applies to all elementary schools under the control of the state or funded by public money. The law is generally understood to require weekly instruction in humane education. New York is the only state to have a penalty; schools that fail to comply with the requirements may lose funding. This penalty provision has never been tested. In fact, it is difficult to determine whether these humane education laws are taken seriously by school administrators or teachers. While these laws are seldom enforced, they do highlight the importance that humane education had at one time. They have also been resurrected in a number of places as part of new developments in education that focus on moral and character education. In an effort to combat the increasing levels of violence seen in schools and communities, educators, parents, and others are introducing new programs to encourage kindness and caring. What they are finding is that George Angell was ahead of them, and that humane education and teaching kindness through proper care of animals can be an important part of the education mainstream once again.

Measuring the impact of humane education has proven difficult. While short-term studies have been able to demonstrate an impact on attitudes, it is more difficult to measure the impact on behavior. Ascione (1992) examined the effect of a humane education program on 32 first-, second-, fourth-, and fifth-grade classrooms. The humane education intervention consisted of a curriculum developed by the National Association for Humane and Environmental Education (NAHEE—a branch of the Humane Society of the United States), and the *Kind News* classroom newspaper distributed by the organization. Students were tested before and after the education program for their humane attitudes. Fourth- and fifth-grade students showed significant improvements in their attitudes toward animals, with a generalized effect on human-related attitudes as well. The younger children did not show significant improvement. Ascione and Weber (1996) followed up one year later with the students who had been in fourth grade and were now in fifth grade. Their results showed that students who had received the humane education program retained higher levels of positive attitudes toward animals than students who had not received humane education training.

A long-term study is now underway at a charter school in Harmony, FL. The **Harmony Neighborhood Charter School** was conceived to focus on humane education linking all elements of the curriculum. The principal and teachers have worked with outside consultants from the Jane Goodall Institute, Albert Schweitzer Institute, and ASPCA to incorporate humane education themes into math, science, language arts, and other subject areas. Each year students are tested on their knowledge and attitudes toward animals and the environment. While this is a small-scale effort in a special setting, it may prove to provide some of the most helpful long-term data on the impact of humane education on children.

Harmony, FL

Located on 11,000 acres between Orlando and Melbourne beaches in Florida, a remarkable community is taking shape (www.harmonyfl.com). Harmony, FL, will eventually be home to about 18,000 residents in a mix of single-family homes, townhouses, and apartment units. Harmony is a dream becoming real under the leadership of Jim and Martha Lentz. The unique vision of Harmony is its effort to be animal friendly. Over 70% of the land will be left undeveloped as a refuge for native wildlife. Even those areas that are developed will be done in a way that is friendly to the animals and to the environment. No homes will be built on the shores of the two 500-acre natural lakes in the community. Shorelines and wetlands around the lakes will be preserved and protected. While boating on the lakes will be permitted, only electric motors will be allowed. Companion animals are part of the plan, including a beautiful dog park for safe off-leash play (Figure 9-3). Integral to the planning of Harmony and its programs and amenities to benefit the human–animal bond is the Harmony Institute (www.harmonyinstitute.org). Composed of leading academic authorities on animals

FIGURE 9-3

The dog park at Harmony features a beautiful lawn and agility equipment for dogs and people to have a great time. Small dogs have their own separate area where they can play without being bullied by bigger dogs. *(Courtesy Dr. Stephen Zawistowski)*

and their interactions with humans, the Harmony Institute Campus Advisory Board (HICAB) provides advice for the community and its development. Members of HICAB will also help to design and carry out research related to human–animal relations in the community. Among some of the planned Harmony Institute Programs are the following:

- Pet Concierge™—A program designed to maximize the benefits of the human–animal bond and intervene in emergency situations that might result in animal relinquishment. Efforts here might include providing advice on solving behavior problems or providing emergency foster care or other services that can help to keep pets in a home.
- WildSide Walk™—A blueprint for integrating natural habitat into communities.
- Center for Community Health—A collaborative and multi-disciplinary center to further human–animal and human–nature interactions as health enhancers.
- Homeowner Documents—Guidelines relating to animals, urban wildlife, and the environment that help to prevent and/or resolve potential issues and concerns.
- Petlife Home Environments—Special design features that make homes more suitable for animal companions and make caring for them easier.
- Community Design—Land design, pet parks, environmentally sensitive plantings and gardens, and lake management. This design includes lighting systems to limit the amount of ambient light that "escapes" into the night as part of the Dark Sky effort to prevent light pollution (www.darksky.org).
- Living in Harmony, A City for People and Animals—a two-year documentary conceived by HICAB board member Jennifer Wolch in collaboration with filmmaker Michael Tobias and Fitzgerald Productions.

CHILDREN AND PETS

While the influence of humane education on children may be difficult to measure, popular culture has seemingly embraced the concept that animals are good for children (Figure 9-4). It is something of a truism that having a pet can help teach a child responsibility. Melson and Fogel (1996) examined this concept by asking parents of pre-school, second-, and fifth-grade students about their children and their care-giving behaviors. They found that caregiving increases with age, and that the impact of pets may be gender neutral. They conclude that pet care may be particularly important to encourage the development of nurturing behaviors in boys. Daly and Morton (2003), on the other hand, did not find that pet ownership was associated with higher levels of empathy in children. They surveyed students from the fourth to eighth grades and found that students with pets scored no higher than students without pets on empathy. The degree of attachment that a child showed toward their pet was also not associated with higher scores on empathy. One difference that they did note was that children with dogs showed higher empathy than children with cats. The Daly and Morton study differed from the Melson and Fogel study in that they directly surveyed the children. Melson and Fogel, on the other hand, surveyed the parents. It may be that parents who observe their children with animals assume a positive, nurturing relationship between the child and pet. Paul and Serpell (1993) surveyed college students and found that concern for animal welfare was positively correlated with companion animal involvement in childhood. The number of pets that the students had while young and how important those pets were to them were positively correlated with humane attitudes as adults and their attitudes toward having pets. These humane attitudes carried over into concern for the welfare of laboratory and farm animals and wildlife.

FIGURE 9-4

Pets and kids seem to be such a natural combination. If parents take the time to train the children and the animals, everyone can have a great time. *(Courtesy Dr. Steve Hansen)*

It is abundantly clear that children who grow up with companion animals will have a greater interest in pets as adults, and will be more likely to have pets as adults (Serpell 1981). Individuals who have a pet earlier in childhood will have a more positive attitude toward having pets as adults than those who had their first pet later in childhood (Poresky et al., 1988). As in many other areas, parents also have a strong influence on their children's interest and attitudes toward pets. Children with parents who have a strong attachment to pets were more likely to report interest in pets and activities associated with pets (Kidd & Kidd, 1990a). Moving back to the discussion on humane education, it is interesting to note that children with pets were also more likely to read stories about pets, and when given a choice, they will do schoolwork assignments on animal-related themes (Kidd & Kidd, 1990b).

It is important to recognize that children of different ages will be at different stages of cognitive development and, as a result, will be capable of performing more complicated care routines only when they are older. Table 9-2 provides a general guide to the different stages that a child will pass through and types of companion animals most appropriate for them, and the range of tasks they can be asked to perform in the care of the pets. Different children will mature at different rates and some will be able to take on more extensive responsibilities at earlier ages. This is especially true in households where companion animals may have preceded the child's appearance in the home. Care must be taken to ensure the safety and welfare of both the child and the pet. In some cases, the animal itself might pose a risk to the child. Dogs may bite and cats may scratch a child who is unable to control his or her excitement and plays too rough or disturbs the pet while it is eating or sleeping. Zoonotic diseases are also a concern, especially for young children who are still working on their hygiene skills. In some cases the equipment associated with the pet can pose a risk. This is especially true for aquariums and terrariums. The electrical cables pose a tangle risk, and

TABLE 9-2

A guide for the interaction of children and pets *(Stephen Zawistowski)*

CHILD'S AGE	PRIMARY ISSUES	RECOMMEND	TASKS FOR THE CHILD	OTHER OBSERVATIONS
Infant	Introduction to current pets	N/A	N/A	Resident dogs and cats need gradual, supervised introduction to infants.
Toddler	Curiosity; pulling, touching, etc.	N/A	N/A	Care must be taken with dog food dishes, toys; litter boxes for cats; aquarium wires.
3–5 years	Learning about contact, empathy	Guinea pigs	Filling water bottle and food dish	Guinea pigs like to be held, seldom bite, and will whistle when excited or happy.
5–10 years	Attention span is variable	Shelf pets, goldfish	Clean cages with adult help; supervised play with dogs/cats	Adults should always check to ensure that pets have food/water and cages are secured.
10–13 years	Greater interest in pets and capacity for responsibility	Dogs, cats, rabbits	Feed pet; walk dog; clean rabbit cage; clean cat litter	Children this age can be reliable, but adults should always check on food/water, etc. Participation in dog training classes is an excellent learning opportunity for the children.
14–17 years	Competition for time and attention (i.e., sports, clubs, etc.)	Birds, aquariums	Most tasks; use allowance to buy treats, etc.	Developing interest as a fancier, more likely to do research or read about the species. Parents should note that dogs and cats acquired at this age will probably stay in the home when the child leaves for college, etc.

electrocution is a danger, especially with water around. Aquariums, terrariums, bird cages, and other enclosures may be tipped over by overly curious children. Children will also need instruction and training in how to approach, pick up, carry, and otherwise interact with companion animals. Small animals can be injured by children who hold them too tightly; they may also bite to defend themselves. While caring for a pet can be an excellent demonstration of responsibility by a child, it remains paramount that an adult take ultimate responsibility for the health and welfare of companion animals in a home.

ANIMALS AND HUMAN HEALTH

There is a substantial literature related to the health benefits of having companion animals. In one of the classic studies, at the University of Maryland, researchers examined the influence of a variety of factors on the survival of heart attack patients after discharge from a hospital coronary unit (Friedman et al., 1980). They investigated the role of income, neighborhoods, and social contacts, such as friends or relatives. Ninety-two patients were enrolled in the study and 14 died in the first year after being hospitalized. When they analyzed the data, they found that the initial social variables that they had set out to study were associated with higher survival rates. However, they were surprised to find that patients with pets at home were three times more likely to survive than patients without pets. The influence of having pets was equivalent to the other social support variables that were studied. The authors of that study point out that pets are not a panacea, but can be part of an overall healthy lifestyle that supports survival after a heart attack. The mechanism of this enhanced survival is still undetermined. However, other studies have shown that animals can have profound effects on cardiovascular function. Katcher et al. (1983) found that simply observing an aquarium with fish in it could reduce an individual's physiological response to stressful situations. Additional studies summarized by Friedman, Thomas, and Eddy (2000) show that having a dog in a room, talking to a dog, or petting a dog can have a beneficial impact on blood pressure during some types of stressful tasks. There are different responses to dogs versus cats, familiar dogs versus unfamiliar dogs, and also differences in response between men and women, with a greater influence on men. The impacts that companion animals have on health are likely complicated and will require additional research to elucidate the mechanisms.

A recent large-scale study in Finland failed to find a positive benefit of pet ownership on health (Koivusilta & Ojanlatva, 2006). Working with a cross section of over 21,000 working-aged Finns, stratified by age groups, they found that pet ownership was associated with poor perceived health. They conclude that pet ownership may have been associated with older people who were more set in their ways and less active. A direct comparison to work done in the United States is difficult because pet ownership patterns may be quite different between the United States and Finland. Surveys of Americans tend to show that pet ownership is higher in families with children, typically at the midlife stage. There are also differences between the overall health delivery systems of the United States and Finland. This study does highlight that the role of companion animals must be considered within the context of the wide range of lifestyle and social factors known to have an influence on health.

COMPANION ANIMAL DEATH AND GRIEF

The level of attachment that people have to their companion animals may best be measured by the grief that they experience when the animal dies. For many, this is the most difficult part of having a pet. It is not made any easier by friends and others who might question their emotional response to the loss of a pet dog or cat (Quackenbush & Graveline, 1985). The past two decades have seen a substantial amount of change in this area and grief **counseling** for pet loss is developing as a special type of counseling. Those working in the field are now able to work within the framework developed by Kübler-Ross (1969) to understand and deal with pet loss. In her classic book *On Death and Dying*, Kübler-Ross described five stages in how people cope with the death of a loved one. While these stages may not follow a clear linear process, they are common to most cases of loss:

1. Denial—Unwilling to accept what has happened. It is almost as if our mind goes into a protective mode, so we do not need to absorb the loss all at once. A sudden death may make this stage more intense. Denial may erode slowly.

2. Anger—Why did this happen? Couldn't someone have done something more? Were all medical options explored?

3. Bargaining—This may happen with an unfavorable diagnosis or poor prognosis. "If she lives I'll never leave her alone again." Guilt may follow—What did I fail to do?

4. Depression—A general lack of interest or motivation. "What's the use if he is not here to walk or play with me."

5. Acceptance—There is new energy and you are able to enjoy life and memories of the one you lost.

Wallace Sife, founder of the Association for Pet Loss and Bereavement, has helped to develop special training programs and professional support for pet loss counselors. He also provides some practical suggestions (Sife, 1998) for dealing with pet loss.

- First and foremost is to let your feelings out, not to hold back, and acknowledge that you have suffered a loss.
- You may want to make a memorial donation in your pet's memory, to a humane organization or a veterinary college.
- Keep a list of things that you enjoyed doing with your pet, and things that you really liked about your pet. This is something that you can continue to add to over time. It may be written or it could be an audio recording.
- Establish new routines. Walk different paths to work or other places.
- Visit other pet owners and their pets. Don't be a hermit.
- Talk to a pet bereavement counselor or attend a pet loss support group.
- Treat yourself to things you would have liked to do, but could not.
- It may help to put away any remaining pet toys and reminders of your pet if they upset you when you see them.
- Talk to your veterinarian and ask the questions that you may not have thought of asking, or were unable to ask when your pet was dying.
- Try to avoid euthanizing a pet on holidays or other special calendar days.
- Understand and respect your grief. If your grief is intense, take some time off from work if you can.
- Don't rush into getting another pet. Visit an animal shelter to see how you feel. Do not be impulsive. If you go back a second time and you still feel like getting another pet, it may be the right time.
- Hold some sort of private memorial service.

Dealing with the death of a pet can be difficult, and made even more awkward by how people respond and react to the situation. It is generally recommended to keep interaction with someone grieving simple and honest. The best response is often, "I am sorry for your loss; is there something that I can do to help?" The grieving person may just need someone to talk to, so be ready to listen. Don't say "I know just how you feel." Feelings of loss are private and unique. There is no way to really know how they feel.

Pet loss can be especially difficult for children (Greene & Landis, 2002). It may be their first experience with death. It is important to be honest. Stay away

FIGURE 9-5

Hartsdale Pet Cemetery features imaginative and touching tributes to pets that have passed away. Whether the memorial is large and formal, a small painted rock in the backyard, or a pretty box with the ashes of a dog or cat, it is part of the process that brings closure and peace to people mourning the loss of a beloved companion. *(Courtesy Hartsdale Pet Cemetery)*

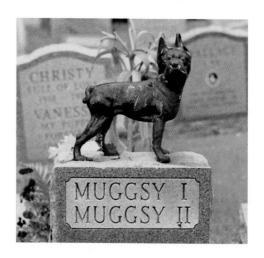

from euphemisms such as "put to sleep" or "the dog ran away." After all, what would have caused the pet to run away? Also, going to sleep can be traumatic for the child. Ask for questions; ask how they feel. Let them know that being sad or angry is normal. Parents can help a child create a scrapbook of their favorite things about their pet. They can also help create a small memorial by decorating a rock or gathering flowers, and making the child a part of a memorial service for the pet. No time limits should be put on how long the child grieves, but if the child continues for three to six months it might be advisable to consult a counselor.

Strong feelings about the loss of a pet are not a recent development. In 1896, Dr. Samuel Johnson, a veterinarian practicing in New York City and at the ASPCA, would sometimes make arrangements for some of his clients to bury pets that had died on his farm in upstate New York. As word of his service became well known, more and more people approached him about burying their pets on his farm. Eventually, Dr. Johnson's apple orchard became Hartsdale Pet Cemetery, the first such dedicated facility in the country (Figure 9-5). Nearly 70,000 pets have found a final resting place there. Hartsdale is also home to a memorial to dogs who have served in war.

Pet owners continue to be strongly influenced by the treatment that they receive from their veterinarian and clinic staff when it comes time to euthanize a companion animal (Adams, Bonnett, & Meek, 1999; Hart, Hart, & Mader, 1990; Martin et al., 2004). Clients are particularly interested in a private place for the procedure, and limited contact with other clients and pets before and afterward. For this reason, it is often recommended that planned euthanasia be performed at home, or at the very beginning or end of office hours at the clinic. Many people are now asking to be present during the euthanasia procedure. This puts additional stress on the staff because procedural errors that result in the pet struggling or appearing to be in pain or discomfort are very upsetting to the pet owner. For this reason it is critical that all staff involved be well trained and prepared to carry out the procedure in a dignified and professional manner. Whenever possible, all paperwork for the procedure should be completed before the euthanasia is performed. This can often be done when the pet is being prepared. This would include sedating the pet, shaving the fur wherever needed, and insertion of a catheter into a vein. Given this preparation, the pet owner and other family members, if desired, can enter the room to find the pet at rest. The pet owner should be fully briefed on what will happen and how the process will be completed. They should be informed that the pet may make a few whimpers, twitch a little, and urinate or defecate as he dies. The actual procedure will be simplified by now needing to inject the euthanasia fluid into the catheter. This will spare the veterinarian the difficult task of "hitting a vein" in a sick or older pet while the owner observes, helping to make the entire experience more controlled and compassionate.

CONCLUSIONS

E. O. Wilson (1984) has presented the hypothesis that humans have evolved deeply imbedded in nature, and that we retain our attachments to nature in our genome. He coined the term **biophilia** to describe this relationship. Evidence of the bond that we have with nature and animals is certainly embedded in human cultures. Art, legends, religion, and economics are bound up with the animals we hunted, herded, bought and sold, worshipped, and reviled. There was a time when most of humanity required some sort of daily contact with animals and nature to earn a living and survive. Technology has "freed" most of us from this requirement. Relatively few people in developed societies need to hunt or raise animals to make a living or earn a meal. It is certainly tempting to say that keeping pets is a way to satisfy an inborn need to have regular contact with animals and through them a more tangible contact with the natural world. While this is a tempting corollary to the biophilia hypothesis, we will need more data to confirm it as a scientific principle. We already have some data in hand. How else can we explain the remarkable influence that watching and touching animals has on basic physiological functions such as blood pressure and heart rate? It would be relatively simple to write about companion animals and extol only wonderful or benign effects. Companion animals are not a panacea. They are much more a Rorschach test of our humanity. What we see in them and how we treat them says much more about us than it does about them. Roger Caras was fond of telling me that "Animals bring out the best in people, and the worst." Going forward, it will be important to use the knowledge we gain about companion animals to ensure that we can be the best as often as possible.

DISCUSSION QUESTIONS

1. What is the most common source of cats and dogs in American homes? What effect does this have on the opportunity to provide education to the new pet owner?
2. What age group is overrepresented among pet owners who relinquish their pets to an animal shelter?
3. What are designer dogs?
4. What were the Bands of Mercy?
5. In which states is teaching humane education mandatory? Is federal funding provided for this purpose?

6. Humane education is being incorporated into programs that focus on character education. What facet of American society has prompted this development?
7. What level of interaction with the family dog is recommended for a toddler? For a 6-year-old? For a 15-year-old?
8. In the study by Friedman et al. (1980), how much more likely to survive were patients with pets at home versus patients without pets?
9. When a companion animal is being euthanized, describe techniques that veterinary staff can apply in order to minimize stress on the animal's owner.

REFERENCES

Adams, C. L., Bonnett, B. N., & Meek, A. H. (1999). Owner response to companion animal death: Development of a theory and practical implications. *Canadian Veterinary Journal, 40*, 33–39.

American Pet Products Manufacturers Association, Inc. (APPMA). (2005). *2005–2006 APPMA national pet owner's survey*. Greenwich, CT: APPMA.

American Veterinary Medical Association. (2002). *U. S. pet ownership and demographic sourcebook*. Schaumberg, IL: American Veterinary Medical Association.

Angell, G. T. (1884). *Autobiographical sketches and personal recollections*. Boston: Franklin Press, Rand Avery.

Antoncic, L. S. (2003). A new era in humane education: How troubling youth trends and a call for character education are breathing new life into efforts to educate our youth about the value of all life. *Animal Law, 9*, 183–213.

Ascione, F. R. (1992). Enhancing children's attitudes about the humane treatment of animals: Generalization to human-directed empathy. *Anthrozoos, 5*(3), 176–191.

Ascione, F. R., & Weber, C. V. (1996). Children's attitudes about the humane treatment of animals and empathy: One year follow-up of a school-based intervention. *Anthrozoos, 9*(4), 188–195.

Cavalli-Sforza, L. L., & Feldman, M. W. (1981). *Cultural transmission and evolution*. Princeton, NJ: Princeton University Press.

Daly, B., & Morton, L. L. (2003). Children with pets do not show higher empathy: A challenge to current views. *Anthrozoos, 16*(4), 298–314.

Friedman, E., Katcher, A. H., Lynch, J. J., & Thomas, S. A. (1980). Animal companions and one-year survival of patients after discharge from a coronary unit. *Public Health Reports, 95*, 307–312.

Friedman, E., Katcher, A., Eaton, M., & Berger, B. (1984). Pet ownership and psychological status. In R. A. Anderson, B. Hart, & L. Hart (Eds.). *The Pet connection: Its influence on our health and quality of life*. Minneapolis: Center to Study Human Animal Relationships and Environments (CENSHARE), University of Minnesota, Minneapolis.

Friedman, E., Thomas, S. A., & Eddy, T. J. (2000). Companion animals and human health: Physical and cardiovascular influences. In A. L. Podberscek, E. Paul, & J. A. Serpell (Eds.), *Companion animals and us: Exploring the relationships between people and pets*. New York: Cambridge University Press.

Good, H. G. (1956). *A history of American education*. New York: Macmillan.

Greene, L. A., & Landis, J. (2002). *Saying good-bye to the pet you love*. Oakland, CA: New Hallinger Publications.

Hart, L. A., Hart, B. L., & Mader, B. (1990). Humane euthanasia and companion animal death: Caring for the animal, the client, and the veterinarian. *Journal of the American Veterinary Medical Association, 197*, 1292–1299.

Herzog, H. A., Bently, R. A., & Hahn, M. W. (2004). Random drift and large shifts in popularity of dog breeds. *Proceedings of the Royal Society London, B, Biological Sciences, 27*(1), S1–S4.

Herzog, H. A., & Elias, S. M. (2004). Effects of winning the Westminster Kennel Club Dog Show on breed popularity. *Journal of the American Veterinary Medical Association, 225*(3), 365–367.

Hirschman, E. (1994). Consumers and their animal companions. *Journal of Consumer Research, 20*, 616–632.

Irvine, L. (2004). *If you tame me: Understanding our connection with animals.* Philadelphia: Temple University Press.

Katcher, A. H., Friedman, E., Beck, A. M., & Lynch, J. J. (1983). Talking, looking, and blood pressure: Physiological consequences of interaction with the living environment. In A. H. Katcher & A. M. Beck (Eds.), *New perspectives on our lives with companion animals* (pp. 351–359). Philadelphia: University of Pennsylvania Press.

Kelly, H. (2006, April). Bespoke bowwows. *Best Life*, 34.

Kidd, A. H., & Kidd, R. M. (1990a). Factors in children's attitudes toward pets. *Psychological Reports, 66*, 775–786.

Kidd, A. H., & Kidd, R. M. (1990b). Social and environmental influences on children's attitudes toward pets. *Psychological Reports, 67*, 807–818.

Koivusilta, L. K., & Ojanlatva, A. (2006). To have or not to have a pet for better health? *Public Library of Science (PLoS) ONE*. Retrieved August 21, 2007, from www.plosone.org1(1):e109.doi:10.1371/journal.pone.0000109.

Kübler-Ross, E. (1969). *On death and dying.* New York: Macmillan.

Martin, F., Ruby, K. L., Deking, T. M., & Taunton, A. E. (2004). Factors associated with client, staff, and student satisfaction regarding small-animal euthanasia procedures at a veterinary teaching hospital. *Journal of the American Veterinary Medical Association, 224*, 1774–1779.

Melson, G. F., & Fogel, A. (1996). Parental perceptions of their children's involvement with household pets: A test of a specificity model of nurturance. *Anthrozoos, 9*, 95–105.

New, J. C. Jr., Salman, M. D., Scarlett, J. M., Kass, P. H., Vaughn, J. A., Scherr, S., et al. (1999). Moving: Characteristics of dogs and cats and those relinquishing them to 12 U.S. animal shelters. *Journal of Applied Animal Welfare Science, 2*(2), 83–96.

New, J. C. Jr., Kelch, W. J., Hutchinson, J. M., Salman, M. D., King, M., Scarlett, J. M., et al. (2004). Birth and death estimates of cats and dogs in U.S. households and related factors. *Journal of Applied Animal Welfare Science, 7*(4), 227–241.

O'Farrell, V. (1997). Owner attitudes and dog behavior problems. *Applied Animal Behaviour Science, 52*, 205–213.

Patronek, G. J., Glickman, L. T., Beck, A. M., McCabe, G. P., & Ecker, C. (1996a). Risk factors for relinquishment of dogs to an animal shelter. *Journal of the American Veterinary Medical Association, 209*, 572–581.

Patronek, G. J., Glickman, L. T., Beck, A. M., McCabe, G. P., & Ecker, C. (1996b). Risk factors for relinquishment of cats to an animal shelter. *Journal of the American Veterinary Medical Association, 209*, 582–588.

Paul, E. S., & Serpell, J. A. (1993). Childhood pet keeping and humane attitudes in young adulthood. *Animal Welfare, 2*, 321–337.

Podberscek, A. L., & Serpell, J. A. (1997). Aggressive behaviour in English cocker spaniels and the personality of their owners. *Veterinary Records, 141*, 73–76.

Poresky, R. H., Hendrick, C., Mosier, J. E., & Samuelson, M. L. (1988). Young children's companion animal bonding and adults' pet attitudes: A retrospective study. *Psychological Reports, 62*, 419–425.

Quackenbush, J., & Graveline, D. (1985). *When your pet dies: How to cope with your feelings.* New York: Simon and Schuster.

Salman, M. D., New, J. C. Jr., Scarlett, J. M., Kass, P. H., Ruch-Gallie, R., & Hetts, S. (1998). Human and animal factors related to the relinquishment of dogs and cats in

12 selected animal shelters in the United States. *Journal of Applied Animal Welfare Science, 1*(3), 207–226.

Salman, M. D., Hutchinson, J., Ruch-Gallie, R., Kogan, L., New, J. C. Jr., Kass, P. H., et al. (2000). Behavioral reasons for relinquishment of dogs and cats to 12 shelters. *Journal of Applied Animal Welfare Science, 3*(2), 93–106.

Scarlett, J. M., Salman, M. D., New, J. C. Jr., & Kass, P. H. (1999). Reasons for relinquishment of companion animals in U.S. animal shelters: Selected health and personal issues. *Journal of Applied Animal Welfare Science, 2*(1), 41–57.

Serpell, J. A. (1981). Childhood pets and their influence on adults' attitudes. *Psychological Reports, 49*, 651–654.

Serpell, J. A. (1996). Evidence for an association between pet behaviour and owner attachment levels. *Applied Animal Behaviour Science, 47*, 49–60.

Sife, W. (1998). *The loss of a pet*. New York: Macmillan.

Spring, J. (1985). *The American school 1642–1985*. New York: Longman.

U.S. Department of Commerce. (1995). *Geographical mobility: March 1993 to March 1994* (Current population reports, population characteristics, pp. 20–485, pp. vii–xvi, 3). Washington, DC: U.S. Government Printing Office.

Wilson, E. O. (1984). *Biophilia*. Cambridge, MA: Harvard University Press.

Zawistowski, S. (1998). Humane education. In M. Bekoff (Ed.), *Encyclopedia of animal rights and animal welfare* (pp. 189–191). Westport, CT: Greenwood Press.

CHAPTER 10

Hot Button Issues

KEY TERMS

ownership	cropping	puppy mills
declawing	free-roaming cats	commercial breeders
docking	feral cats	outdoor cats

While pets or companion animals have a lengthy history living with humans, aspects of this relationship are still evolving. There are a number of circumstances where different groups of people differ in their opinions of how animals should be treated, the type of care they should receive, and what types of animals can or should be kept as companions. These "hot button" issues are debated in the media, in the courts, at the federal level, and in local communities. They may be a point of contention among groups that would normally be allies, or find common ground among groups that differ on other topics or questions. This section will examine a handful of these topics, presenting a summary of the issue, a selection of references, and a list of points from each side of the issue. There are other examples to consider, and new ones are coming up all the time as we continue to redefine our relationship with the animals that share our homes.

OWNER OR GUARDIAN?

Animals are defined as property under law. This made sense at a time when livestock was among a family's most significant economic possessions. A family would buy or sell animals as needed. The animals were needed to provide food, fiber, and labor. The economy of the household and the community depended on this clearly defined relationship. If an animal was lost, killed, or stolen, the family or owner could seek compensation if another person was at fault. Things have changed. Animals are now kept in houses not for their economic value but for companionship. How should this relationship be recognized in our language and the law? Organizations such as In

Defense of Animals are leading an effort to change the title of "owner" to "guardian." It is intended to change the way in which we view our relationship with animals, and our duties for their care. As a guardian, an individual living with an animal will have the responsibility to protect and nurture the animal, in terms of the animal's interests, not the person's.

Pro-Guardian

- People will be required to think about the best interests of an animal in their care.
- It is a proactive and positive term for raising awareness about animals and their treatment.
- The change in language will help to shape attitudes about animals.
- The emotional value of companion animals will be recognized. People who have a pet harmed or killed by accident or intentionally can be compensated for the loss of their companion above and beyond the economic value of the animal.

Pro-Owner

- If there are multiple guardians in a home and they have different opinions, how will a decision be reached regarding the treatment or care of a companion animal?
- People are currently responsible for damage or problems caused by animals that they "own"; how will this be influenced if the people are now guardians?
- This will change the legal status of animals, and may require a substantial change in the infrastructure society has in place to deal with animals and associated issues.
- Introducing the guardian terminology is a stealth tactic to change the legal status of animals without directly addressing the issue.

REFERENCES

Chapman, T. (2005, March/April). Owner or guardian? *Trends*, 31–34.
Katz, J. (2004). Guarding the guard dog? Are you a dog "owner"—or a dog "guardian?" Retrieved March 10, 1997, from http://www.slate.com/id/2096577.
The Guardian Campaign. Retrieved August 15, 2007, from http://www.guardiancampaign.com/article_060524.html.

EXOTIC OR WILD ANIMALS AS PETS?

In addition to many species that have been domesticated and kept as pets for hundreds of years, there are other species that some people enjoy and desire to keep as pets. These include monkeys, large wild cats such as tigers and lions, wolves and wolf–dog hybrids, various large reptiles, and various other species.

Pro-Exotic Animals

- Americans should be free to own any animal that they desire as long as it does not pose a public danger or nuisance.

Anti-Exotic Animals

- Exotic animals pose a significant risk to the owners and the public, not just through injuries and deaths by attacks, but also through exposure to zoonotic diseases.

- The danger of exotic animals is exaggerated. Experienced exotic pet owners are able to provide care for their pets and protect members of the public.
- Exotic pet owners can help to breed endangered animals in captivity to protect them from extinction.

- Many exotic pets are captured from the wild, depleting natural populations of the species.
- Animals bred as pets are not helpful in organized conservation efforts since there is not enough information available on their genetic heritage and breeding pedigrees.
- The specialized needs of exotic animals can seldom be met outside of zoos or other institutions with the experience and resources needed to provide appropriate care. As a result, most of the exotic animals kept as pets suffer poor health and welfare.

REFERENCES

Farintino, R. The whims and dangers of the exotic pets market. Retrieved August 15, 2007, from http://www.hsus.org/wildlife/issues_facing_wildlife/should_wild_animals_be_kept_as_pets/the_whims_and_dangers_of_the_exotic_pets_market.html.

United States Department of Agriculture, Animal and Plant Health Inspection Service. (2000). Animal Care Position Statement: Large wild and exotic cats make dangerous pets. Retrieved August 15, 2007, from http://www.aphis.usda.gov/ac/position.html.

General information on wolf-dog ownership. Retrieved August 15, 2007, from http://www.floridalupine.org.

BREED-SPECIFIC LEGISLATION (BSL)

The fear of dog attacks has resulted in some communities passing, or attempting to pass, laws that would prohibit or restrict the **ownership** of particular breeds thought to be more dangerous than other breeds. Previous generations feared German shepherd dogs and Doberman pinschers. The breed most frequently targeted by current legislation is pit bulls (or the variety of names that dogs in this grouping are known by).

Pro-BSL

- Pit bulls and similar breeds account for a disproportionate number of serious attacks on people and other dogs.
- Pit bulls were bred to be aggressive as fighting dogs, so it should be no surprise that they would be more aggressive, and as a result they are not appropriate companion animals.

Anti-BSL

- The response to pit bull attacks is driven by media frenzy that gives a biased sense of the risk.
- Effective dangerous dog regulation should be breed-neutral and focus on the behavior of individual dogs and their owners.
- Families may be required to relinquish gentle and beloved pit bulls

- Many other Western countries have already recognized the danger that pit bulls represent, and prohibit their ownership.

- kept as companions, which have done no harm when communities pass restrictions.
- Breed-specific legislation has been ineffective as a method to reduce the number of dog attacks in a community.

REFERENCES

Bradley, J. (2005). *Dogs bite: But balloons and slippers are more dangerous.* Berkeley, CA: James and Kenneth Publishers.
Rollin, B. E. (2007, January). Profiling: Two sides of the issue. *Veterinary Forum,* 64–67.

DECLAWING CATS

Cats scratch for a variety of reasons including conditioning their claws, marking by leaving a physical sign (scratch marks) and scent from their paws on objects, and to stretch and exercise. Scratching and damaging furniture is a commonly reported behavior problem among cat owners. Some owners who are elderly or have other health problems can face additional risks if they are accidentally scratched by their cat. The actual procedure involves the amputation of the last bone of each toe to prevent regrowth of the nail, which is attached to the bone.

Pro-Declaw

- The procedure employed has evolved over the years and is performed under anesthesia with post-operative analgesics, resulting in limited pain or discomfort to the cat.
- If it is a choice between someone relinquishing his or her cat due to damage that it causes and **declawing**, it is more humane to declaw the cat so that it can remain in its home.
- The declawing procedure is an important source of income for veterinary practices.

Anti-Declaw

- Declawing is a procedure that has no medical benefits for the cat, and is done entirely for the convenience of the owner.
- It is a form of mutilation that should not be tolerated as a "regular practice" and should only be used when absolutely necessary to keep a cat in a home.
- It deprives the cat of an important form of defense.
- Regular trimming of nails and providing appropriate scratching posts will generally prevent destructive scratching.

REFERENCES

Grier, K. C., & Peterson, N. (2005). Indoor cats, scratching, and the debate over declawing: When normal pet behavior becomes a problem. In D. J. Salem & A. N. Rowan (Eds.), *The state of the animals III* (pp. 27–41). Washington, DC: Humane Society Press.
Landsberg, G. (1991). Cat owners' attitudes toward declawing. *Anthrozoös, 4*(3), 192–197.

Patronek, G., & Dodman, D. (1999). Attitudes, procedures, and delivery of behavior services by veterinarians in small animal practice. *Journal of the American Veterinary Medical Association, 215*, 1606–1611.

DOCKING AND CROPPING OF DOGS

Docking is when a dog's tail is cut short, and **cropping** is when a dog's ears are cut and shaped to stand erect. There are a variety of dog breeds for which these are standard practices. These procedures are usually carried out by a veterinarian under anesthesia, though it is not uncommon for amateurs to also perform the procedures. In many parts of Europe, these procedures are not permitted, even by a veterinarian, unless it is for the health and benefit of the individual animal.

Pro-Docking/Cropping

- This is an important tradition associated with certain breeds, and the essential character and popularity of the breed would be harmed if the practice was banned.
- Some hunting breeds benefit from docking since the tail is prone to injury when the dog is working in the field.
- The procedure is not harmful to the dog when performed by an experienced veterinarian.

Anti-Docking/Cropping

- Breeds have not lost popularity in parts of the world where the practice is banned.
- The claims of benefits for some working breeds are disingenuous since other breeds doing the same work are not docked.
- Since the procedure is generally not for the benefit of the individual dog, any risk of possible complications during surgery is unwarranted.

REFERENCES

Association of Veterinarians for Animal Rights. Retrieved March 29, 2007, from www.avar.org/publications_position.asp#p7.

National Animal Interest Alliance. Retrieved March 29, 2007, from www.naiaonline.org/about/policy_husbandry.htm.

OUTDOOR CATS

Cats at one time were working animals, kept to manage mouse and rat populations around farms and homes. They often spent their entire lives living outside, finding what shelter they could in barns and under porches. That is no longer the case for most cats kept as companions.

Pro-Outdoor

- It is natural for cats to go outdoors to get exercise and enjoy themselves.
- Cats allowed access to the outdoors have fewer behavior problems.

Anti-Outdoor

- **Free-roaming cats**, home-owned, and **feral cats** kill large numbers of small animals and birds. In addition to the harm to these animals, the cats also compete with native predators for prey/food.

- Trap-neuter-release (TNR) programs can effectively deal with populations of free-roaming cats without depending on euthanasia as the primary control practice.
- Their impact on wildlife is overestimated since they usually kill small animals that are in great abundance.

- Free-roaming cats are exposed to a wide variety of dangers, including car accidents, poisons, the elements, fights with other cats and animals, and mistreatment by people.
- Cats kept indoors are healthier and do not risk injury.
- It is more humane to catch and euthanize free-roaming cats than to let them suffer from hunger and injury.

REFERENCES

Fitzgerald, B. M., & Turner, D.C. (2000). Hunting behaviour of domestic cats and their impact on prey populations. In D. C. Turner & P. Bateson (Eds.), *The domestic cat: The biology of its behaviour* (pp. 151–175). New York: Cambridge University Press.

Patronek, G. (1998). Free-roaming and feral cats—their impact on wildlife and human beings. *Journal of the American Veterinary Medical Association, 212,* 218–226.

Slater, M., & Shain, S. (2005). Feral cats: An overview. In D. J. Salem & A. N. Rowan (Eds.), *The state of the animals III* (pp. 43–53). Washington, DC: Humane Society Press.

HORSE SLAUGHTER

When dogs and cats get old, sick, or are no longer wanted, they may be euthanized by a veterinarian or taken to an animal shelter. The same has not been true for horses. For many owners, a horse remains a valuable investment and they do not wish to simply relinquish or discard the horse without some form of compensation. In addition, the size of horses makes disposal of the remains a more difficult or expensive process. Many of these horses may be sold at auctions, or to brokers who buy horses. Some of these horses are kept, resold, and used for riding or other purposes. However, each year as many as 90,000 horses are slaughtered for human consumption. While horsemeat is not usually eaten in the United States, there is a thriving overseas market and most of the meat is shipped to Europe and Japan. For the past several years there have been extensive efforts to prohibit the practice of slaughtering horses for human consumption in the United States.

Pro-Slaughter

- Owners deserve the opportunity to recover their investment in their horses by selling them at auctions, even if the horses may go to slaughter.
- Humans have eaten horses for as long as we have historical records. It is a part of our history, and even if

Anti-Slaughter

- Horses have a unique role in our nation's history and deserve a more dignified fate than being slaughtered for food.
- Horses transported for slaughter are often moved in trucks that do not separate them by age, size, or sex,

we would not eat horsemeat we should recognize that others may still do so.

- Ending the slaughter of horses for human consumption would set a precedent for raising and slaughtering other animals for food.
- There is no adequate infrastructure available to absorb the numbers of horses that would have normally gone to slaughter.
- Slaughter is a more humane option than allowing an unwanted horse to suffer from mistreatment or abandonment.

resulting in injury and suffering during the trip.

- Slaughter plants have not been designed to meet the specific behavioral and anatomical requirements of horses, again resulting in unnecessary injury and pain.
- The availability of the slaughter option allows the horse industry to continue overbreeding of unneeded or unwanted additional horses.

REFERENCES

American Association of Equine Practitioners. Retrieved March 29, 2007, from www.aaep.org.

Veterinarians for Equine Welfare. Retrieved March 29, 2007, from www.vetsforequinewelfare.org.

PET STORE SALES OF DOGS

About 7% of pet dogs are acquired from pet stores (approximately four to five hundred thousand puppies each year). These stores are typically supplied by commercial breeding establishments licensed by the United States Department of Agriculture. Investigations have documented poor conditions at many of these facilities, typically called **puppy mills**. Attention here is directed toward two primary concerns. The first concern is the conditions for the breeding stock, and the second is the health and transport of the puppies that are produced and sold. Breeding dogs may spend their entire lives in cages with limited or no opportunity for exercise. Puppies are often shipped at the age when it is most important for them to socialize with people. There is also concern regarding the overall health and genetic fitness of puppies produced by **commercial breeders**, when compared with those bred by breed fanciers/enthusiasts.

Pro-Sales

- There is a huge demand for puppies and breeds that are not readily available from animal shelters or breed fanciers. The public deserves the opportunity to get the kind of dog they want.

Anti-Sales

- The dogs at puppy mills are in terrible conditions, and the USDA has failed to provide the resources or effort needed to inspect the facilities and enforce the regulations.
- Pet store puppies are frequently sick and subject to inherited diseases.

- Pet stores that sell dogs can also provide purchasers with appropriate information on care and diet.
- If pet stores did not sell dogs, then breeding and selling dogs would become a "backyard" business with no oversight or regulation.
- The inability to sell dogs would be a significant blow to the pet industry, which is a growing part of the community.
- Commercial breeders pay little or no attention to the breed standards; as a result the purchaser may not get what they are looking for when they buy a puppy of a particular breed from a pet store.
- Pet store purchases help contribute to pet overpopulation and the number of homeless dogs.
- Large and successful pet store chains have decided not to sell puppies, and instead have assisted shelters and rescue groups to find homes for dogs in need.

REFERENCES

American Society for the Prevention of Cruelty to Animals (ASPCA). Retrieved March 29, 2007, from www.aspca.org/site/PageServer?pagename=cruelty_puppymills.

Bedwell-Wilson, W. (2007) Pets: To sell or not to sell. *Pet Product News.* Retrieved March 29, 2007, from www.petproductnews.com/top_stories/20070101-pet-stores.aspx.

Humane Society of the United States (HSUS). Retrieved March 29, 2007, from www.hsus.org/pet/issues_affecting_our_pets/get_the_facts_on_puppy_mills/.

National Animal Interest Alliance (NAIA). Retrieved March 29, 2007, from www.naiaonline.org/about/policy_dogs.htm.

DISCUSSION QUESTIONS

1. What is the motivation behind changing the title of pet "owner" to "guardian"?
2. In terms of species conservation, provide an argument for and against keeping exotic animals as pets.
3. What is breed-specific legislation? Name some commonly affected breeds.
4. What are some reasons that cats may engage in scratching?
5. What does the declawing procedure involve?
6. What is the historical purpose behind docking and cropping?
7. What are the arguments for and against **outdoor cats** in terms of their impact on wildlife?
8. What outcomes exist for horses that are no longer wanted by their owners?
9. What are the primary concerns with commercial breeding facilities?

Companion Animal Care Guides

CHAPTER 11

KEY TERMS

house training	parasites	cages
vaccinations	nutrition	habitat
spaying	aquarium	rodents
neutering	diet	filters

DOGS

Species

The domestic dog (*Canis lupus familiaris*) is a carnivore related to wolves and coyotes. It was one of the earliest species domesticated by humans more than 12,000 years ago. There are around 400 different breeds of dogs around the world. Nearly all breeds were developed with a particular work function to perform. While some dogs do perform their original tasks as herders, hunters, or guardians, the vast majority of dogs in the United States are kept as companions. More American households have dogs as pets than any other companion animal.

Behavior

Social Behavior. Dog behavior has evolved from the social structure of wild canids. It functions to maintain social contact with other individuals of various ages and both sexes. Subtle changes in body posture, ear and tail positions, and other signals are signs of arousal, threat, submission, or play (Figures 11-1 to 11-10).

FIGURE 11-1

Neutral Relaxed. *(Courtesy ASPCA)*

FIGURE 11-2

Arousal. The dog has been stimulated by something in his environment. When the dog is excited by something pleasurable, the hackles will be down and the tail will be carried a little lower and will loosely wag. The muzzle will be relaxed and the tongue may be seen. This posture may be displayed to subordinates in order to express higher ranking pack position. *(Courtesy ASPCA)*

FIGURE 11-3

Aggressive Attack. This threatening posture is used to chase another away or, if need be, to attack in order to protect possessions, pack, or self. *(Courtesy ASPCA)*

FIGURE 11-4

Active Submission. This pacifying posture is used when a dog acknowledges another dog or human's higher social ranking, or to inhibit another's aggression. *(Courtesy ASPCA)*

FIGURE 11-5

Passive Submission. Bellying up indicates surrender, a pacifying gesture offered to a more dominant or aggressive individual. *(Courtesy ASPCA)*

FIGURE 11-6

Defensive Aggression. When fearful, a dog will give warning signals to indicate he does not wish to be approached. If unheeded, he will bite to protect himself. *(Courtesy ASPCA)*

House Training. Effective **house training** is based on consistency. Feeding the dog at the same time each day and walking the dog immediately after it arises from sleep and 20–30 minutes after feeding will take advantage of the dog's natural biological need to relieve itself after waking and eating. The dog should be taken outside (or if paper training, to the location in the home where paper has been placed) and allowed to sniff and relieve itself. Praise and a treat can be used to reinforce the desired behavior. During periods when the dogs or puppies cannot be observed in the home it can be helpful to confine

FIGURE 11-7

Maternal Correction. A mother dog will discipline a pup with a quick muzzle grasp. The pup learns to offer submissive body postures. *(Courtesy ASPCA)*

FIGURE 11-8

Play Solicitation. The play bow is a combination of dominant and submissive gestures. It is offered to invite another to play or as part of courtship behavior. *(Courtesy ASPCA)*

FIGURE 11-9

Greeting Behavior. A submissive dog may greet a more dominant dog with a muzzle nudge as an appeasement (pacifying) gesture. *(Courtesy ASPCA)*

FIGURE 11-10

Greeting Posture. Dogs sniff each other's genital region when greeting to gather information on sexual status. *(Courtesy ASPCA)*

them to a crate or kennel in the home. Dogs are comfortable in a den, and will generally avoid soiling their sleeping area in a den. The crate should be large enough for the dog to stand up, turn around, and be comfortable, but not large enough to eliminate and lie down in a separate part of the crate. As a rule of thumb, puppies can *hold themselves* for about one hour for each month of age. That is, a 2-month-old dog can typically go for about two hours before needing to eliminate, and a 7-month-old dog for seven hours. If an older dog is brought into a home it is a good idea to spend a couple of weeks on remedial training to help get the dog accustomed to the daily schedule of the new home.

Obedience Training. All dogs should receive some basic training. This will help to ensure that they are easy and safe to walk, and can be controlled in the home and around company or when traveling. Some basic commands would include *sit*, *stay*, *come down*, and *no*. The *no* command would be a general communication to the dog to stop whatever it is doing at the moment and look at the owner for further instructions. It can be used to stop a dog from running into the street, picking up some garbage, or jumping on a guest or furniture. Teaching responses to commands should be done with patience and reinforcement for desired behavior. There is no need for yelling or striking the dog. All members of a family should be part of training. Attending a group obedience class will have several benefits. An experienced instructor will provide guidance and ensure that proper techniques are being used. Group classes are also a valuable opportunity to socialize a dog to other people and dogs.

General Care

Diet. Dogs should have fresh, clean water at all times. Puppies should be fed high-quality puppy food. Specialized puppy foods are now available for large-breed dogs to help control their growth rate. Puppies between 8 and 12 weeks should be given four meals per day. From 3 to 6 months, they should receive three meals per day, and then from 6 to 12 months, two meals per day. After one year of age, small- to medium-size dogs can have just one meal per day. Large-breed dogs, and others that are prone to bloat, should be fed twice a day. Table foods and treats can be given, but should account for no more than 10 percent of the daily calorie consumption. Care needs to be taken with a variety of foods that humans commonly eat but are dangerous for dogs. Chocolate, raisins, and onions are all toxic. The ASPCA Animal Poison Control Center (<www.aspca.org>) maintains a comprehensive list of foods, plants, and other products that are toxic to dogs and other animals.

Health Care

Vaccinations. Puppies should be vaccinated with a combination vaccination (called a 5 in 1) at 2, 3, and 4 months of age and then revaccinated as recommended by a veterinarian. These **vaccinations** protect against distemper, hepatitis, leptospirosis, parvovirus, and parainfluenza. Dogs older than 4 to 5 months will receive two vaccinations 2 to 3 weeks apart. Most communities also require that dogs be vaccinated for rabies. This will normally be done after 3 months of age, with a follow-up vaccination a year later, and then renewed every 3 years after that. Vaccinations are available for kennel cough, Lyme disease, and other illnesses and should be considered upon consultation with a veterinarian based on potential risk factors for the dog.

Parasites. Dogs in nearly all regions and locations are at risk for infestation by worms. For the most part, these are contracted through worm eggs that are present in the environment, deposited in the feces of other dogs and animals. The presence of an infection will usually be diagnosed through a fecal examination. Medications are available to treat dogs having worms, and prevent subsequent infestations. Heartworms are spread by mosquitoes and can be fatal. It can most commonly be prevented with a monthly pill.

Fleas can be detected by combing a dog over a piece of white paper. Actual fleas may be seen, or the presence of flea feces will appear as small dark spots on the paper. Adult fleas can be removed with a flea comb. Ticks will be found attached to the dog's skin, and can be removed carefully with tweezers, making sure that the head and mouthparts are not left embedded in the skin. A variety of flea and tick products are available to prevent or control infestations.

Neutering. In addition to the benefits of preventing unwanted pregnancies and contributing to the homeless animal population, **spaying** females and **neutering** males have health benefits as well. Spaying before maturity (before the first heat is best) significantly reduces the risk of breast cancer and eliminates the risk of pyometra, an infection of the uterus. Neutering eliminates testicular diseases and reduces the likelihood of prostrate problems.

CATS

Species

Cats (*Felis domesticus*) descended from small wild cats that lived in North Africa. There are probably around 100 different breeds of cats. Cats do have a wide variety of colors and coat types, some variation in body type, head shape, and ears; they do not show the extreme variation that is seen in different breeds of dogs.

Behavior

Cat behavior is largely adapted to manage the distance and spacing with conspecifics. Cats are one of the few domesticated species that were derived from a wild ancestor that did not live in social groups. While many people do not think of cats as being social, the cat has evolved as a companion animal that is far more social than its ancestors and adapts well to living with people and other animals. A key element of the social nature of an individual cat is its exposure and experience with other cats, animals, and people at a young age. Kittens who remain with their litter until fully weaned (8 to 12 weeks) and are given exposure to people and dogs will be the most social when adults.

Social Behavior. Cats will signal their interest in being approached or avoided by other cats and people primarily by the position of their tails and ears (Figures 11-11 to 11-18).

FIGURE 11-11

Confident Cat. The confident cat purposefully moves through space, standing straight and tall with tail erect. He is ready to explore his environment and engage those he meets along the way. His upright tail signifies his friendly intentions, while his ears are forward and erect, adding to the cat's alert expression. *(Courtesy ASPCA)*

FIGURE 11-12

Confident Cat at Ease. When relaxed, a confident cat stretches out on his side or lies on his back exposing his belly. He is in a calm but alert state and accepts being approached. His entire posture is open and at ease; but beware, not every cat that exposes his abdomen will respond well to a belly rub. Some will grasp your hand with their front paws, rake your forearm with their hind feet, and bite your hand. *(Courtesy ASPCA)*

FIGURE 11-13

Distance-Reducing Behaviors. Distance-reducing behaviors encourage approach and social interaction and are meant to telegraph to others that the cat means no harm. The act of rubbing against a person's hand or another cat (scent marking) to distribute glandular facial pheromones from the forehead, chin, or whisker bed is calming and seems to guarantee friendly interaction immediately afterward. The tail is usually held erect while the cat is scent-rubbing. *(Courtesy ASPCA)*

FIGURE 11-14

Distance-Increasing Behaviors. The goal of distance-increasing behaviors is to keep others from coming closer. Aggressive interactions are avoided when the warnings are heeded. Conflicted cats lack the confidence to stare down and charge others. Instead, they assume a defensive threat posture, warning others away by appearing as formidable as possible by arching their backs, swishing their tails, and standing sideways and as tall as possible. Arousal and fear causes their fur to stand on end (pilo-erection) and pupils to dilate. *(Courtesy ASPCA)*

FIGURE 11-15

Anxious Cat. When a cat becomes anxious, he crouches into a ball, making himself appear smaller than usual. Muscles are tensed and the cat is poised to flee if necessary. The tail is held close to the body, sometimes wrapped around the feet. The head is held down and pulled into the shoulders. *(Courtesy ASPCA)*

FIGURE 11-16

Defensive Aggression. The pariah threat is another distance-reducing posture. When a cat determines that he cannot escape an unwanted interaction with a more dominant animal, he prepares to defend himself. The ears are pulled back and nearly flat against the head for protection and the head and neck are pulled in tight against the body. Facial muscles tense, displaying one weapon—the teeth. The cat rolls slightly over to one side in order to expose the rest of his arsenal—the claws. He is now ready to protect himself. *(Courtesy ASPCA)*

FIGURE 11-17

The Predator. Even when fed two meals a day, cats are still predators. The predatory sequence is stalk, pounce, kill, remove, and eat. When stalking prey, a cat may stealthily move forward or lie in wait, shifting his weight between his hind feet. When movement is detected, the cat pounces on his prey and delivers a killing bite. He may then take the fresh-killed prey to a quiet place to eat, or a female may take it to her kittens. Even cats that don't hunt for their meals still enjoy chasing moving objects, including toys and, in some cases, human body parts. *(Courtesy ASPCA)*

Litter Training. Cats will preferentially eliminate in a place where they can cover their waste. Kittens can be placed in a litter box several times a day and after feeding. They will normally take to using the litter box quite naturally. A new cat in the home can be confined for a short period of time in the room or area where the litter box is located until the cat has used the litter box several times. The litter box should be placed in a quiet area of the home, outside of busy traffic. It is generally best to place it in a location separate from where the cat is fed. In large homes or multilevel homes it can be helpful to have several litter boxes. Multi-cat homes will also need additional litter boxes, usually at least one for each cat in the home. Individual cats will show preferences for different types of cat litters, with many showing a preference for the various clumping litters. Litter should be cleaned frequently since dirty litter boxes are one of the most common causes of problems with the use of the litter box.

General Care

Diet. Fresh, clean water should be available at all times. Kittens should be fed high-quality kitten food—two to three times per day. A dry kibble food is most convenient and can be mixed as needed with canned food as a treat or to encourage eating. Kibble can also be softened with meat stock. Adults can be fed one to two meals per day. If canned food is fed it is important to remove uneaten food before it has a chance to spoil.

Health Care

Vaccinations. Kittens should be vaccinated with a combination vaccination (called a 3 in 1) at 2, 3, and 4 months of age and then renewed as recommended by a veterinarian based on anticipated risk of exposure of the individual cat. This vaccination protects against panleukopenia (also called feline distemper), calicivirus, and rhinotracheitis. An older, unvaccinated cat should receive two

FIGURE 11-18

The Groomer. Cats spend 30–50% of their waking time grooming. Backward-facing barbs on the tongue act as a comb to loosen tangles and remove some **parasites**. Beyond maintaining the cat's coat, grooming also relieves tension and promotes comfort. Licking also facilitates cooling off in warm weather. *(Courtesy ASPCA)*

vaccinations 2 to 3 weeks apart, and then renewed as needed. Based on the municipality, region, and possible exposure risk to a cat (depending on whether they are strictly indoor or spend time outdoors as well), rabies vaccination may be considered as well. There is also vaccination available for feline leukemia virus (FeLV). This is one of two immune system viruses that infect cats. The other is feline immunodeficiency virus (FIV). There is no vaccine available for FIV.

Parasites. Cats, especially those that go outdoors, can be infected with several types of worms. Fecal examinations are required to confirm an infection. Deworming should be done under the supervision of a veterinarian. Cats can be examined for fleas and ticks in the same manner as dogs. Preventive treatments are available for cats. Care needs to be taken to use only those products specifically approved for use with cats. Pyrethrin-based products used for dogs are extremely toxic to cats.

Neutering. As with dogs, neutering male cats and spaying females have advantages in addition to preventing unwanted pregnancies. Neutered males are less likely to spray to mark territory in the home, and will be less likely to roam in search of mates, when they are subject to bites and scratches in fights with other male cats. Spaying helps to prevent breast cancer and eliminates the possibility of pyometra. The procedures should be performed before 6 months of age, and can be done as early as 3 months.

 FERRETS

Species

Ferrets (*Mustela putorius furo*) are likely the descendents from a species of stoat. They are about 38 to 40 cm long and live to about 8 years of age. (Figure 11-19). In addition to their natural coloration, there are several different color patterns available. Some locations (e.g., California, New York City) prohibit ownership of pet ferrets, so it is important to check with local authorities regarding relevant laws or restrictions.

FIGURE 11-19

A common Sable ferret. *(Photo by Isabelle Francais)*

Behavior

The normal behavior of ferrets is playful and mischievous. They are quick and able to sneak into small spaces, so care is needed when a ferret is allowed to roam through a home. They are active predators and will play with toys, but care must be taken if other small animals such as mice or hamsters live in the home. Periods of intense activity are balanced against sleeping up to 18 hours per day. Nipping is a natural behavior and will be used to get attention or when defensive. Gentle handling and training will reduce the likelihood of nipping. Digging is a favorite occupation, so houseplants will need to be protected. They will use a litter box, similar to a cat. Special harnesses are available and ferrets can be trained to walk on a leash.

General Care

Diet. Fresh water and food should be available at all times, with limited treats. A high-quality ferret food will provide the best **nutrition**. Avoid dairy products, chocolates, seeds, and nuts.

Habitat. Ferrets need a large, well-ventilated cage for times when they cannot be supervised. It should be kept out of drafts or direct sunlight. A multi-level cage can be outfitted with a litter box, toys, and hammock for sleeping. A bedding of either hardwood shavings or recycled paper products can be used to line the cage. The litter box will need to be scooped daily, and the bedding changed weekly or more often as needed.

Health Care

Ferrets should be vaccinated for distemper at 9, 12, and 16 weeks of age, and rabies between 13 and 16 weeks, and renewed as needed under the supervision of a veterinarian. They should be neutered to prevent health complications.

HAMSTERS

Species

Syrian hamster (*Mesocricetus auratus*). This hamster is also known as the golden hamster because of its golden brown coat and is the most common hamster kept as a pet. They were first captured from the wild in Syria in 1930. They are 15 to 17 cm long, and will typically live for 2 to 3 years. There are several different color versions and hair coat types.

Dwarf Campbell's Russian hamster (*Phodopus campbelli*). This hamster is the most common *dwarf hamster* available in pet stores. Their natural color is gray, though there are some different colored varieties. They are 10 to 12 cm in length and round and pudgy in appearance. Their typical lifespan is 1 to 2 years.

Dwarf winter white Russian hamster (*Phodopus sungorus*). This hamster is also known as the Siberian, or Djungarian hamster. They are not very common in pet stores. Generally grayish in color, they are called winter white because their coat may turn all white during the shorter daylight hours of winter. They are 8 to 10 cm in length and will usually live 1 to 2 years.

Roborovski hamster (*Phodopus roborovski*). This is the smallest of the hamsters (4 to 5 cm in length), and is seldom seen in pet stores. They are usually available only from specialty breeders. They will live 3 to 4 years.

Chinese hamster (*Cricetulus griseus*). Chinese hamsters are also very rare. They are 10 to12 cm in length and will live 2 to 3 years.

Behavior

Hamsters are nocturnal by nature, so they will usually be active during the evening and at night. Place the cage in a room where they will not disturb people with their nighttime activity. During the day your hamster will enjoy sleeping in a nest that it makes. Care should be taken if you try to pick the hamster up while it is sleeping. Hamsters will nip or bite if they are handled roughly or disturbed while sleeping. Hamsters also have large pouches in their cheeks. When they are fed, they will pack their cheeks with food and carry it back to store in their nests.

Syrian hamsters are solitary by nature. After 10 weeks of age it is best to keep them in a cage by themselves. Putting two together could result in serious fighting.

Dwarf hamsters are social if they have been introduced to other hamsters at a young age. Do not try to put a new hamster in with either a group that has been living together, or a dwarf hamster that has been living alone as an adult.

You can tame your hamster by taking it out and holding it gently in your hands. Start for just brief periods of time, which can be prolonged gradually over time. Hand feeding your hamster small treats will help them get used to you.

Interaction Level. Hamsters can be temperamental and are best left at the observational level. They are very entertaining if they are housed in an interesting environment. This can be a commercial setup with tubes and slides, etc., or in a large (20 to 30 gallon) **aquarium** with cardboard boxes, paper tubes, and other places to climb and hide.

General Care

Diet. In the wild, hamsters primarily eat seeds in addition to some fruits and occasionally insects. As pets, hamsters do well on a basic **diet** of hamster mix available from pet stores. This will usually have a selection of seeds, grains, cracked corn, and rodent pellets. You can supplement this with bits of fresh vegetables or fruits. Care needs to be taken when feeding because hamsters tend to store large amounts of food in their nests or in the corners of their **cages**. Fresh food can spoil if the hamster does not eat it right away. Fresh, clean water should be available at all times in a drip tube.

Habitat. There are a wide variety of cages available for hamsters. Wire cages with plastic bottoms provide a jungle gym for the hamster to climb on. The plastic base is easy to clean and you can fill it with bedding so that they can dig and tunnel. The wire should be small enough to prevent the hamster from escaping. Cages made for Syrian hamsters will have spaces too large for dwarf hamsters, and they may escape. Dwarf hamsters can be kept in cages made for mice. Bedding should be made of aspen or hardwood shavings, or recycled paper pellets that are available at pet supply stores. Do not use cedar or pine as these can cause health problems for the hamsters. You can put the cardboard tubes from paper towels in the cage for them to use as tunnels.

The fancy hamster environments with tunnels, hideaways, and other options are also good. These are more expensive, however, and can be hard to clean. If you have dwarf hamsters, they will sometimes have trouble climbing up and down the tubes. You can put a thin branch in the tube so that they can climb up and down.

A 10-gallon aquarium with a wire cover also makes a good home for a hamster (Figure 11-20). You can add cardboard tubes and a small flowerpot for the hamster to use as a cave.

Hamster do enjoy running on an exercise wheel (Figure 11-21).

The cage should be cleaned weekly. The bedding should be cleaned out, and any leftover food stored by the hamster should be thrown out.

Hamsters need to chew to keep their teeth sharp and also to wear down the growing teeth. Provide pieces of wood that have not been painted or treated with chemicals, or pieces of a dog biscuit for them to chew on.

Provide small pieces of paper towel for the hamster to shred and make a nest.

FIGURE 11-20

Aquariums make good homes for hamsters.
(Courtesy of Carolyn Miller)

FIGURE 11-21

A hamster running on an exercise wheel.
(Courtesy of PhotoDisc)

Special Concerns

- Hamsters are nocturnal, and will not be very active during the day. They may bite if disturbed while sleeping.

Maintenance.

- Daily—food/water
- Weekly—clean litter
- Monthly—scrub cage
  - Note: The more complicated hamster gyms with tubes and chambers are interesting, but they are much more difficult to clean. They will require dismantling to clean and scrub.
- Annual—veterinary care

Costs.

- Acquisition
  - Hamster—$5 to $10
  - Cage and accessories—$25 to well over $100 for a very sophisticated **habitat**
- Maintenance
  - Litter—$10 to $15/month
  - Food—$5/month

Health Care

Zoonoses.

- Rare concerns, possible Salmonella

Factoids

Hamster comes from a German word that means *to hoard*.

 GERBIL

Species

Mongolian gerbil (*Meriones unguiculatus*) has been kept as a pet for about 30 years. They are approximately 10 cm long with a tail of about the same length. Unlike rats and mice, gerbils have hair on their tails. Their typical life span is 3 to 5 years. Their natural color is brownish gray to blend into the color of

FIGURE 11-22

A gerbil's natural color is brownish gray. *(Photo by Isabelle Francais)*

desert sand (Figure 11-22). Pet gerbils are available in a variety of different colors, including white/albino and black.

Behavior

Gerbils that have been introduced to other gerbils while young will be very social as a group. However, strange adults may fight when introduced. It is best to keep gerbils separated by sex because they will breed and reproduce very quickly.

Gerbils are active throughout much of the day, looking for food and digging.

Gerbils seldom bite, and can be picked up by scooping them into your hand. Do not lift them by their tails since this can cause injury.

Interaction Level.

- Can be observed, and under supervision can be petted and held

General Care

Diet. Feed a basic rodent mix with seeds, grains, and pellets, supplemented with small pieces of fresh fruit and vegetables. Gerbils will select sunflower seeds from a mixture first. Do not provide extra sunflower seeds because they alone do not provide adequate nutrition. Fresh, clean water should be provided with a drinking tube.

Habitat. A pair of gerbils will do very well in a 10-gallon aquarium with a wire cover. Gerbils love to dig, so provide extra bedding of aspen or hardwood shavings. Cardboard tubes from paper towels can be used as tunnels.

Gerbils will also do well in plastic habitats with connecting tubes. However, they may scratch the tubes and sides of the habitat with their nails, so it can be difficult to see them. For this reason, glass tanks are typically preferred.

Because gerbils are adapted for desert living, they produce very little urine; as a result their cages tend to have very little odor. Weekly cleaning is required to remove leftover food and other waste.

Maintenance.

- Daily—feeding/water
- Weekly—clean cage litter
- Monthly—scrub cage
- Annual—veterinary care

Costs.

- Acquisition
  - Gerbils—$5 to $10/ea.
  - Cage—$50
  - Accessories—$25
- Maintenance
  - Food—$5/month
  - Litter—$15/month
  - Veterinary care

Health Care

Zoonoses.

- Rare cases of Salmonella

Factoid

When gerbils are alarmed, they may stamp their feet as a warning to the other gerbils in their group.

GUINEA PIG

Species

Guinea pig (*Cavia porcellus*). The guinea pig, or cavy, comes in three common breeds: the smooth coated, the Abyssinian that has curly hair or rosettes, and the Peruvian that has long hair (Figure 11-23). Males weigh about 800 to 1200 g, while females are smaller (250 to 350 g). Their typical life span is 5 to 7 years.

Behavior

Guinea pigs are very social. Groups of females do especially well together. If you keep males and females together you should have a veterinarian neuter the males and spay the females to prevent them from breeding.

Guinea pigs are one of the few small pets to make noises. They will whistle and grunt when they are excited. When you come to feed or play with them, they will often greet you with a chorus of loud excited whistles.

Interaction Level.

- Guinea pigs can have a high level of social interaction. They seldom bite, do not run fast, or jump, so they can be placed in someone's lap. They are also vocal and their frequent grunts, whistles, and other sounds usually delight people. Their lack of a long tail tends to protect them against anti-rodent prejudices.

FIGURE 11-23

The guinea pig comes in three common breeds: the American (A); Abyssinian (B); and Peruvian (C) *(Photos by Isabelle Francais)*

(A)

(B)

(C)

General Care

Diet. Guinea pigs need vitamin C in their diet. Feed them a mix of food pellets made for guinea pigs, a variety of fresh fruits (especially citrus fruits) and vegetables, and a grass hay such as timothy.

A guinea pig's front teeth continue to grow, so it is important to give them a piece of log or wood to chew on to wear their teeth down.

Habitat. Guinea pigs should be kept in a solid-bottom cage with bedding of aspen or hardwood shavings or grass hay. The minimum size for the cage is 2 square feet for each guinea pig. Guinea pigs are not good climbers, but a sturdy cover will be needed to stop other family pets from bothering them. Plastic-bottom *tub-cages* with a wire top will work very well. Do not use wire-floor cages as this irritates their feet.

Provide empty oatmeal boxes, medium-size flower pots, plastic pipes, and other places for them to run through and play in.

Each day guinea pigs can be allowed to run around in your house for extra exercise. However, make sure that they do not get stuck behind furniture or chew electrical wires. If you are patient, some guinea pigs can be trained to use a litter box.

Maintenance.

- Daily—food/water/vitamin C
- Weekly—clean cage litter
- Monthly—scrub cage, clip toenails
- Annual—veterinary care

Costs.

- Acquisition
  - Guinea pigs—$10 to $25
  - Cage and accessories—$75 to $100
- Maintenance
  - Litter—$15/month
  - Food—$5 to $10/month

Health Care

Zoonoses.

- Mange mites

Factoid

Long-haired guinea pigs will need to have their coats combed or brushed frequently to prevent tangles.

FIGURE 11-24

An albino domestic mouse. *(Getty Images, Inc.)*

MICE

Species

Mouse (*Mus musculus*). The domestic mouse has been kept as a pet for many generations. It is about 5 cm long with a tail that is about another 5 cm. They will typically live 1 to 3 years. They tend to be wonderful, active pets. While white (albino) mice are most common in pet stores, fancy mice come in a wide variety of colors and color combinations and coat types (Figure 11-24).

Behavior

Mice are curious and playful and will be active at various times during the day. Females will do especially well in a group. You can keep three to four in a 10-gallon aquarium. Males are social if they are introduced at a young age. Strange males will fight. If you keep several males together, be certain to provide enough room and hiding places so that they can get away from one another if needed. Do not keep males and females together since they breed quickly, resulting in large litters.

Mice are very good jumpers, so you will need to be careful when you take them out of their cages. You can scoop them up in your hand or in a paper cup to get them out of the cage. Do not grab them by the middle or end of the tail since they can be hurt. If you need to catch a mouse quickly, you can grasp them at the base of their tail and lift them up, and then cup them in your hand. If you take time to feed your mice by hand while holding them gently, you can tame them to sit calmly in your hand or on your shoulder.

Interaction Level.

- Mice can be difficult to hold, so it is best to keep interactions at the observation/feeding level.

General Care

Diet. Mice will do well on a good, basic rodent mix of seeds, grains, and some pellets. You can include small amounts of fresh fruit and vegetables as treats. Always provide fresh, clean water using a drinking tube.

As with all **rodents**, a mouse's front teeth grow continuously. Provide unpainted, untreated pieces of wood or pieces of a dog biscuit for your mice to chew and wear down their teeth.

Habitat. Mice probably do best in a 10-gallon aquarium with a wire cover. There should be several inches of bedding. Use either aspen or hardwood shavings or reprocessed paper products. Do not use cedar or pine shavings as these may cause health problems. Provide your mice with small boxes or flowerpots to hide in, and cardboard tubes to chew and run through. You can also add a tree branch for them to climb on.

Plastic habitats will also work for mice. Since mice are smaller than hamsters, you may need to put small branches in the tubes so that the mice can climb up and down. These cages can be more difficult to clean.

Wire cages made for hamsters may have bars spaced too far apart to keep mice inside. If a person can stick his fingers through the bars, a young mouse could probably sneak out as well.

Mice will enjoy running on an exercise wheel. Make sure that it has a solid surface with no wire rungs so that its tail does not get caught while running.

Clean the cage every week, replacing the dirty bedding and wiping down the rest of the cage. Male mice will tend to produce more odor than females, so their cages need to be cleaned more often.

Maintenance.

- Daily—food/water
- Weekly—clean litter
- Monthly—scrub cage
  - Note: The more complicated environments with tubes and chambers are interesting, but they are much more difficult to clean. They will require dismantling to clean and scrub.
- Annual—veterinary care

Costs.

- Acquisition
  - Mice—$2 to $10
  - Cage and accessories—$25 to well over $100 for a very sophisticated habitat
- Maintenance
  - Litter—$10 to $15/month
  - Food—$5/month

Health Care

Zoonoses.

- Rare allergies to mouse dander or urine, potential for Salmonella

Factoids

Fancy *show mice* may be up to twice the size of regular pet mice.

RATS

Species

The domestic rat is a descendant of the wild brown rat (*Rattus norvegicus*) and has been bred as a pet for about a hundred years. Domestic rats come in a variety of colors and color combinations, and are less fearful than their wild cousins (Figure 11-25). When handled gently, they quickly learn to enjoy riding on people's shoulders or napping in their laps.

Baby rats can be removed from their litters at about 6 weeks of age. They are able to reproduce at this time, so it is recommended that males and females be kept separate. Rats are fully grown at 6 months of age. Males, called bucks, generally weigh 400 to 700 g, while females, called does, are smaller, weighing 200 to 500 g. A typical life span will be 2 to 3 years.

General Care

Diet. Rats will do well on a high-quality pellet chow formulated for rodents. This can be supplemented with some fresh fruits, vegetables, and occasional table scraps. Treats need to be limited to prevent obesity. Fresh, clean water should be made available at all times using a drinking bottle with a tube.

Housing. Rats do best in wire cages because they enjoy climbing, and there is good ventilation too. A cage that is 60 cm × 60 cm × 60 cm will generally do for a pair of rats. The floor should be solid, and a bedding of aspen or pelleted recycled paper should be provided. Do not use pine or cedar shavings because this may be harmful to your pets. A large, multi-level ferret cage is an especially good home. Provide PVC tubes and boxes for rats to play in, and a variety of toys to play with. Rats can be prone to colds, so be sure to keep the cage out of drafts. Intense direct sunlight should also be avoided.

A large aquarium can also be used as a home for rats, which will require a screen cover to provide ventilation. It will probably need cleaning more often than a wire cage to keep odor problems under control.

FIGURE 11-25

A rat with Berkshire markings. *(Courtesy of Brian Yacur)*

General Care

Rats are social animals and should be kept in pairs at least. Males will do well together if they are introduced when young. Females will be more friendly when introduced later in life.

The average rat litter is 12 and has been known to be as high as 20, so care needs to be taken to separate males and females when they are 6 weeks of age.

Rats love to play with toys. You can provide many of the toys that are enjoyed by parrots, including swings and ropes to climb. They also love to chew, so untreated wood helps to keep their teeth in shape.

Care.

- Daily—food/water
- Weekly—clean litter
- Monthly—scrub cage
  - Note: Rats do well in the large, multi-level wire cages designed for ferrets.
- Annual—veterinary care

Costs.

- Acquisition
  - Rats—$5 to $10
  - Cage and accessories—$100 to $200
- Maintenance
  - Litter—$10 to $15/month
  - Food—$5/month

Health Care

Zoonoses.

- Potential allergies to dander and urine

Special Concerns

- Rats may cause discomfort in some people. Choosing rats with piebald or other marked coat patterns can help to ease some of these concerns because they look more like dogs or cats in color patterns than wild rats.

FIGURE 11-26

(A) The Rhinelander rabbit and (B) the American rabbit. *(Photo by Isabelle Francais)*

(A) (B)

RABBITS

Species

Domestic rabbits (*Oryctolagus cuniculus*) are social animals that share a combination of behavior patterns with dogs, cats, and guinea pigs. There are more than 60 different breeds that vary in size from 1 kg to over 6 kg. In addition to size variation, there are a number of different color and coat pattern variants, and long- and short-hair types. There are two common ear positions. The normal ear position for rabbits is upright, though size will vary, while lop-eared breeds have ears that hang pendant, similar to a basset hound (Figure 11-26).

Baby rabbits are known as kittens or bunnies, and can be weaned at 4 to 6 weeks of age. A well-cared-for indoor rabbit has a life span of 7 to 10 years. Males (bucks) should be neutered and females (does) should be spayed to prevent unwanted breeding and to control some behavior problems, especially in males. Rabbits are lagomorphs, not rodents.

Behavior

Rabbits are social and tend to enjoy being around people. However, they vary in their tolerance to being picked up. Care must be taken to support the hind legs fully. When scared, a rabbit can kick out and cause severe scratches and cuts. Also, if unsupported when lifted up, a kicking rabbit can badly injure itself, including breaking its back. A common and easy way to transport a rabbit is by placing it in a picnic basket. It can then be carried safely and introduced to people for petting.

Because rabbits are social, care must be taken while introducing two rabbits. They are territorial and will fight when meeting for the first time. Initial introductions should take place in a neutral territory where there is room for one or the other to run away as needed. They should be under observation at all times until they start interacting peacefully on a regular basis.

A common behavior that is observed in rabbits is *chinning*. Rabbits have scent glands in their chins, and they rub their chins on various surfaces to mark their territories. It is not uncommon for them to mark human companions in the same way as cats do when they rub their faces against people.

Another frequent behavior is *thumping*. This is a form of communication used most often to signal agitation or send a warning.

Interaction Level.

- Rabbits can be highly interactive; they provide opportunities for observation and petting if they are allowed to roam about a room. In addition, if trained to walk on a leash and harness, they can be walked outside or from room to room. Fruit and vegetable portions of their diet can be given as treats throughout the day to encourage interactions.

General Care

Diet. The primary diet for a rabbit should be unlimited access to grass hay (i.e., timothy hay). Fresh commercial rabbit pellets should be provided at a rate of about ¼ cup/day/3 kg body weight. About a cup of dark green or yellow vegetables (sprouts, carrots and carrot tops, cabbage, squash, or collard greens) should be provided per 3 kg body weight per day, and 1 to 2 tablespoons of fruit such as apple, peach, melon, strawberry, etc.

Rabbits are obligate coprophages. When they first pass feces it is soft and they will eat it whole. This helps them to fully process the food they eat. The second form of feces is a small, hard round pellet.

Clean, fresh water should be available at all times.

Habitat. Rabbits should be kept indoors in a secure cage. The cage should be at least four times the size of the rabbit and should have a solid bottom. Do not use wire-bottomed cages since the wire is harmful to the rabbit's feet. Hardwood shavings, pelleted recycled paper products, straw, or hay can be used as bedding. Rabbits enjoy having a nest box in their cage, and will often sit on top of the box.

Rabbits will typically choose one corner of their cage for elimination. This habit makes it possible to litter box train a rabbit. You can put a litter box in the corner of the cage, and when the rabbit starts using it on a regular basis, it can be placed in a room. Unless you have a very large cage, you will need to provide exercise for the rabbit each day, outside of the cage. This can be accomplished by letting it run around a room or by training the rabbit to walk on a leash and harness. If the rabbit is allowed to roam freely in a room, care must be taken to ensure that it cannot escape and that it cannot reach electrical wires. Rabbits can get tangled in wires behind computers, televisions, and stereos, and also have the habit of chewing on electrical insulation on wires. Spraying the wires with a chew inhibitor such as *bitter apple* can help to prevent this.

Care.

- Daily—food/water, brushing and petting, cleaning litter box
- Weekly—cage cleaning
- Monthly—cage scrubbing, basic physical/wellness exam, nail clipping
- Annual—veterinary care

Costs.

- Acquisition
  - Rabbit—$25 to $50
  - Cage and accessories—$75 to $150
- Maintenance
  - Food—$20 to $30/month
  - Litter—$20 to $30/month

Health Care

Zoonoses.

- Campylobacter
- Tularemia
- Salmonella
- Pasteurella
- Bacterial infections from scratches

CHINCHILLAS

Species

Chinchillas (*Chinchilla laniger*) are agile and, at times, humorous rodents. Chinchillas are nocturnal, though they may be active during the day. They are from high, dry areas in the Andean mountains and do best in low humidity (30 to 40%) and at cool temperatures (10 to 15 degrees Celsius). This can make it a challenge to maintain a healthy environment in a residential facility. Their weight ranges from 400 to 800 g, with females typically larger than males (Figure 11-27). Their average life span is 9 to 17 years.

FIGURE 11-27

The average lifespan of a chinchilla is 9–17 years. *(Courtesy of Brian Yacur and Guilderland Animal Hospital)*

Behavior

Chinchillas rarely bite and do seem to enjoy being held. Their luxurious coat can provide a pleasant tactile experience for someone petting them. Care must be taken when lifting them and holding them. Their hair will come off in clumps (*slipping*) if they are held too tightly or improperly. They can be safely picked up and carried by grasping the base of the tail, scooping and supporting the body, and then placing it into the hollow of the arm, against the body, or in a basket.

They will learn to urinate in a specific spot in their cage, or will use a litter box. Defecation is more often random as they will leave small, hard pellets here and there.

Interaction Level.

- Petting, feeding, and observation

General Care

Diet.
Chinchillas should be provided with unlimited timothy hay and 20 to 30 g commercial chinchilla pellets per day. They will enjoy small daily treats of fresh carrots, green vegetables, or sunflower seeds, not to exceed one teaspoon per day. Fresh, clean water should be available at all times. A drinking bottle is the best choice since they will dump water bowls.

Habitat.
Chinchillas enjoy climbing, so a multi-level cage is a good choice. A wire, multi-level cage designed for ferrets, with a solid bottom, would be a reasonable choice. Along with a bedding of shavings or pellets, they should be provided with a dustpan (there is commercial chinchilla dust available) for 10 to 15 minutes per day. A wooden or cardboard nest box or PVC tube for sleeping is a welcome addition.

Maintenance.

- Daily—food/water, clean litter pan if used
- Weekly—clean cage
- Monthly—scrub cage; check toenails
- Annual—veterinary care

Costs.

- Acquisition
  - Chinchillas—$75 and much higher for rare colors
  - Cage and accessories—$100 to $200
- Maintenance
  - Litter—$10 to $15/month
  - Food—$5/month

Health Care

Zoonoses.

- Listeria
- Bacterial infection of bites (rare) or scratches

BIRDS

Recommended Species

Parakeet (budgerigar or budgie; *Melopsittacus undulatus*). Parakeets may live 5 to 8 years. The bird's original color was green with black markings, but they now come in a wide variety of colors including blue, white, yellow, and various other combinations (Figure 11-28).

Zebra finch (*Taeniopygia guttata*). These finches may live for 7 to 10 years. The grayish males have black/white stripes on the breast (hence the name *zebra*) and orange cheek patches. While females are mostly gray, there is a variety of color forms including fawn, white, and others.

Canary (*Serinus canarius*). Canaries may live 7 to 10 years. Their original color was greenish-yellow, but now most people are familiar with all-yellow varieties. There are breeds with red as well, and some have a *cap* of feathers.

Cockatiel (*Nymphicus hollandicus*). Cockatiels may live 12 to 15 years and can be 13 inches long. Their original color was gray with a yellow face and orange cheek patches, but there are now other color varieties as well.

Behavior

Budgies are quite social; they can be kept together to play and keep each other's company.

Zebra finches are also social. If you have more than one male, however, the cage will need to be large enough to let them set up separate territories.

Cockatiels can also be kept together.

Canaries do best if kept as singles. While two or more females may do well together, male canaries will fight. Only the male canary sings.

Cockatiels and budgies both enjoy climbing and playing with a variety of swings and hanging toys. If you are patient, cockatiels and budgies can be taught to talk. Canaries and zebra finches can be very active. They like to have several perches at the same level in the cage so that they can hop back and forth.

Most birds love to bathe, so provide a separate water source for bathing. There are special covered *bird baths* that fit into the cage door. They prevent the birds from splashing all over and let you watch your birds enjoy themselves. You can also give your birds a shower by using a spray bottle with warm water. Do not put soap in the water for their bath or shower.

Interaction Level

Birds provide both visual and auditory stimulation.

FIGURE 11-28

Budgerigar. *(Photo by Isabelle Francais)*

Diet. There are pelleted foods available for small birds. These foods provide good, balanced nutrition and have the advantage that there are no seed hulls left after the birds eat. However, not all birds like to eat the food pellets, especially if they have been on a seed-based diet. You can convert this diet to pellets by mixing small amounts of the pellets with the seed mixture. Over 1 to 2 weeks, gradually decrease the amount of seeds and increase the amount of pellets. You will be able to tell whether your birds are eating the pellets when there are none left in the seed cup. This can be supplemented with fresh fruits and vegetables. Small amounts of hard-boiled egg are also good. Any fresh food, especially egg, needs to be removed from the cage before it can spoil.

Food and water should be provided in cups that attach to the side of the cage. Use separate containers for the seed mixture and any fresh foods provided.

All birds should also be given cuttlebone to provide calcium.

Habitat. All birds require a cage large enough to let them extend their wings and fly over short distances. In addition, there should be enough room for a variety of perches that will let them climb or hop back and forth. Provide perches of different diameters for exercise of the bird's feet. Do not use perches that are wrapped with sandpaper as this will irritate the feet of the birds.

A typical cage should be 65 cm × 65 cm × 65 cm with bars spaced 10 mm apart. Spacing between bars greater than this may allow birds to escape, or get stuck between bars.

Parakeets and cockatiels prefer cages that are taller with horizontal bars since they climb up and down for play and exercise. Finches and canaries prefer cages that are longer since they fly or hop back and forth.

The cage bottom should be covered with paper. Newspaper can be used if it is printed with nontoxic soy-based ink. Remove the dirty paper and replace with a clean paper each day. Once a week, or as needed, clean the cage more thoroughly with warm soapy water.

Keep the cage in a warm part of the house, out of drafts. Do not place it in direct sunlight because this can cause overheating.

There are a variety of new cages on the market that are constructed of Plexiglas. These help to control dust, droppings, and other mess in the environment. Some are well equipped with air-filter systems and built-in lights.

The kitchen is not a good place for a bird. Birds are very sensitive to fumes. Self-cleaning ovens and some nonstick cookware give off fumes that are dangerous to birds.

Care.

- Daily—feeding, cage cleaning (removing droppings, paper)
- Weekly—cleaning perches, cage bars, etc.
- Monthly—checking and clipping toenails
- Annual—veterinary care

Costs.

- Acquisition
  - Birds—$12 to $100 (depending on species)
  - Cages—$75 to $500 (depending on size and construction)

Special Concerns

- Birds are attractive and active additions to a room. However, they are sensitive to a wide range of environmental changes including temperature, fumes, drafts, and intense sunlight.

- Maintenance
  - Food—$1 to $10/week
  - Bedding—$10/month

Health Care

Zoonoses.

- Campylobacter
- *Escherichia coli*
- Listeria

Factoids

Male budgies can be identified by the blue color of their cere—the area above the beak where their nostrils are located.

Canaries are named after the Canary Islands where they were found. However, the islands were named after dogs. Roman sailors found that the people of the Canary Islands raised large, fierce dogs. The sailors named the islands after the Latin word for dog, *Canis*.

Capture escaped birds by tossing a light towel over the bird

Birds will molt at least once each year. At this time they will lose many of their feathers, and new ones will grow in.

Maintaining red/orange feather colors in canaries may require foods high in carotenoids.

FISH

Recommended Species

Cold Water. Goldfish come in a variety of breeds that differ in color, fin length, and type as well as unusual head shapes (Figure 11-29). The common goldfish can live 10 to 20 years and grow to over 30 cm in length. Some of the other types of goldfish are as follows:

- Comet: A bit slimmer in body with longer tail fins than the common goldfish.
- Shubunkin: Similar in body shape to the comet but with a splotchy orange and white color pattern.

FIGURE 11-29

Goldfish come in a variety of breeds that differ in color, fin length, and type. *(Photo by Aaron Norman)*

- Fantail: Rounded body with two tail fins.
- Veiltail: Similar to the fantail with very long fins.
- Moor: Similar to the fantail but black with enlarged eyes.

Weather loaches are another hardy and common coldwater aquarium fish. They are brown/bronze with brown markings. They have a long snake-like body and will grow to 7 to 10 cm.

Warm Water. There are hundreds, if not thousands, of different warm-water tropical fish species available for aquariums. They vary greatly in color, activity, and size. It is important to choose types of fish that will do well in the same type of water conditions, and with the other types of fish chosen. Some common types of fish that will generally do well in a community aquarium include the following:

- Tetras: Generally small (2.5 to 5 cm) and quick moving. These include the brightly colored neon and cardinal tetras. Tetras like to form schools, so it is best to have four to eight of the same type. They will swim as a group back and forth across your aquarium flashing their colors.
- Danios: Generally small (2.5 to 5 cm) and quick moving as well. Zebra danios and other members of this group are also schooling fish and will form groups in your aquarium.
- Barbs: Active colorful fish that include the aptly named orange- and black-striped tiger barbs. Larger barbs (5 to10 cm) may nip the fins of slower moving aquarium residents with long fins.
- Angelfish: Among the most elegant of all aquarium fish, they will grow to 12 to 16 cm or larger in a large enough aquarium. They have a flattened, triangular body shape with trailing fins. The original coloration pattern was a silver body with dramatic vertical black stripes. Breeders have

FIGURE 11-30

Whiptail catfish. *(Photo by Aaron Norman)*

developed other versions of angels that are all black, marble colored, and pink, as well as veiltail varieties with extremely long fins. Larger angels will eat smaller fish, so care needs to be taken in matching sizes. They are also prone to having their fins nipped by more aggressive fish.

- Platys: Hardy fish that come in a variety of oranges, yellows, and other colors. They are 2.5 to 5 cm in size, and are good community fish.

- Swordtails: Males sport a long sword-like extension of their tail fin giving this group of fish their name. Common colors for swordtails include reds and oranges. Care must be taken to ensure that the aquarium has a secure cover since they are champion jumpers and will leap out of a tank.

- Catfish: There are a number of different sorts of tropical catfish that are attractive in an aquarium (Figure 11-30). Some remain small (2.5 to 5 cm) while others may grow to over 15 cm. Those of the Corydorus group are hardy and effective in cleaning up food that has fallen to the bottom of the tank. Catfish may be striking or comical with their whiskers that they use to find food.

Behavior

Interaction Level.

- Fish are colorful and attractive. Their presence is soothing, and research data indicates that observation can lower blood pressure. Interaction can be through viewing or assisting with feeding. Food should be measured in small cups to prevent dumping/overfeeding from the food container. There are also magnetic algae scrapers with a piece inside and a part outside the aquarium. Moving the outside part drags the inside scraper across the glass. Residents would be able to help with this.

General Care

Adding Fish.
Start your aquarium with a few hardy fish such as platys about a week after the tank has been set up. Do not add more fish for another week. These first fish will provide the bacteria needed to have a healthy tank.

After that, add a couple of fish at weekly intervals until your tank is fully stocked. An old rule of thumb is to have about one 2.5-cm fish for each gallon of water. That means you could have about ten 2.5-cm fish in a 10-gallon aquarium. Modern **filters** let you increase this number, but it is good to keep in mind that your aquarium should not be overcrowded. When you bring fish home from the store, they will be in plastic bags. Float these bags in the aquarium for at least 15–30 minutes to equalize the water temperature in the bag and in the aquarium. Open the bags carefully and allow the fish to swim out.

Habitat. It is best to start with a 10- or 20-gallon aquarium. These sizes are large enough to accommodate a nice selection of fish, and will fit in most areas of a home. The tank should be located somewhere close to a source of water to simplify filling the tank, cleaning, and water change. It should not be in direct sunlight as this makes temperature control difficult, and may promote an overgrowth of algae. A 10-gallon aquarium, filled with water and gravel, will weigh over 45 kg (and a 20-gallon tank over 90 kg), so it is important that the table or counter it is on is sturdy. You will not be able to move the tank once it is set up, so take your time choosing the location.

Water. Water is the home for your fish. The water you start with should be clean and free of chemicals as much as possible. When you fill your aquarium, let the water stand for several days to allow it to *age* and let some of the chemicals found in tap water to evaporate. You can also buy a chemical neutralizer from the pet store to add to the water in your aquarium. Most fish do well in water that is near to neutral, neither alkaline nor acidic. You can test your water using a kit from the pet supply store. It should have a pH around 7.0. Some fish require water that is more or less acidic or alkaline in nature. You should check with someone at the store about this, and how to adjust the pH of the water in your aquarium. You should remove ten liters every week or two, and replace it with clean, aged water. This will help to remove chemicals that can build up in the aquarium and are not eliminated by evaporation or filtration.

Temperature. Tropical fish require water temperatures of 78 to 82 degrees Fahrenheit. You will need an aquarium heater to keep the water warm. Buy a heater appropriate for the size of aquarium you have and a thermometer that you can put in the water. Once your aquarium is set up and filled with water, put the heater in the tank and plug it in. Start at a low temperature, and gradually increase the setting over a couple of days until the heater maintains the water at the desired temperature.

Filters. The water quality of your aquarium will be maintained by one or more filters. Filters clean the water by physical, chemical, and biological means. Physical filtration removes large particles of waste from the water. The water will usually be passed through some sort of fiber or sponge where these particles will be trapped. The filter medium, the fiber or sponge, will need to be replaced on a regular basis. Chemical filtration will typically use activated charcoal to absorb chemical impurities as the water is passed through it. Biological filtration requires a healthy bacterial culture in the aquarium and filter medium. These bacteria then act on impurities in the water, changing them to less harmful or beneficial chemicals.

The most common types of filters available are as follows.

Box Filters. As the name suggests, these filters are usually box-like in design. They will be filled with activated charcoal and a filter fiber. They are often placed in the corner of an aquarium, or may be attached to the inside wall of the tank. Water is drawn through the filter by either air pressure from an air pump or a mechanical pump. These are effective for tanks up to about 10 gallons, and are most practical for temporary aquarium setups, isolation tanks for sick fish, or nursery tanks for baby fish. They are easy to remove and clean, but detract from the appearance of the aquarium.

Undergravel Filters. These are flat, plastic *platforms* that are placed on the bottom of the aquarium and then covered with gravel. There are slots in the plastic platforms, and water is drawn through the gravel and under the filter platform and then up through tubes at the corners of the filter. Air pressure or mechanical pumps (called powerheads) provide the power to move the water through the system. Solid waste is captured in the gravel and under the filter platform. Bacterial cultures in the gravel then act on the waste. These filters are effective in 10- to 20-gallon aquariums if there is adequate water flow through the system. They have limited impact on the visual appeal of the aquarium since the filter platform is covered by the gravel, and the only visible elements are the *lift tubes* in the corners of the aquarium. An added advantage is that the waste trapped in the gravel provides food for plants that are rooted in the gravel. A disadvantage is that cleaning under the filter platform requires a complete breakdown of the aquarium.

Outside Filters. These filters usually hang on the side or back of the aquarium. Water is drawn through a tube and into a filter box that contains filter medium and activated charcoal, and then is returned to the tank. The filter medium traps waste particles and provides a surface for bacteria that perform biological filtration. The activated charcoal removes additional chemical wastes. These filters are highly efficient and easy to clean. It is important to select a filter that provides adequate water flow for the size of aquarium that you have. These filters usually have information on the box that indicates the size of the aquarium for which they are appropriate.

Gravel. The gravel in your aquarium serves several functions. It provides a place for helpful bacteria to live and carry out biological filtration, and a place for plants to root; it also adds to the beauty of your aquatic landscape. Gravel should be thoroughly rinsed with running water to remove dust and small particles before placing it in the aquarium. The gravel texture will depend on the type of filter you use, and whether you are planting live plants. Coarser gravel works best with undergravel filters since it allows a better water flow through the system. Finer gravel works well with outside filters since it tends to prevent large amounts of waste being trapped in between the gravel. Very fine gravel may pack too tightly for plant roots to grow and spread. You will usually need about .5 kg of gravel for every gallon of water your aquarium holds (e.g., 5 kg of gravel is usually about right for a 10-gallon aquarium).

Light. Your aquarium will require a light source. This is best provided with a combination aquarium cover and light fixture. These are specifically designed to fit aquariums of a particular size. It will center the light over the aquarium, and provide a cover that limits water evaporation, things from falling into the tank, and fish from jumping out. A fluorescent fixture that provides full-spectrum lighting, similar to all the wavelengths of sunlight, will best show the colors of your fish and provide proper light to support plant growth. Care must be taken with incandescent light fixtures because the heat generated by the bulbs can make it difficult to keep a constant temperature in the aquarium. The light should usually be on a schedule of 12 hours on and 12 hours off.

Plants. Plants in your aquarium add to its beauty. In addition, plants provide places for your fish to hide and play a role in the biological cycle of the aquarium. There are a wide variety of plants to choose from at most stores that sell fish. Use several different kinds to provide your tank with variety in appearance.

Diet. There are a number of very good commercial fish foods available. Dried flake foods provide a good balanced diet for fish. In addition, dried tubifex worms and bloodworms are available. Fresh foods such as live brine shrimp, tubifex worms, and bloodworms provide variety. Live foods are frequently needed to stimulate breeding behavior.

The most important rule when it comes to feeding fish is not to overfeed. Excess food will fall to the bottom of the tank and spoil. Even with filtration, excess food will result in reduced water quality. It is best to feed several small meals each day. Feed just enough so that your fish eat everything before it ends up at the bottom of the tank. If you are going on vacation, have someone feed your fish while you are gone, or you can get an automatic feeder that will drop small amounts of food into the tank at measured intervals.

While you want to scrape any algae that grow on the front glass of your tank so that you can see your fish, let the algae grow on one end or corner. Your fish will enjoy snacking on the bits of algae that grow there.

Care.

- Daily—turning on lights (could be put on a timer), monitoring water temperature, feeding
- Weekly—water changes, scraping algae, water testing
- Monthly—vacuum gravel, prune plants, clean filter/replace charcoal and filter pads
- Annual—full system check and partial breakdown and cleaning if needed
- Options—services available to maintain aquaria for businesses and other facilities

Special Concerns

Electricity and water can be a dangerous combination. Never use aquarium equipment with frayed electrical lines. Unplug aquarium equipment before you put your hands in the water to adjust plants or decorations. Always have an adult inspect the aquarium setup.

Costs.

- Acquisition—20-gallon tank, filters, stand, light, heater, air pump, gravel, plants, fish, food—$300
- Maintenance—very low feeding costs; electricity consumption

Health Care

Zoonoses.

- Possible Salmonella risk

American Dog Agility Organizations

APPENDIX

American Kennel Club (AKC)
5580 Centerview Drive
Raleigh, NC 27606-3390
Phone: (919) 233-9767
E-mail: info@akc.org

American Mixed Breed Obedience Registration (AMBOR)
179 Niblick Road #113
Paso Robles, CA 93446
Phone: (805) 226-9275
E-mail: ambor@amborusa.org

Australian Shepherd Club of America (ASCA)
6091 East State Highway 21
Bryan, TX 77803-9652
Phone: (978) 778-1082
E-mail: agility@asca.org

Canine Performance Events (CPE)
PO Box 805
South Lyon, MI 48178
(No phone number listed)
E-mail: cpe@charter.net

Dogs on Course in North America (DOCNA)
PO Box 83238
Phoenix, AZ 85071-3238
Phone: (602) 375-0385
E-mail: info@docna.com

Just for Fun (JFF)
8658 Slocum Road
Ostrander, OH 43061
Phone: (740) 666-2018
E-mail: dogwood1@earthlink.com

North American Dog Agility Council (NADAC)
11550 South Highway 3
Cataldo, ID 83810
Phone: (208) 689-3803
E-mail: NADACK9@aol.com

North American Dog Racing Association (NADRA)
PO Box 84
Fenton, MI 48430
Phone: (810) 210-5315
E-mail: HIJUMPNSAMS@aol.com

Teacup Dogs Agility Association
PO Box 158
Maroa, IL 61756
Phone: (217) 521-7955
E-mail: www.k9tdaa.com

United Kennel Club (UKC)
100 East Kilgore Road
Kalamazoo, MI 49001-5593
Phone: (616) 343-9020
E-mail: registration@ukcdogs.com

United States Dog Agility Association (USDAA)
PO Box 850995
Richardson, TX 75085-0955
Phone: (972) 231-9700
E-mail: info@usdaa.com

GLOSSARY

A

Abuse—Intentional acts that cause unwarranted or unnecessary pain or suffering to an animal.

Acquisition—The act of acquiring possession of pets.

Affiliative—The tendency to associate with other individuals.

Agility—A competitive sport emphasizing teamwork between the handler and dog as they negotiate a timed series of obstacles.

American Boarding Kennel Association (ABKA)—Represents the profession and promotes training and standards of practice.

American Fancy Rat and Mouse Association (AFRMA)—Was founded in 1983 to promote and encourage the breeding and exhibition of fancy rats and mice for shows and pets.

American Kennel Club (AKC)—The oldest and largest purebred dog registry in the United States, founded in 1884.

American Pet Product Manufacturers Association (APPMA)—Was founded in 1958 (<www.appma.org>). Its 830 members include product manufacturers, importers, and livestock suppliers.

Americans with Disabilities Act of 1990 (ADA)—Refers to the U.S. public law that prohibits, under certain circumstances, discrimination based on disability.

Angell, George—Founder of the Massachusetts Society for the Prevention of Cruelty to Animals.

Animal-assisted activities (AAA)—Provides opportunities for motivational, educational, recreational, and/or therapeutic benefits to enhance quality of life.

Animal-assisted therapy (AAT)—The incorporation of trained and certified animals into well-defined therapeutic interventions for humans.

Animal control—The enforcement of laws and regulations related to the seizure, capture, impoundment and disposition of animals in a community.

Animal fighting—The practice of allowing animals to fight. They may be animals of different species such as dogs and bears or bulls, or more frequently, animals of the same species. Dogfighting and cockfighting are among the most common manifestations of this practice, though fish (Siamese fighting fish) and crickets are among other species that are also bred and allowed to fight.

Animal protection—A general term used to describe activities to ensure that animals are treated responsibly. It is a compromise term used to avoid potential negative associations with either animal rights or animal welfare.

Animal rights—The proposition that animals have rights regarding their treatment by humans regardless of their economic value or other human defined conditions

Animal sheltering—The practice of providing homeless, lost, and abandoned animals with housing, including food, water, and veterinary care.

Animal welfare—The position that humans should not mistreat animals or cause them unnecessary pain or suffering. Human use of animals is justified if unnecessary pain and suffering is prevented or kept to a minimum.

Animal Welfare Act (AWA)—The most significant federal law related to companion animals.

Applied animal behaviorists—Professionals who apply a wide range of behavior theory to the management and treatment of animals. They commonly work with companion animals, but also with animals in zoo, farm, and laboratory settings.

Aquarium—Generally constructed of glass, lexan, or other transparent material, it is used to keep animals that live in water. Reptiles and small mammals may also be kept in an aquarium.

Asilomar Accords—A document for saving healthy, treatable companion animals in the United States. It is accepted and utilized by reputable animal-welfare organizations, shelters, and rescue groups across the United States.

Assistance Dogs International, Inc. (ADI)—An organization focusing on setting standards for the assistance dog industry since 1987.

Assistance dogs—These by partnering with a person with disabilities provide a specific service as well as provide their handler with greater independence and companionship.

Association of American Feed Control Officials (AAFCO)—A nonregulatory, nongovernmental group that conducts feeding trials to verify the nutritional claims made by pet food companies.

Association of Pet Dog Trainers (APDT)—A professional organization that provides dog trainers with continuing education and professional services.

B

Bands of Mercy—After-school clubs that encouraged children to learn about animals and how to care for them.

Behavior evaluation—The evaluation of animals, such as dogs, in animal shelters to check whether they are appropriate for placement in new homes.

Bestiality—Is sexual contact with animals.

Biophilia—A term coined by E.O. Wilson to describe the relationship between human beings and nature.

Birth rate—The number of progeny born per one thousand individuals in a population. Combined with the death rate it will provide an estimate of population growth.

Boarding kennels—A temporary place for dogs, cats, or other animals to stay.

Breed—A particular type or form of a domesticated species. Members of a breed will share common characteristics, and when mated with one another will "breed true" meaning that they will produce progeny with similar characteristics.

Breed-specific bans—Under these laws, municipalities may preemptively ban specific breeds of dogs as a prophylactic effort to protect public safety.

267

Breed standard—A set of guidelines that determine and define the appearance, and sometimes behavior of a breed of domestic species. For a species such as dogs, the breed standard will define the shape and set of the ears, acceptable colors, overall size, and other characteristics. In show competitions individual animals are judged on how closely they match the standard for their breed.

C

Cages—At one time people who kept pets often needed to build their own enclosures or cages to keep their pets. These would often be made with wood frames and wire fencing, or woven from sticks and reeds. Modern cages are often fabricated from combinations of stainless steel wire and plastic, simplifying cleaning and helping to ensure that the pet is unable to escape.

Canine Freestyle—A choreographed musical program performed by handlers and their dogs. The object of musical freestyle is to display the dog and handler in a creative, innovative, and original dance, using music and intricate movements to showcase teamwork, artistry, costuming, athleticism, and style in interpreting the theme of the music.

Canine Good Citizen (CGC)—Established in 1989 by the American Kennel Club, it promotes responsible dog ownership and training dogs to be well-mannered in public. Dogs and their owners are required to take a short behavior evaluation with items that include accepting a friendly stranger, walking through a crowd, and coming when called. Dogs do not need to be registered with the AKC to earn their CGC.

Cat Fanciers' Association (CFA)—One of the oldest and largest registries for purebred cats, it was founded in 1906.

Cockapoos—A mix of cocker spaniel and poodle.

Commercial breeders—People who breed animals with the purpose of selling them.

Companion animal—Generally thought to be a more respectful term for "pet." Both terms refer to domestic or tame animals that live with humans, usually in the home, for companionship rather providing work, food, or fiber.

Competitions—People with companion animals compete in a wide variety of events to see whose animals are the best looking, fastest, or most accomplished in one activity or another. Some competitions are strictly for fun, others are very organized, and animals are able to earn points towards championships or other titles.

Conformation—The physical appearance of animals. When judged in competitions the conformation of different breeds of animals are defined by the breed standard.

Conservation dogs—Offers assistance around the world in a variety of conservation efforts.

Convention on International Trade in Endangered Species of Wild Fauna and Flora (CITES)—Has over 150 member countries or parties, and regulates the capture, killing, confinement, or possession of wildlife designated as endangered, whether alive or dead, whole or parts.

Counseling—Something that offers direction or advice to a course of action.

Cropping—When a dog's ears are cut and shaped to stand erect.

Cruelty—The legal definition of cruelty may vary from state to state, but its commonly accepted definition is to cause unnecessary suffering or pain to an animal.

D

Dealers—People who sell animals as part of a business or commercial enterprise. They may breed the animals themselves, or buy them from others and resell them to retailers or the public.

Death rate—The number of individuals per one thousand in a population that die each year. Combined with the birth rate it will provide an estimate of population growth.

Declawing—A procedure involving the amputation of the last bone of each toe to prevent regrowth of the nail, which is attached to the bone. The procedure is most commonly performed on domestic cats.

Decompression chambers—A pressure chamber for euthanizing animals in shelters; generally not accepted as a humane practice now.

Delta Society—An organization best known for its work pertaining to the human-animal bond, supporting research and formalizing training and ethics surrounding service animals.

Designer dogs—Intentional hybrids of two different purebred dogs.

Diet—The amount and balance of nutrients provided for an individual.

Disasters—A disaster is a natural or man-made event that negatively affects industry, property, life, or livelihood, often resulting in permanent changes to human societies, ecosystems, environment, and animal shelters.

Docking—When a dog's tail is cut short.

Dog training—Began in Germany in the early 1900s. It was based on the practices developed for training dogs for German military and police applications.

Domestication—The process by which a wild species is adapted to live with humans.

E

Earthdog—A competition for terriers or other dog breeds that were originally bred to pursue and catch rodents, badgers, and other animals underground.

Enrichment—Providing animals under managed care with environmental stimulation to improve the animals' quality of life. The goal of enrichment is typically to increase physical activity, stimulate natural behaviors, and reduce the likelihood of repetitive or stereotypical behaviors.

Euthanasia—The practice of ending the life of a terminally ill animal in a painless or minimally painful way.

Evidence dogs—Trained to use scent to alert to the presence of a variety of controlled or contraband substances or products.

Evolution—The change in inherited traits in a population from generation to generation.

Exotic animals—The capture, ownership, and transport of these animals are limited by several different federal laws.

Exotic pets—Animals other than native wildlife or common domestic pets. Examples would be monkeys, tropical birds, and snakes.

F

Fancy—The hobby of keeping, breeding, and exhibiting a domestic species.

Federal laws—Those laws passed by Congress and signed by the President.

They apply to all 50 states. There are limited federal laws related to the care and treatment of companion animals.

Feral—Members of a domestic species that are living in a wild state.

Feral cats—Refers to cats living and breeding entirely in the wild.

Filters—Used to maintain water quality in aquariums.

Flyball—It is a team sport, matching two teams of four dogs each that run in relays over a series of jumps to retrieve balls that they release from a box opposite the starting point.

Free-roaming cats—See outdoor cats.

G

Genome—The complete DNA sequence of an organism.

Genotype—The genetic constitution of an individual.

Green Chimneys Children's Services and School of Little Folk—Offers care for children who cannot live at home because of their mental or physical disabilities.

Grooming—The care typically provided to keep an animal's fur or hair trimmed or cut as needed, combed or brushed to remain free of tangles. It will also include keeping the nails or claws trimmed.

H

Habitat—The spatial area where an animal lives. Humans provide the habitat for companion animals. The habitats for many companion animals, especially reptiles and rodents, have become more complex and sophisticated as people have learned more about their needs for temperature, humidity, and enrichment.

Harmony Neighborhood Charter School—Conceived to focus on humane education linking all elements of the curriculum.

Hearing dogs—These are a kind of assistance dogs that are specially chosen and trained to help people who are deaf or hearing impaired.

Herriot, James—The pen name of James Alfred Wight, a British veterinary surgeon. He wrote a beloved series of books about his work as a veterinarian between the first and second World Wars in the Yorkshire region of Britain.

Hoarding—Keeping more than the typical number of companion animals, failing to provide even minimal standards of nutrition, sanitation, shelter, and veterinary care, with this neglect often resulting in starvation, illness, and death; and denial of the inability to provide this minimum care and the impact of that failure on the animals, the household, and human occupants of the dwelling.

Hoarding of Animals Research Consortium—Recommends a task-force approach that includes the following:

- Animal control
- Public health
- Mental health
- Child and adult protective services
- Zoning boards
- Fire prevention
- Veterinary assistance

Hobbyist—An amateur who participates in an activity for enjoyment rather than financial reward. Many people who breed, raise, and exhibit various types of pets do so for the enjoyment and entertainment that it brings to their lives.

House training—Training a dog to defecate or urinate outside the house. Is based on consistency. Feeding the dog at the same time each day, and walking the dog immediately after it arises from sleep and 20–30 minutes after feeding will take advantage of the dog's natural biological need to relieve itself after waking and eating.

Household (HH)—All of the people who occupy an individual housing unit. The approximate U.S. household size is 2.6 people.

Humane education movement—A movement promoted by George Angell to prevent cruelty against animals, under which children were taught to show respect and kindness toward animals.

Humane movement—Beginning with the formation of the Society for the Prevention of Cruelty to Animals (SPCA—now Royal SPCA) in 1824, it is the formal organized effort to prevent the mistreatment of animals.

Humane societies—Organizations providing services for both animals and children.

Humane Society of the United States (HSUS)—A Washington, D.C.-based animal welfare organization. The HSUS provides many forms of assistance to local animal shelters, operates a number of direct animal care programs, and is a leader in promoting best practices for local animal shelters throughout the country.

Hurricanes—A tropical storm, with winds over 74 mph, that usually also has rain, thunder, and lightning, and common along the east coast of the United States and the Gulf of Mexico. People need to evacuate areas in the path of the storm and are often confronted with the difficulty of what to do with their companion animals.

I

Iditarod—Best known sled dog endurance race held in Alaska. It is a highly popular event, but does attract the criticism of animal rights groups regarding the treatment of the dogs.

Impartial (Pet adoption)—These people have generally had animals in the past but have few explicit preferences. They are open to whatever might be available, just looking for something that will be a good pet and member of the family.

Internet—An electronic network providing access to millions of resources worldwide. It has become an important part of pet owning by providing easy to access information, on-line shopping for products, supplies, and other services. It is also used in promoting and supporting pet adoptions.

Insurance—Companion animals, their owners, and other people that come in contact with the animals will also be affected by laws and regulations that govern liability and insurance practice.

K

Kennel club—An organization associated with breeding, registering, and exhibiting dogs in competition.

Kitty litter—Granulated mineral clay.

L

Labradoodles—A mix of Labrador retriever and poodle.

Law—A system of rules implemented by institutions.

Leader Dogs—A nonprofit organization that provides all training, expenses, and the dogs free of charge to the visually impaired.

Legislation—A law that has been promulgated by a legislature or representative body.

Levinson, Boris—A psychologist who theorized and demonstrated that using animals as part of the treatment process had benefits for emotionally disturbed children.

Liability—When an individual is responsible for damages caused by their actions or behavior.

Licensing—Providing the legal authority to practice a particular profession or operate a business or service. This is done to ensure that professional standards are met to protect the public.

Livestock—Typically used to describe domestic animals raised for production. In the pet industry the term will be used in reference to animals that are bred and sold.

Local laws—Are the main source of regulations and legal oversight regarding animals in the United States.

Lure coursing—A sport for dogs that involves chasing a mechanically operated lure. The competition is most common for sight hounds, or those breeds originally developed to hunt by visually tracking their prey as opposed to using scent.

M

Maddie's Fund—Dave and Cheryl Duffield founded this charitable foundation in the name of Maddie, a miniature schnauzer, to help the nation's most needy dogs and cats.

Media—Companion animals have appeared in the primary media of almost every era.

Messenger dogs—These dogs have been used in war to carry messages between different units in combat. These dogs are known for their loyalty, and they learn to travel silently and take advantage of natural cover when moving between the handlers.

Microchips—An integrated circuit that is placed under the skin of a dog, cat, or other animal. It is about the size of a rice grain. When "read" with a scanner it will provide a digital identification number. When these numbers are registered with an ownership database it is possible to link lost animals with their owners and arrange for their return.

Mine dogs—Also known as M-dogs or mine detection dogs that are trained to find trip wires, booby traps, metallic and nonmetallic mines.

Mitochondrial DNA (mtDNA)—DNA found in mitochondria, the cellular organelles associated with metabolism. This DNA is inherited separately from nuclear DNA and can be tracked through the female line of a population.

Morris, Mark—Had a profound impact on the science of pet nutrition and on the veterinary profession. He produced the first *prescription diet* for a specific condition.

N

Narcotics dogs—Refer to evidence dogs.

National Animal Control Association (NACA)—An organization to support the movement toward more professional animal control management. The NACA also offers training certification in defensive driving, chemical immobilization, euthanasia, and the use of a bite stick (used for protection against an aggressive dog).

National Council on Pet Population Study and Policy (NCPPSP)—Formed in 1993, it is a coalition of companion animal groups that sponsors research related to the reasons that dogs and cats end up in animal shelters, how many pass through shelters and, how to reduce that number.

National Dog Groomers Association of America (NDGAA)—Was formed in 1969 to promote excellence in professional standards. It offers a variety of accredited training workshops to enhance an individual's skills.

Natural history—The study of animals and plants in their natural environments, including their description, life history, and relationships with other plants, animals, and the environment.

Neglect—This is a category of maltreatment. It is generally associated with the failure to provide adequate care rather than the intentional mistreatment of an animal.

Neoteny—The retention of juvenile characteristics in adults. Many domestic species kept as companions show neoteny, retaining physical traits, such as round heads, and behavioral traits such as submission and attention seeking as adults.

Neutering—Rendering an animal incapable of reproduction through surgery in addition to the benefits of preventing unwanted pregnancies, thus contributing to the homeless animal population, spaying females and neutering males have health benefits as well.

New welfarism—The position that favors working to reduce animal suffering as they are used for food and other purposes as an intermediate step, until the use of animals is completely eliminated.

No-kill—A premise that no healthy, behaviorally sound dog or cat should be killed simply for the convenience of eliminating an unwanted animal.

Nonprofit—Sector supported by charitable contributions. Their size ranges from the Humane Society of the United States with a budget of over $100 million and a staff of hundreds to small local organizations that are organized and run by volunteers with little or no budget.

Nuclear DNA—DNA found in the nucleus of a cell.

Nutrition—See diet.

O

Obedience—Many clubs arrange and host dog shows, field trials, and obedience events.

Obedience training—Teaching basic commands to a dog that would include *sit*, *stay*, *come*, *down*, and *no*. This helps to ensure that they are easy and safe to walk, and can be controlled in the home and around company or when traveling. People may pursue more sophisticated training with their dogs and enter obedience competitions.

Outdoor cats—Spend their entire lives living outside, finding what shelter they can in barns and under porches.

Ownership—The exclusive right or control over property. Under most current laws, animals are considered property and the person in possession is considered the owner.

P

Parasites—Organisms that survive by deriving their sustenance from another organism, frequently compromising the health of the host.

Pet—Generally thought to be derived from the French "petit," meaning small or little, it was first applied to spoiled or indulged children. Eventually it was used to describe domestic or tamed animals that families kept for pleasure.

Patrol dogs—Can detect human scent in the air and pursue that odor to find anyone hiding in a confined area such as a building.

Pet food industry—Was started by James Spratt in the 1860s. Purina revolutionized the pet food industry in the 1950s when they introduced Purina Dog Chow, the first pet food product produced by high-pressure extrusion.

Pet Food Institute (PFI)—Was founded in 1958 and represents about 97% of the industry. It provides education and consumer research for its member companies. It also lobbies legislative bodies on issues of importance to the industry.

Pet industry—It encompasses the broad range of services and products that are provided to care for pets.

Pet Industry Joint Advisory Council (PIJAC)—Is a second major organization in the field and includes pet stores and provides training and certification programs for animal care and husbandry in pet stores.

Pet sitting—Pet owners will arrange to have someone come into their home and care for their pets when they are away.

Petfinder—A website (<www.Petfinder.com>) to promote and support pet adoptions. Now, it includes over 9,000 shelters and rescue groups that post pets available for adoption.

Phenotype—The measurable traits or characteristics or an organism.

Planners (Pet adoption)—These are people who know what they are looking for in terms of breed, size, and color. They may be trying to replace a companion animal from the past, or continuing a family history with a particular type of pet.

Pocket pets—Generally refers to small mammals such as mice, hamsters, gerbils, and others.

Poundmaster—In the early days of animal shelters, the location where the animals were held was called the impound (later shortened to pound). The man in charge of the impound was called the pound master.

Pound seizure—A term commonly used to describe the practice of using lost, homeless, and abandoned dogs and cats from municipally funded pounds for use in experimentation (research, teaching and testing).

Pounds—Old term for animal shelter, derived from the term "impound." It is used less frequently now and, more often than not, as a perjorative term.

Puggles—A mix of pug and beagle.

Puppy mills—Large-scale breeding establishments for dogs that have substandard conditions for the dogs.

Pure bred—An individual that is the result of mating two parents of the same breed or type.

R

Radio Frequency Identification (RIFD)—see Microchip.

Rare breed—Most often used to describe uncommon breeds of domestic livestock.

Regulations—Restrictions or rules promulgated by an administrative agency rather than a legislative body. Agencies within a city government may regulate pet ownership by requiring that dogs be kept on a leash or limit the number of pets that are kept in a dwelling.

Relinquishment—When a person is unable or unwilling to keep a pet and brings their pet to an animal shelter.

Rescues—Generally small groups that work to care for homeless animals and work to find them new homes. Some will focus on particular breeds of dog or cat (i.e., breed rescue); others may work in cooperation with larger animal shelters to remove animals with special needs to provide care.

Rodents—An order of mammals characterized by two continuously growing incisors in the upper and lower jaws. Many small rodents have been domesticated and kept as pets, such as hamsters, gerbils, rats, and mice.

S

Schutzhund—Training and testing of dogs or various breeds for the traits needed for police-type work.

Search and rescue (SAR)—It refers to a technique under which dogs are trained to find lost people whether they are living or dead.

Seeing Eye—The name of only one of many guide dog training schools.

Selective breeding—The deliberate effort to mate specific individuals of a species or breed to produce progeny of a desired type. It has played an important role in the development of many domestic species.

Self—An animal of a solid color such as an all-black rat or mouse.

Sentry dogs—Worked on a short leash and taught to give warning by growling, alerting, or barking.

Service animals—These are legally defined and are trained to meet the disability-related needs of their handlers.

Service dogs—These help people with disabilities to overcome problems in their environments and achieve greater independence.

Shelf pet—Similar to pocket pet. It generally refers to small animals that can live in cages or habitats that would fit on a book shelf.

Shelter design—The first animal shelters were often adapted from stables and barns that had been used to house livestock. Over the past 50 years many developments have helped to improve the materials and plans for building animal shelters to make them more sanitary and appropriate for maintaining the physical and psychological well-being of the animals, as well as being more welcoming to the public.

Shelter programs—Many animal shelters are now becoming important resources in their communities for people and their companion animals. These programs may include humane education, dog training classes, and other community activities.

Shows—Organized events where people can show their farm or domestic animals for fun or as a way to enhance their value as breeders of their species.

Smitten (Pet adoption)—People in this group describe an irresistible pull to a particular animal.

Social/therapy animals—Animals that did not complete service animal or service dog training due to health, disposition, trainability, or other factors, and are made available to people with disabilities to have as pets.

Socialization—The process by which young animals become accustomed to the presence, proximity, and behavior of their own species. Young animals may have a brief period when they are most receptive. Domestic animals generally have longer receptive periods than their wild ancestors, and will also socialize to humans and other species if exposed when they are young.

Society for the Prevention of Cruelty to Animals (SPCA)—A generic term that refers to organizations that work to prevent the mistreatment of animals. SPCAs in the United States are not organized under a single administrative structure.

Society of Animal Welfare Administrators—Founded in 1970, it is a professional organization of executives who oversee the work of humane organizations. It provides professional training and resources to encourage high standards of performance. It sponsors an accreditation program for Certified Animal Welfare Administrators (CAWA).

Spaying—Surgical removal of a female's ovaries and uterus. In addition to the benefits of preventing unwanted pregnancies, thus contributing to the homeless animal population, spaying females has health benefits as well.

Species—A basic term of biological classification, it refers to a group of organisms capable of breeding and producing viable offspring.

State laws—Most laws related to the treatment of companion animals are passed and enforced at the state level.

Survey—A research method where a portion or sample of a population may be contacted to provide information. When proper statistical methods are employed to select the sample the results can be used to infer how the entire population may have responded to the questions.

T

Tame—Used to describe an individual animal with a reduced tendency to flee or be defensive around humans. Selective breeding of the tamest animals in a population can lead to domestication.

Technology—Used by many sheltering and rescue groups to assist in their work.

Temperament tests—A test to determine the personality of your puppy or dog.

The Link—Used to describe the observation that cruelty to animals is often associated with, or linked to, other forms of violent behavior such as child and spousal abuse.

Therapy dogs—Dogs trained to interact with people for therapeutic purposes. Therapy dogs and their human partners may visit hospitals, homes for seniors, or other facilities.

Tracking—One of the ways for a dog to follow the trail of an animal or person. In this the dog follows the exact path the person has taken.

Trade—The exchange of goods and services.

Trailing—One of the ways for a dog to follow the trail of a person. In this the dog can pick up the scent, even when it is dispersed by wind, rain, and temperature and the actual path is unclear.

Trait—A single feature of an organism such as hair color or size.

Trap-Neuter-Release (TNR)—The practice of live-trapping feral cats, sterilizing them through surgery, and then releasing them back to where they were trapped for population control.

Trials—A form of competition for dogs that typically involves some form of hunting.

V

Vaccinations—A preventive treatment to limit the likelihood that an individual will contract an infectious disease.

Veterinary behaviorists—Practitioners providing behavior services for pets.

Veterinary medicine—The profession of providing medical care for animals.

Violence—Physical, sexual, or psychological abuse directed towards an individual.

W

Water rescue—When dogs are trained to assist people in water.

Weight pulling—Competitions where animals pull carts or sleds with weights on them.

White, Carolyn Earle—One of the early leaders of the humane movement in the United States. She founded the Pennsylvania Society for the Prevention of Cruelty to Animals, but because she was a woman she was unable to take an active role in its activities. She eventually formed the Women's Auxiliary of the PennSPCA, which eventually became the Women's Humane Society.

Wildlife—Non-domestic animals that are native to a region or area.

Working Dogs for Conservation Foundation (WDCF)—An organization having dogs to help locate a variety of species.

Wright, Phyllis—In the 1970s and 80s, while at the Humane Society of the United States, she was a primary force in improving the conditions at animals shelters and raising the profile of the field of animal sheltering.

Z

Zoonotic—Any infectious disease that can be transmitted from other animals, both wild and domestic, to humans.

Zoophilia—The sexual attraction of a human to a non-human animal.

INDEX

Note: Entries followed with f indicate reference is to be found in a figure. Specific animal breeds are found under dog breeds, cat breeds, etc.

A

Acquisition of pets, 207, 207f
Acupuncture charts, 144
Adoption, 79
Affiliative behaviors of cats, 33
Agell, George, 59
Aggressive dogs, 208
Agility courses and trials, 165, 167, 170–172
Agility Stakes competition, 170
AHBA Herding Trial Program, 175
Air scent, 198
AKC Champion of Record, 168
AKC registration, 167, 209
Albert Schweitzer Institute, 213
Alert, 199
Alexander the Great, 5
All-breed shows, 167
All Creatures Great and Small (Herriot), 143
Allen, Karen, 187
Alliance for the Contraception of Cats and Dogs (ACCD), 86
All Things Wise and Wonderful (Herriot), 143
American Animal Hospital Association (AAHA), 87, 148
American Association of Veterinary State Boards, 120
American Bar Association, 122
American Boarding Kennel Association (ABKA), 154
American Bred, 168
American Cat Association (ACA), 37
American College of Veterinary Behaviorists (ACVB), 153
American Fancy Rat and Mouse Association (AFRMA), 181
American Herding Breed Association (AHBA), 175
American Humane Association (AHA), 61, 74, 84, 87, 121
American Humane Association Annual Conference, 92
American Humane Education Society, 211–212
American Kennel Club (AKC), 18, 33, 65, 87, 100, 151, 166, 170, 202, 265
American Livestock Breeds Conservancy (ALBC), 47–48
American Mixed Breed Obedience Registration (AMBOR), 265
American Pet Products Manufacturers Association (APPMA), 8, 87, 137–138
American Psychiatric Association, 129
American Psychological Association, 195
American Sighthound Field Association (ASFA), 174
American Society for the Prevention of Cruelty to Animals (ASPCA), 31, 57–61, 72, 82, 85, 87–88, 113, 160, 207–208, 213, 219
Americans with Disabilities Act of 1990 (ADA), 118–119, 185, 192, 193
American Veterinary Medical Association (AVMA), 8, 65, 79, 87, 122, 145, 148
Amory, Cleveland, 63
Amur Tiger Scent Dog Monitoring Project, 204
Angell, George, 75, 211, 212
Animal abuse and neglect, 124–125
Animal and Plant Health Inspection Service (APHIS), 115
Animal-assisted activities (AAA), 186
Animal-assisted therapy (AAT), 186, 196
Animal behavior programs, 93, 149–153
Animal Behavior Society, 153
Animal care and control, 76
Animal Care Expo, 84
"Animal Clinical Psychology: A Modest Proposal" (Tuber, Hothersall, and Voith), 152–153
Animal cruelty, 55, 79, 113, 123–129
Animal fighting, 114, 125, 127
   See also Dogfighting; Bull baiting; Blood sports
Animal hoarding, 127–129
Animal Intelligence (Romanes), 149
Animal Law, 112–113
Animal Law Committee, 122
Animal Legal Defense Fund, 63
Animal Planet, 156, 157f
Animal Precinct, 157f
Animal rescue, 84
Animal rights, protection, and welfare, 53, 62, 65, 66
Animal Rights (Salt), 65
Animals and children, 210–214
Animals and human health, 216–217
Animals as Friends, and How to Keep Them (Saunders and Fisher), 3, 4
Animals as property, 113, 224
Animal sheltering, 61
Animal Sheltering (HSUS), 84
Animal shelter medicine, 92
Animal Shelter Medicine for Veterinarians and Staff (Zawistowski and Miller), xiii
Animal Statistics Table, 102
Animal Welfare Act (AWA), 63, 114, 124
Animal Welfare Institute, 63
Annual Animal Statistics and Live Release Rate formulas, 103f–104f
Annual Live Release Rate formulas, 105
A Pet of Your Own (Zappler and Villiard), 4
Appleton, Mrs. William, 59
Applied animal behaviorists, 151, 152–153
Aquariums, balanced, 5, 260–262
Army, U.S., 151
Artificial hand (Assess-a-Hand), 94
Asilomar Accords, 90, 100–105
ASPCA Animal Poison Control Center, 141, 235
Assistance dogs, 184, 188
Assistance Dogs International, Inc. (ADI), 188–189
Association for Pet Loss and Bereavement, 218
Association for Veterinary Epidemiology and Preventive Medicine, 87
Association of American Feed Control Officials (AAFCO), 141
Association of Pet Dog Trainers (APDT), 152

273

Astor, Jacob, 58
Australian Shepherd Club of
    America (ASCA), 176, 265
Avanzino, Richard, 88
AVMA Task Force on Canine
    Aggression and Human-
    Canine Interactions, 2001, 122
Axelrod, H. R., 5

B

Babylonian Code of Hammurabi, 143
Backyard breeders, 86
Bands of Mercy, 211–212
Banfield, The Pet Hospital, 149
BARF diet, 141
Bark and hold, 200
Barnes, Juliana, 30
Barnum, P. T., 59
Bastet, 35, 36
Beach, Frank, 45
Beautiful Joe (Saunders), 3
Behavior problems and programs,
    93–99, 207–208
Be Kind to Animals Week, 62
Belyaev, Dmitry, 26
Benjamin, Carol, 152
Bentham, Jeremy, 55, 66
Beresford Cat Club of Chicago, 37
Bergh, Christian, 57
Bergh, Henry, 31, 57–61, 67–68,
    72–75, 113
Bering Strait, 20, 22, 40
Bestiality, 129
Biomedical research, 81
Biophilia, 220
Birds, care of
    behavior, 255
    general care, 256–257
    recommended species, 255
Birth rates, 10–11
Bite stick, 86
Black Beauty (Sewell), 55, 211f
Black cats, 36–37
Blascovich, Jim, 187
Blood sports, 75
Boarding kennels, 154
Boke of St. Albans (Barnes), 30
Bomb-sniffing dogs, 199–200
Bones and raw foods diet (BARF), 141
Borchelt, Peter, 153
Bred-by-Exhibitor, 168
Breed clubs, 86
Breeders, commercial, 230–231
Breeding, 5
Breeds, 17, 28
    See also Cat breeds; Dog breeds
Breed-specific legislation (BSL), 122,
    226–227
Breed standard, 168
British Kennel Club, 170

Broome, Arthur, 57
Buddy and Socks, 7
Buffon, Georges, 36–37
Bull baiting, 57
Bulldog Club of America, 167
Bureau of Census, 8
Bureau of Transportaion Statistics
    (BTS), 116
Burial sites, 19, 20f
Burns, Kit, 59
Bush family, 7
Bustad, Leo K., 186
Byrd, Admiral, 138

C

Cadaver search dogs, 199
Call of the Wild, The (London), 151
Canary Islands, 29
Cancer detection, 200
Canine age-related cognitive
    disorder, 153
Canineality/personality of dogs, 95–96
Canine archaeology, 19–20
Canine film stars, 151
Canine Freestyle Federation, Inc.
    (CFF), 179–180
Canine genetics and evolution,
    20–25
Canine Good Citizen (CGC), 167
Canine Performance Events
    (CPE), 265
Canis lupus familiaris, 20, 232
Canned pet food, 140
Caras, Roger, 220
Carbon monoxide, 79
Care guides
    birds, 255–257
    cats, 236–239
    dogs, 232–236
    ferrets, 239–240
    fish, 257–263
    gerbils, 243–245
    guinea pigs, 245–246
    hamsters, 240–243
    mice, 247–248
    rabbits, 251–253
    rats, 249–250
Castration, 144
Cat breeds
    Himalayan, 39, 40
    Maine coon cat, 37, 38f
    munchkin cats, 40
    Persian, 39
    Ragdoll, 164
    Scottish fold, 40
    Siamese, 39
Cat Courier, 38, 39f
Cat Fanciers' Association (CFA), 37,
    40, 87, 166, 180
Cat fancy, 37–40

Cat Journal, 38
Cats, care of
    general care, 238–239
    social behavior, 236–238
    species, 236
Cats, declawing, 227
Cats, feral and free-roaming, 15, 228
Cats, mutation of, 40
Cattle, miniature, 47
Cause-marketing agreements, 160
Cave paintings, 41
Cavia porcellus (guinea pig or cavy),
    46, 245–246
Centers for Disease Control, 117
Certification Council of Pet Dog
    Trainers (CCPDT), 152
Certified Animal Health
    Technicians, 121
Certified Applied Animal
    Behaviorists, 153
Certified Pet Dog Trainer
    (CPDT), 152
Certified Veterinary
    Technicians, 121
CGC program, 176
Charities and non-profit
    organizations, 158
Charles II, King, 6
Charlotte's Web (White), 47
Chemical immobilization, 86
Child abuse and labor, 62, 131
Children and pets, 214–216
Children and animals, 210–214
Chinchilla laniger (chinchillas),
    253–255
Chinese hamsters, 241
Chinning, 251
Choosing pets, 208–210
City Refuge for Lost and Suffering
    Animals, 73
Civil Defense, 62
Civilizing influences, 53
Clever Hans, 149
Clicker training, 152
Clinton, William J., 7
Clomipramine, 153
Coalition for Reuniting Pets and
    Families, 100
Coast Guard, 82, 202
Code of Federal Regulation, 116
Combat stress in dogs, 203–204
Commercial breeders, 230–231
Common school philosophy, 211
Companion Animal Recovery
    system (CAR), 100
Companion versus property, 224–225
Competitions, 165
Conditioned reflexes, 150
Conformation events, 167
Connolly, Mary, 68
Conservation, 117–118

C

Conservation dogs, 204
Conservation priority list, 48
Convention on International Trade in Endangered Species of Wild Fauna and Flora (CITES), 117–118
Coolidge, Calvin, 6
Coronado, 41
Corporate veterinary practice, 149
Cows, 36
Cricetulus griseus (Chinese hamster), 46, 241
Cropping, 228
Cruelty. *See* Animal cruelty; Animal abuse and neglect
Cruft, Charles, 138
Crufts dog show, 170
Crystal Palace, 37
Culp, Val, 179
Cyanuric acid, 142
Cycle of Violence, 130
Cyprus, 33

D

Dahmer, Jeffrey, 130
Dancing with dogs, 179–180
Danforth, William, 138
Danger signal, 191
Dark Sky effort, 214
Darwin, Charles, 19, 54, 149–150
Darwin, Erasmus, 54
Dealers, 114
Dear Socks, Dear Buddy (Buddy and Socks), 7
Death rates, 10–11
Decompression chambers, 78, 79
Defense of Animals, 225
Delta Society, 121, 186, 188, 195
Demographics, 8–12
Department of Defense Dog Center, 203
Department of Health and Human Services, 117
Designer dogs, 210
Diplomate of the American College of Veterinar Surgeons (DACVS), 145
Disasters, 81, 83
Disney, Walt, 33
Distemper, 145
Djungarian hamster, 240
DNA analysis, 23f
Docking, 228
Doctorate in veterinary medicine (DVM), 145
DoD Military Working Dog Program, 203
Dog Adopter Survey (ASPCA), 97
Dog breeds
 Afghan hounds, 23, 174
 Akita, 23
 Alaskan malamute, 23, 177, 184
 American Staffordshire bull terriers, 122
 Australian shepherds, 171
 Basenji, 23
 basset hounds, 170
 beagles, 198
 Belgian Malinois, 199
 Belgian sheepdog, 24
 Belgian Tervuren, 24
 Bernese mountain dog, 23
 bloodhounds, 197, 198
 border-border, 171
 Border collies, 171, 175, 175f
 borzoi, 24
 Bouvier des Flandres, 201
 boxers, 23, 190, 201
 bulldogs, 23, 30, 57
 Cavalier King Charles spaniels, 6
 Chihuahua, 17, 18f
 Chinese crested, 18, 25
 Chow chow, 23, 122
 cockapoos, 210
 Collie, 24
 Dalmatians, 209
 designer dogs, 210
 dingo, 24
 Doberman pinschers, 122, 201, 226
 Egyptian sight hounds, 28
 German shepherd, 23, 122, 190, 199, 200, 226
 giant schnauzers, 201
 golden retrievers, 170, 190
 Great Dane, 17, 18f
 greyhound, 24, 30, 174
 hybrid breeds, 171, 210
 Ibizan hound, 24, 174
 Irish wolfhound, 24, 174
 labradoodles, 210
 Labrador retrievers, 190, 198, 199
 Laconians, 29
 Lhasa apso, 23
 lurchers, 31
 Maltese, 30
 Mastiffs, 23, 24, 28, 30, 31, 178
 mixed breeds, 190
 Molossians, 28
 mongrels, 30
 New Guinea singing dog, 24
 Pekingese, 23
 Pharaoh hound, 24
 pit bulls, 122, 125, 226
 pointers, 24
 poodles, 178
 puggles, 210
 puli, 18
 rat terriers, 170
 retrievers, 24, 171
 Rottweilers, 23, 122, 201
 Saint Bernard, 24
 Saluki, 23, 28
 Samoyed, 23, 177
 scent hounds, 24
 shar-pei, 23
 shepherds, 198
 Shetland sheepdog, 24, 171
 Shiba Inu, 23
 shih tzu, 23
 Siberian husky, 23, 177, 184
 sight hounds, 28, 174
 spaniels, 24, 30
 Staffordshire bull terriers, 122
 Tibetan terrier, 23
 vizlas, 199
 whippets, 174
 Xoloitcuintli (Mexican hairless), 25
 Yorkshire terriers, 183
Dog-Dog aggression test, 94
Dogfighting, 59, 125, 127
Doggie daycare, 155
Dog grooming, 121, 153–154
Dog personality/canineality, 95–96
Dog pound, 78
Dogs, care of
 general care, 235–236
 house training, 233–235
 obedience training, 235
 social behavior, 232–234
 species, 232–236
Dogs, dancing with, 179–180
Dogs, dangerous, 121–122
Dogs, pet store sales of, 230–231
Dogs, social history of. *See* Social history of dogs
Dogs for Defense, 202
Dogs on Course in North America (DOCNA), 265
Dog sports, 165
Dog Star, 29
Dog training, 121, 151
Dog walks, 159, 160f
Domestication
 of cats, 33–40
 of companion animals, 1–2, 14–15, 17–49, 66–67
 of dogs, 17–33, 25–28
 domesticated elite, 26
Dominion: The Power of Man, the Suffering of Animals, and the Call to Mercy (Scully), 67
Dressage, 179
Drugs, 199
Duffield, Dave, 89, 159

Dunbar, Ian, 152
The Duty of Mercy and the Sin of Cruelty to Brute Animals (Primatt), 55

E

Ear cropping, 63, 64f
Earthdog trials, 165, 172–173
Egyptian *Book of the Dead*, 35
Egyptians, 28–29, 33, 34, 36
Enrichment, 98–99
Eohippus, 40
Equine abuse, 84
Equus caballus (modern horse), 40
Equus przewalski, 41
Escherichia coli, 141
Eustis, Dorothy Harrison, 189
Euthanasia, 39, 74, 79, 80f, 84, 86, 88–89, 219
Euthanized for Humane Reasons (EHR), 79
Evidence dogs, 199
Evolution, 54, 149
Exotic animals, 117, 121, 225–226
Exploitation of animals, 67
Expression of the Emotions in Man and Animals, The (Darwin), 149

F

Facility licensing, 119–121
Factory farming, 64
Fair chase philosophy, 64
Falcons, 30
Fancy, 31, 32
Fargo, William, 59
Farm Bill, 116
Favre, David, 113
FBI Behavioral Sciences Unit, 130
Federal Aviation Administration (FAA), 203
Federal Emergency Management Agency (FEMA), 83
Federal laws, 113, 114–119
Federal wildlife regulations, 117–118
Feeders, 45
Feline domestication. *See* Domestication
Felis catus (domestic cat), 33
Felis domesticus, 236
Felis sylvestris lybica (African wildcat), 33, 34
Felis sylvestris sylvestris (European wildcats), 34
Ferrets, 42–42, 44
Ferrets, care of
  behavior, 240
  general and health care, 240
  prohibitions against, 239
Field and Fancy, 38
Field trials, 167
Fillmore, Millard, 59
Films and television, animals in, 84
Fish, care of
  aquariums, 260–262
  behavior, 259
  general care, 259–263
  recommended species, 257–259, 257–263
Fisher, James, 3
Fitzgerald Productions, 214
Flyball, 165, 173–174
Food aggression test, 94
Food and Drug Administration (FDA), 117, 153
Foods, toxic, 140–141
Four Stages of Cruelty (Hogarth), 55
Francione, Gary, 66
Friedman, Carl, 88
Friends of Animals, 63
Frisch, Karl von, 151
Fund for Animals, 63
Funding for animal shelters, 76–77, 159
Fur and Feather, 38

G

Gambian giant pouched rat, 117
Gas chambers, 75f
Geisler, Max, 136
Genetic variation, 16
Genomes, 19
Gerbils, care of
  behavior, 244
  general care, 44f, 243–245
German Shepherd Dog Club (Shaeferhund Verein), 201
Gerry, Elbridge T., 68
Glossary for annual animal statistics, 105–110
Great Race of Mercy to Nome, 177
Greeley, Horace, 58
Green Chimneys Children's Services and School of Little Folk, 195–196
Greyfriars Bobby, 32
Greyhound racing, 175
Grief, stages of, 217–218
Grier, K. C., 3, 5
Grogan, John, 156
Grooming, 153
Group competitions and shows, 168–169
Guardians, 225
Guide dogs, 118, 188–191
Guide for the Care and Use of Laboratory Animals, 98
Guidestar, 159
Guinea pigs, care of
  behavior, 245
  general care, 246

H

Habeus Corpus Act, 68
Haggerty, Captain Arthur, 152
Hale, Doug, 139
Hamsters, 44, 46–47, 240
Hamsters, care of
  behavior, 241
  general care, 241–243
  species, 240–241
Handbook of Tropical Aquarium Fishes (Axelrod and Schultz), 5
Handi-Dandi Dancers, 180
Harmony, FL, 213–214
Harmony Institute, 213, 214
Harrowsby, Lord, 58
Hartsdale Pet Cemetery, 219, 219f
Hartz Mountain, 136
Health benefits of animals, 216–217
Health insurance, veterinary, 149
Hearing dogs, 188, 191–192
Hearing Dogs for Deaf People, 191
Heating, ventilation, and air conditioning systems (HVAC), 91
Heelwork-to-Music, 179
Herding, 175–176
Herriot, James, 143
Herzog, Hal, 209
Hill Packing Company, 148
Hippocrates, 144
History of Pets (Grier), 3
HMS *Beagle*, 54
Hoarding, animal, 124, 127–129
Hoarding of Animals Research Consortium (HARC), 128–129
Hobbling, 31
Hogarth, William, 55–57
Homeowner's insurance, 123
Homer, 29
Homo erectus, 27
Homologous sequences, 20
Horsemeat, 229
Horses
  development of, 40–42
  miniature, 47
  as service animals, 185
  slaughter of, 229–230
Hot button issues, 224–231
  breed-specific legislation (BSL), 226–227
  cats, declawing, 227
  docking and cropping dogs, 228

exotic pets, 225–226
horse slaughter, 229–230
outdoor cats, 228–229
ownership versus guardianship, 224–225
pet store sales of dogs, 230–231
Hothersall, D., 152–153
Households (HHs), 8
House of Lords, 6
House training, 233–234
Houston SPCA, 82
Humane, defined, 74
Humane attitudes, 214
Humane education, 84, 211, 212
Humane movement, 62
Humane shelters, 59, 73
Humane Societies, 73, 75
Humane Society of the United States (HSUS), 62, 79, 84, 87, 158, 212
Humanitarian League, 65
Humor, 158f
Hunting, 28, 64, 75
Hurricanes, 81, 82–83, 116
Huxley, Julian, 4
Hybridization of breeds, 40, 210
Hyde, James T., 37

I

Iams, 139, 140, 142, 160
Iams, Paul, 139
Iditarod, 176–178
Illinois Humane Society, 61
Ill-Treatment of Cattle Act, 57
Impartial choosers, 208
Impound, 71
Imprinting, 15
Inbreeding, 16
Insurance, 119, 123, 149
Insurance Information Institute (III), 122
Intelligent disobedience, 190
International Horse Show at Olympia, 170
Internet, 99, 129, 156–157
Irvine, Leslie, 208

J

Jacob sheep, 48
Jane Goodall Institute, 213
Jingles Award, 195
Johnson, Lyndon B., 7
Johnson, Samuel, 219
Journal of Applied Animal Welfare (JAWWS), xiii
Judging, 168–169
Julius Caesar, 29
Jumpers course, 172
Just for Fun (JFF), 265

K

Kansas Humane Society, 93
Kant, Immanuel, 53, 66
Kennel clubs, 31, 32
Kennel runs, 90–91
Kilcommons, Brian, 152
Kind News, 212
Koehler, William, 151
Kübler-Ross, Elizabeth, 217
K-9 units, 202

L

Lady and the Tramp, 33
Lagomorphs, 43, 251
Land Grant Colleges, 145
Law, liability, tort, and insurance, 112, 113, 122–123
Law of Effect, 150
Leader dogs, 190
Leader Dogs for the Blind, 190
Leech, 144
Lentz, Jim and Martha, 213
Levinson, Boris M., 194–195
Liability, tort, and insurance law, 119, 122–129
Licensed Veterinary Technicians (LVT), 121
Licensing, 72, 74, 115
"The Life, Adventures, and Vicissitudes of a Tabby Cat," 210
Limited admission, 88
Lincoln, Abraham, 71
Lions Club, 190
Listeria, 141
Litter training, 39, 142, 238
Little, Clarence Cook, 45
Live Release Rate, 90, 101
Livestock, 42, 47–49, 75
Locke, John, 55
London, Jack, 151
The Lord God Made Them All (Herriot), 143
Lord, Marjorie, 59
Lorenz, Konrad, 19, 151
Louisiana SPCA, 82
Louis XV, King, 145
Loving Paws Assistance Dogs, 185
Lowe, Ed, 142
Lure coursing, 167, 174–175

M

Maddie's Fund, 159
Madison Square Garden, 31, 37
Mann, Horace, 211
Marine mammal trainers, 151
Marley and Me (Grogan), 156
Martin, Richard, 57
Massachusetts Sate Police Canine Explosive Detection Program, 200
Massachusetts Society for the Prevention of Cruelty to Animals, 87, 211
McCulloch, Michael, 186
McGuffey's Newly Revised Eclectic Reader, 211
M-dogs. *See* mine dogs
Mead, Margaret, 130
Meanwell, Peter, 170
Media coverage, 156–158
Meet Your Match Canineality Adoption Program, 95–98, 160
Melamine, 142
Melopsittacus undulatus (parakeets), 255
Memorials, 219
Mendel, Gregor, 45
Menu Foods, 142
Meriones unguiculatus (gerbils), 44, 243
Merychippus, 40
Mesocricetus auratus (Syrian or golden hamster), 46, 240
Mesohippus, 40
Messenger dogs, 202
Mice, care of
  behavior, 247
  general care, 247–248
Microchips, 99–100, 116
Military dogs, 201–204
Military Working Dog School, 202–203
Miller, Lila, xiii
Miller, Polk, 136
Mine dogs, 202–203
Mink, 139
Mitochondrial DNA (mtDNA), 21
Mobility dogs, 185
Model State Veterinary Practice Act, 120
Mongolian gerbil, 243
Monkey pox, 117
Monodactyl, 40
Moral rights, 66
Morrill Land Grant Act of 1862, 145
Morris, Mark, 148
Morris Animal Foundation, 148
Mus domesticus (house mouse), 44
Mushers, 176–178
Musical Canine Freestyle, 179–180
Musical Dog Sport Association, 180
Mus musculus (mouse), 247
Mustela putorius, 42
Mustela putorius furo, 239
Myers, Fred, 62
My Pets, Real Happenings in My Aviary (Saunders), 3

N

NACA Animal Control Training Manual, 86
NACA News, 86
NACA 100 Training Academy, 86
Narcotics dogs, 199
National Animal Control Association (NACA), 79, 86, 87
National Animal Interest Alliance (NAIA), 65
National Association for Biomedical Research (NABR), 65
National Association for Humane and Environmental Education (NAHEE), 212
National Association of Dog Obedience Instructors, 152
National Association of Professional Pet Sitters (NAPPS), 155
National Association of Protection Dogs, 200
National Cat Club (NCC), 37
National Certified Master Groomer, 154
National Council on Pet Population Study and Policy (NCPPSP), xiii, 86, 87
National Dog Groomers Association of America (NDGAA), 153
National Family Opinion (NFO), 8
National Guard, 82
National Humane Society, 62
National Register of Cats, 37
National Research Council, 98
National Rifle Association (NRA), 65
National Service Dog Center, 186
National Shelter Outreach (ASPCA), 85
Natural breeds, 40
Natural selection, 149
Neglect, animal abuse and, 124–125
Neoteny, 17
Neutering, 79, 236, 239
Newcastle, 31
New England Federation of Humane Societies, 86
New welfarism, 66
New York City pound, 72, 74
New York Society for the Prevention of Cruelty to Children (NYSPCC), 74
NFO (National Family Opinion), 8
Nixon, Richard M., 7
Nobel Prize, 150, 151
No-kill movement, 86, 88–90
Nonprofit organizations. *See* Charities and non-profit organizations
North American Dog Agility Council (NADAC), 172, 265
North American Dog Racing Association (NADRA), 265
North American Flyball Association (NAFA), 173–174
North American Veterinary Conference, 92
Norwegian rats, 45
Nuclear DNA, 21
Nymphicus hollandicus (cockatiels), 255

O

Obedience, 151, 167, 176, 179
Obedience training, 235
Obesity, 140
Obligate coprophages, 251
Odyssey (Homer), 29
Old Yeller, 33
Olfaction, 199, 204
Olson, Patricia, 87
On Death and Dying (Kübler-Ross), 217
One-bite rule, 122
101 Dalmations, 209
Origin of Species (Darwin, C.), 54
Oryctolagus cuniculs (rabbits), 43
Oryctolagus cuniculus (domestic rabbits), 251
Our Cats, 38
Outdoor cats, 228–229
Owned strays, 72
Owners and ownership of pets, 66–67, 208, 215, 225–226
Oxen, 41

P

Parakeets (budgerigar, or budgie), 255
Parasites, 236
Parrots as service animals, 185
Passive responder, 200
Patch, John, 156, 157f
Patrol dogs, 198, 200
Pavlov, Ivan, 150–151
Pennsylvania Society for the Prevention of Cruelty to Animals (PSPCA), 73, 74
People for the Ethical Treatment of Animals (PETA), 63, 66
PeopleSoft Company, 159
Pet care booklets, 78
Pet care industry, 134–135
PETCO, 135, 160
Pet Concierge™, 214
Pet death and grief, 217–220
Pet Evacuation and Transportation Standards Act (PETS Act), 83
Petfinder, 99
Pet food industry, 138, 142
Pet Food Institute, 141
Pet Industry Joint Advisory Council (PIJAC), 138
Pet-oriented psychotherapy, 194–195
Pet overpopulation, 87
Pet products, 135–136
Pets, 2, 185
Pets for Life program (HSUS), 85
Pet sitting, 155
PetSmart, 135, 149, 160
PetSmart Charities, 81f
Pet Stock News, 38
Pet store sales of dogs, 230–231
Pet Talk, 156, 157f
Pfungst, Oskar, 150
Phenotypes, 15
Phodopus campbelli (Russian dwarf hamster), 46, 240
Phodopus roborovskii (Roborovski hamster), 46, 240
Phodus sungorus (dwarf winter white Russian hamster), 46, 240
Pigeons, 16
Pinch test, 94
Planners, 208
Pliohippus, 40
Pocket pets, 42
Police dogs. *See* Patrol dogs
Polk Miller Drug Company, 136
Poodle Club of America, 167
Positioning, 168
Poundmaster, 71–72
Pound seizure, 81
Prairie States, 86
Precocial, 15
Prescription Diet k/d®, 148
Primatt, Humphrey, 55
Primum non nocere (Above all, do no harm), 144
Proctor and Gamble, 140
Professional Standards for Dog Trainers, 186
Protection dogs, 200
Pryor, Karen, 152
Psychoactive drugs, 153
Psychologists for the Ethical Treatment of Animals (PSYeta), xiv
Psychology Today, 2
Pulka, 166
Puppy mills, 33, 115–116, 230–231
Purebred, 22
Purebred dogs, 32, 166
Purina, 138–139

R

Rabbit breeds
  Dutch, 43
  Flemish, 43
  New Zealand white, 43

Rabbits, care of, 251–253
  behavior, 251
  general care, 252–253
  species, 251
Rabies, 145
Radio Frequency Identification (RFID), 99, 116
Raisins, 141
Rally obedience, 176
Ralston Purina, 138
Ramses the Great, 6
Rape kits, 129
Rare breeds, 47, 49
Rats and mice, 44–46, 249–250
Rattus norvegicus (wild brown rat), 44, 249–250
Rattus rattus (black rat), 44
Raw foods, 140
Red Cross emergency shelters, 81
Redemption fees, 71
Red Star Animal Relief, 62
Reflexive behavior, 150
Registered Veterinary Technicians, 121
Regulations, 114
Reid, Pamela, 171f
Relinquishment, 79, 207
Reproductive potention, 15–16
Resale value of animals, 72
Rescue/adoption groups, 79
Rescue Waggin', 81
Return to Owner (RTO), 79
Richmond SPCA, 88
Rin Tin Tin, 33, 156
Roborovski hamsters, 241
Rodents, 34, 247–248, 249–250
Rolenson, George, 138
Romanes, George, 149–150
Romans, 29, 36, 42
Romulus and Remus, 29
Roosevelt, Franklin D., 6–7
Roosevelt, Teddy, 6
Ross, Myra and Sam, 195
Royal Society for the Prevention of Cruelty to Animals (RSPCA), 57, 58, 60
Runts, 47

S

Sadism, 130, 131
SAFER (Safety Assessment for Evaluating Rehoming), 93, 95
Salman, Mo, 87
Salmonella, 117, 141
Salt, Henry Stephens, 65
Samburu-Laikipia Wild Dog Project, 204
San Francisco Animal Care and Control (SF AC&C), 88

San Francisco SPCA (SFSPCA), 88
Santorum, Senator, 116
Sassy Seniors, 180
Saul, Betsy and Jared, 99
Saunders, Blanche, 151
Saunders, Marshall, 3
Save the Tiger Fund, 204
Sayres, Ed, 88
Scala naturae (chain of being), 54
Scent hounds, 24, 29, 197–198
Schultz, Jacque, 157f
Schultz, L. P., 6
Schutzhund, 200–201
Schwerker, Victor, 44
Scout or patrol dogs, 202
Scully, Matthew, 67
Seal of approval, 160
Search and rescue (SAR) dogs, 194, 197
The Seeing Eye, 189–190
Seeing Eye dogs, 47, 189
Seeing Eye, Inc., 189
Selective breeding, 19, 40
Selfs, 45
Senator Vest's Tribute to a Dog, 136, 137
Sensitivity test, 93
Sentry dogs, 202
Separation anxiety, 153
September 11, 2001, attacks, 82, 194, 197
Sergeant's Pet Products, 136
Serinus canarius (canaries), 255
Service animals, 118, 184, 185, 187, 188, 197
Sewell, Anna, 55, 211f
Shaeferhund Verein (German Shepherd Dog Club), 201
Shaw, Margaret, 3
Shelf pets, 42
Shelter design, 85, 90–92
Shelter intake and disposition data, 80f
Shelter medicine, 92
Shibuya station, 32
Shortall, John G., 61
Shows, 165–168, 167f, 169, 170, 209
Siberian hamster, 240
Sife, Wallace, 218
Signal dogs, 118, 191
Simpson, F., 37–39
Sinclair, Upton, 61
Singer, Peter, 66
Sirius, 29
Skinner, B. F., 151
Slaughter, humane, 66
Slaughter of horses, 229–230
Sled dog racing, 166, 177, 184
Small animals, 42–47

Small Ladyes Poppees, 30
Smitten choosers, 209
"The Snark Was a Boojum" (Beach), 45
Social history of dogs, 28–33
Socialization, 15
Social/therapy animals, 185
Society for the Prevention of Cruelty to Animals (SPCA), 55, 57, 60, 75
Society for the Prevention of Cruelty to Children, 68
Society of Animal Welfare Administrators (SAWA), 85, 87
Society of North American Dog Trainers (SNADT), 152
Sodium pentobarbital, 79
Software packages, 99
Sootering, 166
Sopwell Nunnery, 30
Sound alert dogs, 191
Sources of companion animals, 207f
Souvetaurinarii, 144
Spaying and neutering, 79, 236, 239
Speciality shows, 167
Species, 14
Spitz, Carl, 151
Sports Illustrated, 7
Spratt, James, 138
Spratt's Dog Cake, 138
Stacking, 168, 169
Stare test, 93
State and local laws, 113, 119–129
State Board of Veterinary Medicine, 145
Sterilization, 79. *See also* Spaying and neutering
Stevens, Christine, 63
Stockdog, 176
The Story of the Robins (Trimmer), 210f
Support dog, 185
Surveys, 2, 8–10

T

Taeniopygia guttata (Zebra finch), 255
Tag test, 94
Tail docking, 63
Tame, 25
Tanzanie Carnivore Project, 204
Tax Code—Nonprofit and 501(c)(3), 119
T-Bo Act, 123
Teacup Dogs Agility Association, 265
Technology, 99–100
Temple catteries, 36
Terriers, 24, 30

Therapy dogs, 192–196
Thorndike, Edward, 150–151
341st Training Squadron, 203
Tinbergen, Niko, 151
Tischler, Joyce, 63
TNR (trap-neuter-release)
Tobias, Michael, 214
Toxic foods, 140–141, 235
Tracking, 197–198
Trade, 117
Trailing, 197–198
Traits, 16
Trap-neuter-release (TNR), 17
Traps, 172
Treats, 140
Trials, 171
Trimmer, Mrs., 210f
The Truth About Cats & Dogs, 156
Tuber, D. S., 152–153
Tufts Animal Care and Condition (TACC) scales, 124–125, 126
Turner, James, 54
Turnspits, 32

U

Uncle Tom's Cabin, 211f
United Kennel Club (UKC), 266
United States Department of Agriculture (USDA), 115, 116, 198
United States Dog Agility Association (USDAA), 266
United States Fish and Wildlife Service, 204
United States Veterinary Medical Association (USVMA), 145
Urban legends, 156
U.S. Army, 151
Utilitarian position, 66

V

Vaccinations, 90, 235–236
Vegan diets, 66
Veterinarium, 144
Veterinary behaviorists, 151, 153
Veterinary Clinics of America (VCA), 149
Veterinary education, 144–148
Veterinary health insurance, 149
Veterinary Information Network, 92
Veterinary medicine, 120–121, 143–149, 153
Veterinary Practice Act, 120
VetSmart, 149
Victoria, Queen, 57
Victorian era, 53–54
Villages, 25–26
Villiard, Paul, 4
Violence, cycle of, 130–131
Vivisection, 55, 60, 75
Voith, V. L., 152–153
Voyage of the Beagle (Darwin, C.), 54
Vulpes vulpes (silver foxes), 26

W

Walker dog, 185
War Dog Program, 202
Washington, George, 6
Water rescue dogs, 199
Watson, J. B., 151
Weber, Josef, 151
Weight pulling, 166, 178–179
Weir, Harrison, 37
Weiss, Emily, 93, 95
Wells, Henry, 59
Western States, 92
Westminster Kennel Club, 31, 166
Westminster Kennel Club Dog Show, 167f, 169, 209
Wheeler, Marietta "Etta" Angell, 67–68
Where Purity Is Paramount, 138
White, Carolyn Earle, 59, 60f, 73, 74, 75
White, E. B., 47
Whitehouse-Walker, Helen, 151
Wight, James Alfred. *See* Herriot, James
Wild Bird Conservation Act (WBCA), 118
Wildlife, 113, 121
WildSide Walk™, 214
Wilson, E. O., 220
Wilson, Mary Ellen, 61, 67–68, 74
Wolch, Jennifer, 214
Wolves, 25
Women's Auxiliary of the PSPCA, 73–74
Women's Humane Society, 73f, 74
Working Dogs, 179
Working Dogs for Conservation Foundation (WDCF), 204
World Canine Freestyle Association, 179–180
World Trade Center, 82, 197
World War II, 33, 62, 202
Wright, John D., 68
Wright, Phyllis, 79

Z

Zappler, Georg, 4
Zoonomia (Darwin, E.), 54
Zoonotic diseases, 75, 215
Zoophilia, 129